Organotransition-Metal Chemistry

Organotransition-Metal Chemistry

Edited by

Yoshio Ishii
Department of Synthetic Chemistry
Nagoya University
Chikusa, Nagoya

and

Minoru Tsutsui
Department of Chemistry
Texas A&M University
College Station, Texas

PLENUM PRESS · NEW YORK AND LONDON

Library of Congress Cataloging in Publication Data

Main entry under title:

Organotransition-metal chemistry.

Includes bibliographies and index.
1. Organometallic compounds—Congresses. 2. Transition metal compounds—Con-
gresses. I. Ishii, Yoshio, 1914-   ed. II. Tsutsui, Minoru, 1918-   ed. III.
United States. National Science Foundation. IV. Nippon Gakujutsu Shinkōkai.
QD410.P76   547'.05   74-28165
ISBN 0-306-30832-0

Proceedings of the first Japanese—American seminar on "Prospects in Organotransition-Metal Chemistry," held at the University of Hawaii's East—West Center in
Honolulu, May 1—3, 1974 and jointly sponsored by the National Science Foundation
and the Japan Society for the Promotion of Science

© 1975 Plenum Press, New York
A Division of Plenum Publishing Corporation
227 West 17th Street, New York, N.Y. 10011

United Kingdom edition published by Plenum Press, London
A Division of Plenum Publishing Company, Ltd.
4a Lower John Street, London, W1R 3PD, England

Printed in the United States of America

Preface

In every generation the achievements in science have served
mankind. The progress accomplished by one generation stimulates
the next generation to even greater achievements, which may take
the form of increasing, crystallizing, or detailing existing theo-
ries. Other forms, generally resulting from persistence and enlight-
ened fortune, open new areas of investigation previously unimagined
and have an impact that may be felt for many years.

An example of this latter form of achievement was the preparation and elucidation of the structures of dicyclopentadienyliron
(ferrocene, reported in 1951) dibenzenechromium iodide, triphenyl-
chromium tristetrahydrofuranate, and numerous olefin-metal π-complexes which provided an introduction to new types of chemical bonds-
the sigma carbon-transition metal bond and the metal π-complex bond.

Initial progress in the field of organotransition-metal chemistry followed the lines of interest generated separately by organic
and inorganic chemistry. However, it is becoming increasingly clear
that organotransition-metal chemistry is not only bridging these two
fields, but also crosslinking many other fields of science.

The stabilization and isolation of both cyclobutadiene and benzene derivatives of transition-metal organometallics were first a-
chieved because of academic interest. However, industrial processes
such as the Ziegler-Natta olefin polymerization, the Wacker oxidation
of ethylene to aldehydes and ketones, and the hydroformylation of o-
lefins, among others, have provided practical applications for the
achievements in organotransition-metal chemistry. Still to be determined are the exact role of transition-metal organometallics in many
metal biological functions and processes such as nitrogen fixation
by nitrogenase and multifunctions of vitamin B_{12} and functions of or-
ganomercury as a pollutant and catalytic functions in fuel cells.

During the past decade, a marked trend toward research on organometallic chemistry by distinguished Japanese organic chemists
has been observed. As a result, Japanese organotransition-metal
chemists have demonstrated distinctive capabilities in discovering

new reactions and isolating unstable organotransition-metal compounds. On the other hand, many organotransition-metal chemists in
the United States were originally inorganic chemists and have made
outstanding contributions in specific areas of academic interest,
such as the nature of bonding, structures, and mechanisms of reactions of organotransition-metal complexes.

While rapid progress in the field has been made during the past
two decades, detailed mechanisms for only a few reactions of organo-
transition-metals are known. It would seem that the mechanisms of
more reactions could be better evaluated and understood by simultaneously applying the characteristic approaches and research experiences that have been developed in both countries. For these reasons,
it was most timely to hold a seminar on the following specific themes:

a. Factors influencing the stability of transition metal carbon
    bonds and mechanisms of the form and breaking of such bonds.
b. Mechanistic features of reactions involving unstable organotran-
    sition-metal complexes.
c. Application to catalytic processes, for example, carbonylation
    and related processes in which both countries have some scienti-
    fic and technological interest.
d. Application to environmental problems; notably the formation of
    toxic organometallic compounds such as methylmercury.

This bilateral meeting would benefit both countries in furtherance
of their uniqueness of development in specialized areas, and in sti-
mulating active interest in neglected and deficient areas in the
field, which would be of the greatest benefit to science and society.

On May 1-3 the first Japanese-American seminar on "Prospects in
Organotransition Metal Chemistry" was held at the University of Hawaii's East-West Center, in Honolulu, jointly sponsored by the National Science Foundation and the Japanese Society for the Promotion
of Science. The purpose of the seminar was to acquaint participants
with the major trends and direction of research in each other's coun-
try. Attendance was limited to 17 Japanese and 24 Americans in order to encourage informal discussion and exchange of ideas between
the two nationalities.

This monograph includes 26 papers presented by the participants
and nine papers invited by the editors. In order to provide spiri-
tual and scientific feelings of the seminar, the outline of the sessions is described as follows:

There were six sessions of three to four hours length scheduled
for the three days. No less than three full sessions were held on
the first day in order to allow two subsequent afternoons free for
informal socializing on the beach.

Brief addresses by the Japanese and American organizers of the conference, Dr. Minoru Tsutsui and Dr. Yoshio Ishii, opened the first session, devoted to novel organometallic compounds. Two participants from U.C.L.A. dealt with compounds of unusual geometry: Dr. M. F. Hawthorne discussed his recent work in polyhedral expansion and contraction of metallocarboranes, and Dr. H. D. Kaesz described unsaturated ruthenium hydrido-carbonyl cluster complexes and their possible application to catalysis. Presentations of work concerning compounds with novel ligands were represented by K. Itoh (Nagoya University) on the bridging, bi-dentate behavior of benzoyl isocyanates and by Dr. R. B. King concerning complexes of polycyano olefins, specifically comparing the di-cyanovinylidene ligand with carbon monoxide. The only paper on the actinide metals was presented by Dr. Tobin Marks (Northwestern University) on tris(cyclopentadienyl) alkyl complexes.

The largest single topic of interest, covered from many viewpoints, was homogeneous catalysis. While the original controversy over the concerted vs. step-wise mechanism of various metal-catalyzed rearrangements has been largely resolved, four papers on this subject were presented. First and foremost was Dr. R. Pettit's very thorough presentation on the concerted nature of silver ion catalyzed rearrangement of polycyclic hydrocarbons. Dr.'s Frank Mango (Shell Oil) and Robert Grubbs (Michigan State University) both discussed metallocycles as intermediates in these reactions, Dr. Mango in particular questioned whether any reaction proceeding in this way should really be called "concerted". The Japanese contribution to this area was given by Dr. R. Noyori (Nagoya University) on nickel(0) catalyzed rearrangements involving highly strained σ bonds.

Dr. J. Halpern, also active in homogeneous catalysis and a strong participant in the preceding discussion, presented a paper on a different aspect of catalysis - a detailed analysis of the mechanism of catalysis by tris(triphenylphosphine) rhodium chloride. More specific applications of homogeneous catalysis were covered in papers by Dr. Jitsuo Kiji (Kyoto University) on diolefin polymerization catalyzed by Ni(0) complexes and acid, Dr. G. W. Parshall (DuPont) on activation (for exchange with deuterium gas) of aromatic C-H bonds, and Dr. Jiro Tsuji (Tokyo Institute of Technology) on addition reactions of butadiene catalyzed by palladium(II) complexes.

Two homogeneous catalytic reactions applicable to laboratory synthesis were presented: (1) the efficient catalytic coupling of Grignard reagents with aromatic halides by phosphine complexes of nickel (by Dr. M. Kumada, Kyoto University) and (2) the selective reduction of unsaturated ketones via rhodium catalyzed hydrosilylation (by Dr. Iwao Ojima, Sagami Chemicals). The latter synthesis was also reported to have yielded products of up to 50% optical purity from racemic substrate and chiral catalyst.

One most unusual and timely report was that of Northwestern University's Dr. James Ibers, who had observed cationic rhodium-phosphine complexes to homogeneously catalyze the conversion of CO and NO to CO_2 and nitrogen.

In addition to the studies of catalytic reactions mentioned above, the kinetics and mechanism of several other reactions were presented. Dr. J. Osborn (Harvard) analyzed the oxidative addition of alkyl halides to Pt(O), and concluded that such reactions generally occur via radical chain pathways. There were two contributors from Osaka University in this area; Dr. Akira Nakamura who examined the mechanism of a molybdenum dihydride complex with various unsaturated compounds, and Dr. Toshio Tanaka, who had determined the kinetics of TCNE addition to a cationic rhodium complex. Dr. G. Whitesides (MIT) managed to cover two topics in twenty minutes, the first on the decomposition mechanisms of metal alkyls and the second, some results on synthetically useful Li-Hg-C complexes of unknown structure. Further discussions of metal alkyl complexes were heard, one concerning nmr studies of nickel-alkyl phosphine complexes by Dr. Akio Yamamoto (Tokyo Institute of Technology), and a second about decarboxylation of allyloxycarbonyl platinum(II) complexes by Dr. Hideo Kurosawa (Osaka University).

There were only two papers specifically on stereochemistry. Dr. Kazuo Saito (Tohoku University) concentrated on the asymmetric induction arising upon coordination of an olefin to chloro-L-prolinato platinum (II), while Dr. J. Faller (Yale) lectured on the substituents required to prevent palladium-allyl chiral centers from racemizing.

The last session was devoted to biological aspects of organometallic chemistry, dealing chiefly with the metalloporphyrin and -corrin systems. Dr. James Collman (Stanford) discussed a synthetic Iron(II) porphyrin with a hydrophobic "pocket" in which oxygen could be reversibly absorbed. Dr. Minoru Tsutsui of Texas A&M presented his work on some novel bimetallic rhenium porphyrins; Hisanbou Ogoshi (Kyoto University) also talked about noble-metal porphyrins, considering their similarity to vitamin B-12. For the final paper, Dr. John Wood (University of Illinois) presented his work on the mechanism of mercury neurotoxicity, which arises from methylmercury ion catalyzed cleavage of a vinyl ether linkage found only in brain lipids.

The seminar was quite successful, and the organizers hope to repeat it three years hence, although a meeting location has not been selected. In comparing the Japanese and American contributions, the only general observation which might be made is that while the Japanese tended to collect more data, the Americans collected less, but analyzed it more thoroughly.

Heartfelt appreciation is extended to each of the authors and particularly to Jack Halpern, James P. Collman and Seinosuke Otsuka whose help in organizing this symposium was invaluable. Also, we would like to express our gratitude to Mrs. Sally Hayes for an excellent job in the preparation and typing of this monograph.

Yoshio Ishii
Minoru Tsutsui

Contents

BIOLOGICAL AND RELATED ASPECTS

METALLOCARBORANES: PAST, PRESENT, AND FUTURE

Timm E. Paxson, Kenneth P. Callahan, Elvin L. Hoel and
M. Frederick Hawthorne[*]
Department of Chemistry, University of California
Los Angeles, California 90024

The first metallocarborane complex, $[3,3'-Fe^{II}(1,2-C_2B_9H_{11})_2]^{2-}$,[1] the carborane analog of ferrocene, was first synthesized ten years ago. This compound bridged boron hydride and carborane chemistry with that of the transition metals, and heralded the explosive pro-liferation of metallocarborane chemistry.

The rationale behind this combination of diverse specialties was developed from two chemical advances: the synthesis[2] and the molecular orbital description[3] of ferrocene and the discovery[4] of the base degradation of closo[5] icosahedral carboranes to produce anionic species.

The production of the icosahedral carboranes and their degra-dation products from the parent boron hydride, decaborane, $B_{10}H_{14}$, is shown in Figure 1. The bridging hydrogen of the nido-[5] $C_2B_9H_{12}$ anion,[6,8] may be removed by such reagents as NaH,[6,8] alkyllithium rea-gents[7] or by strong aqueous base:[8]

$$7,8-C_2B_9H_{12}^{-} + RLi \longrightarrow 7,8-C_2B_9H_{11}^{2-} + RH + Li^{+}$$

$$7,8-C_2B_9H_{12}^{-} + OH^{-} \longrightarrow 7,8-C_2B_9H_{11}^{2-} + H_2O$$

It was believed that the fully deprotonated $C_2B_9H_{11}^{2-}$ (dicar-bollide) ions should have orbitals of proper symmetry and energy to interact with transition metals in a manner analogous to the cy-clopentadienyl ion. A simplified orbital scheme with lobes point-ing toward the missing vertex is illustrated for $7,8-C_2B_9H_{11}^{2-}$ in Figure 1. The synthesis of the first metallocarborane followed shortly after this realization.

1

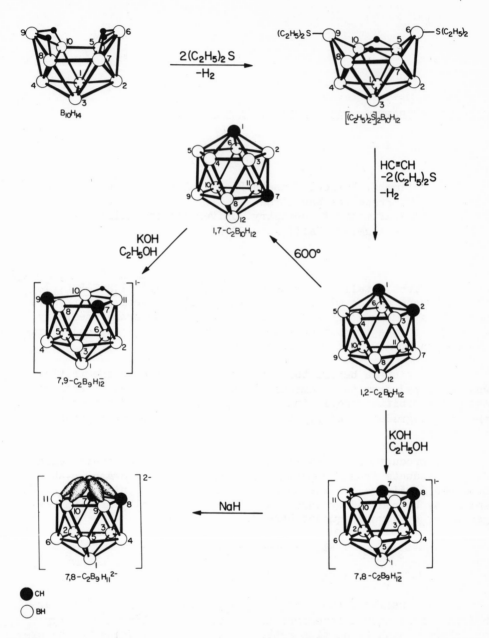

Figure 1. Synthesis of icosahedral carboranes and anionic derivatives. Orbital representation of $7,8$-$C_2B_9H_{11}^{2-}$ (dicarbollide) ion.

Subsequent to this preparation, "mixed" sandwich compounds involving both the dicarbollide and carbonyl or cyclopentadienyl ligands were synthesized,[9],[10] and complexes of other transition metals
including chromium,[11] molybedenum,[8] tungsten,[8] manganese,[8],[9a] rhenium,[8],[9a] cobalt,[8],[12] nickel,[8],[13],[14] palladium,[8],[14] platinum,[14]
copper,[8],[15] and gold[15] were reported. Four examples of the various
metallocarboranes which have been characterized are illustrated in
Figure 2. Complexes utilizing the $C_2B_7H_9^{2-}$ [16] and $C_2B_6H_8^{2-}$ [17],[18]
ligands were also obtained.

The properties of the metallocarboranes thus synthesized indicated that there were important differences between these species
and their bis-cyclopentadienyl analogs, most notably in their thermal and hydrolytic stabilities and in their electronic properties.
The metallocarborane complexes generally prefer high formal oxidation states of the metal,[19] and their thermal stabilities are so
great that the study of thermal isomerization, involving atomic migration across polyhedral surfaces, is an important aspect of metallocarborane chemistry.[20]

Perhaps one of the most versatile of all methods for producing
metallocarboranes is the so called "polyhedral expansion" reaction.
The concept of the expansion reaction is based upon the apparent
similarity of aromaticity in benzenoid hydrocarbons and electron
delocalization in $C_2B_nH_{n+2}$ carboranes. The former species accept
electrons into their nonbonding and lowest energy antibonding molecular orbitals upon chemical reduction, and it was reasoned that a
similar process in carboranes might produce geometrical changes and
orbital populations favorable to metallocarborane formation. This
proved to be true as is outlined below:

$$C_2B_nH_{n+2} + 2Na \longrightarrow 2Na^+ + [C_2B_nH_{n+2}]^{2-}$$

$$[C_2B_nH_{n+2}]^{2-} + CoCl_2 + C_5H_5^- \xrightarrow{[O]} (\eta-C_5H_5)Co^{III}C_2B_nH_{n+2}$$

$$6 \geq n \geq 10.$$

In each case the carborane has been "expanded" to its next higher
homolog with the $\{(\eta-C_5H_5)Co^{III}\}$ moiety residing at one vertex of
the new polyhedron.

The gross geometric structure of some of the <u>closo</u>-carboranes
and metallocarboranes (and the <u>closo</u>-$B_nH_N^{2-}$ boron hydrides) is shown
in Figure 3. Thus, reduction of the 9-vertex tricapped trigonal
prismatic $C_2B_7H_9$ carborane and reaction with $CoCl_2$ and cyclopentadienyl ion would produce the expanded product, a 10-vertex bicapped
square antiprismatic $C_5H_5CoC_2B_7H_9$ metallocarborane.

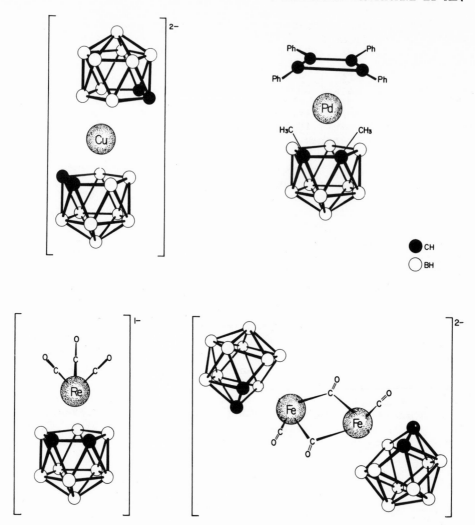

Figure 2.   Four metallocarborane complexes. $[3,3'-Cu(C_2B_9H_{11})_2]^{2-}$ (8,15); $1,2-(Me_2)-3-(\eta-Ph_4C_4)-3,1,2-PdC_2B_9H_9$ (8); $[3,3,3-(CO)_3-3,1,2-ReC_2B_9H_{11}]^{1-}$ (8); and $\{(\mu-3,3'-CO)_2-[3-(CO)-3,1,2-FeC_2B_9H_{11}]_2\}^{2-}$ (9b).

Polyhedral expansion has also been successfully extended to metallocarboranes. The three electron reduction of monometallocarboranes, followed by addition of $CoCl_2$ and $C_5H_5^-$ produced bimetallocarborane complexes:

$$(\eta-C_5H_5)Co^{III}C_2B_nH_{n+2} + 3Na \longrightarrow 3Na^+ + [\eta-C_5H_5)Co^{II}C_2B_{n+2}]^{3-}$$

$$[(\eta-C_5H_5)Co^{II}C_2B_nH_{n+2}]^{3-} + CoCl_2 + C_5H_5^- \xrightarrow{[O]} (\eta-C_5H_5)_2Co_2^{III}C_2B_nH_{n+2}$$

$$6 \geq n \geq 10.$$

The immense scope[21] of this reaction does not permit a complete description of expansion of each carborane and metallocarborane we have examined, but two examples will serve to illustrate the nature of the process.

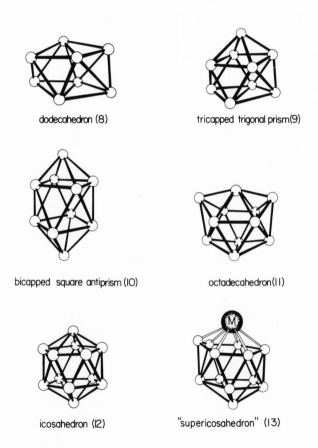

dodecahedron (8)

tricapped trigonal prism (9)

bicapped square antiprism (10)

octadecahedron (11)

icosahedron (12)

"supericosahedron" (13)

Figure 3.   The polyhedral figures containing 8-13 vertices.   The "supericosahedron" has only been observed containing one or two metal vertices.

Figure 4 depicts the polyhedral expansion technique as applied to $1,6-C_2B_8H_{10}$[22] and reveals that the expansion reaction is actually very complicated in nature. Besides the expected $(\eta-C_5H_5)CoC_2B_8H_{10}$ complex, $2,3-(\eta-C_5H_5)_2-2,3,1,7-Co_2C_2B_8H_{10}$, a bimetallic species of icosahedral geometry (the result of a double expansion), and $2-(\eta-C_5H_5)-2,1,6-CoC_2B_7H_9$, a species containing one <u>less</u> {BH} vertex than the starting material were also isolated. In addition $[1,1'-Co^{III}-(2,3-C_2B_8H_{10})_2]^-$, an anionic complex comprised of two octadecahedra fused through a common metal center, and a complex formulated as $(C_2B_8H_9)-(\eta-C_5H_5)CoC_2B_8H_{10}$ were obtained. This latter compound was spectrally identified as a $(\eta-C_5H_5)CoC_2B_8H_{10}$ complex substituted at a polyhedral carbon atom by a $C_2B_8H_9$ carboranyl moiety also bonded through carbon. A further product, isolated from the expansion reaction carried out at $-80°$, was characterized as $[(\eta-C_5H_5)CoC_2B_8H_{10}-CoC_2B_8H_{10}]^-$ and was proposed to contain both a monodentate and a bidentate $C_2B_8H_{10}$ ligand. This structure has since been confirmed by a single crystal x-ray diffraction study[23] and is the last structure shown in Figure 4.

The polyhedral expansion of a monometallocarborane, $2-(\eta-C_5H_5)-2,1,6-CoC_2B_7H_9$,[24] is depicted in Figure 5. The myriad of products isolated again reflects the complexity of the expansion reaction. The compounds characterized included not only three isomeric bimetallic $(\eta-C_5H_5)_2Co_2C_2B_7H_9$ complexes, the expected expansion products, but also a positional isomer of the starting metalloborane, $2-(\eta-C_5H_5)-2,1,10-CoC_2B_7H_9$; a bimetallic product containing one less {BH} vertex than the starting material, $3,7-(\eta-C_5H_5)_2-3,7,4,5-Co_2C_2B_6H_8$; and the trimetallic icosahedral complex $2,3,5-(\eta-C_5H_5)_3-2,3,5,1,7-Co_3C_2B_7H_9$. The last was the first reported trimetallic carborane complex and suggests that the expansion reaction may be extended to bimetallic and higher polymetallic species.

The polyhedral expansion method has also been applied to synthesize mono- and bimetallic iron complexes,[24] and mixed bimetallic iron-cobalt[25] and nickel-cobalt[26] complexes.

Although strictly not metallocarboranes, two <u>closo</u> icosahedral nickelaboranes have been obtained from <u>closo</u>-anionic boranes using a special application of the polyhedral expansion technique. Reaction of $[(\eta-C_5H_5)Ni(CO)]_2$ with $B_{10}H_{10}{}^{2-}$ and $B_{11}H_{11}{}^{2-}$ yielded $1,2-(\eta-C_5H_5)_2-1,2-Ni_2B_{10}H_{10}$ and $[1-(\eta-C_5H_5)-1-NiB_{11}H_{11}]^{1-}$ respectively. Evidently the $\{(\eta-C_5H_5)Ni\}$ moiety provided the electrons necessary for the expansion of the <u>closo</u>-borane starting materials. The reaction sequence and [11]B nmr spectra of the compounds are shown in Figure 6.

Figure 4. The proposed structures of the products from the polyhedral expansion of $1,6-C_2B_8H_{10}$ (22).

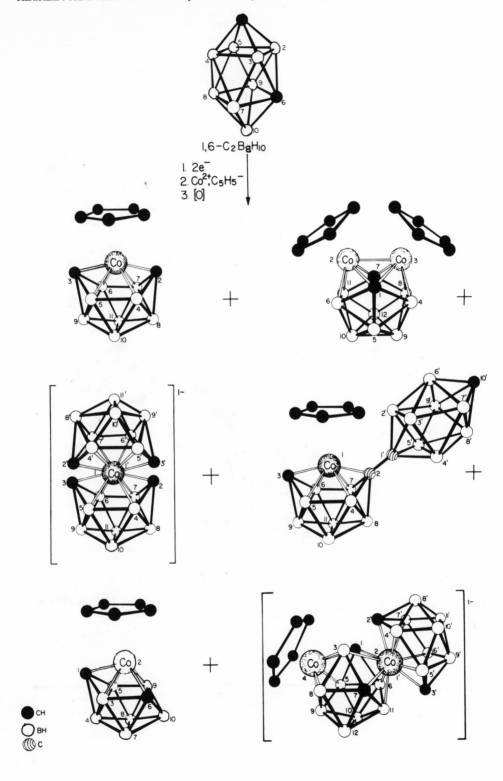

Whereas the polyhedral expansion reaction increases the number
of vertices of a polyhedral carborane or metallocarborane by the addition of $[(\eta-C_5H_5)M](M=CO, Fe, Ni)$ units, another technique has
been developed in which a {BH} unit is removed from one vertex of a
closo-metallocarborane by degradation, producing a nido-complex.
Two electron oxidation of this species provides a closo-metallobor-

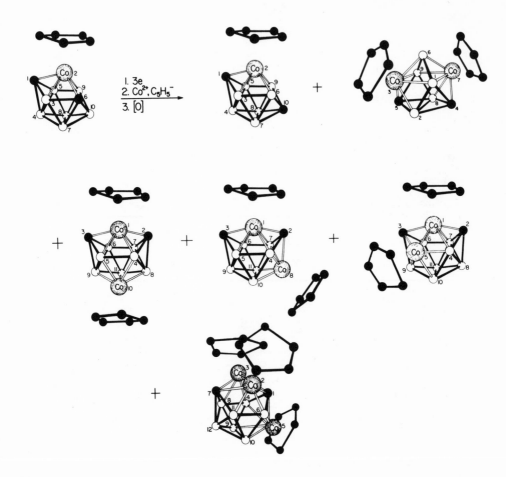

Figure 5. The proposed structures of the products from the polyhedral expansion of $2-(\eta-C_5H_5)-2,1,6-CoC_2B_7H_9$ (24).

Figure 6. Preparation and ^{11}B nmr spectra of $1,2-(\eta-C_5H_5)_2-1,2-$
$Ni_2B_{10}H_{10}$ and $[1-(\eta-C_5H_5)-1-NiB_{11}H_{11}]^{1-}$. The proposed structures
of the products are depicted.

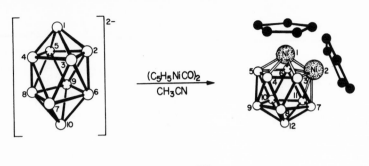

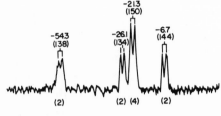

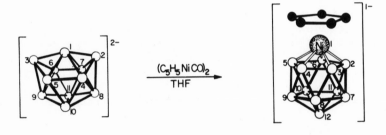

80.5 MHz ^{11}B nmr
(in 1,2-Dimethoxyethane)

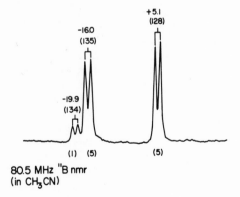

CH
BH

80.5 MHz ^{11}B nmr
(in CH_3CN)

ane containing one less vertex that the starting material (see Figure 4). Not surprisingly, this synthetic method has been dubbed the "polyhedral contraction" reaction.[27]

A good example of this reaction is the sequence depicted in Figure 7. The icosahedral complex, $3-(\eta-C_5H_5)-3,1,2-CoC_2B_9H_{11}$ was degraded[27,28] to the intermediate nido complex, $[9-(\eta-C_5H_5)-9,7,8-CoC_2B_8H_{11}]^-$, which has been isolated and characterized.[29] The two electron oxidation of this intermediate produced a high yield of the octadecahedral complex $1-(\eta-C_5H_5)-1,2,4-CoC_2-B_8H_{10}$. This was in turn degraded[28] to yield nido-$8-(\eta-C_5H_5)-8,6,7-CoC_2B_7H_{11}$, which has also been isolated.[29] The structure of this complex, including the stereochemistry of the briding hydrogens, was unambiguously determined by an x-ray diffraction study carried out at $-160°$.[30] The metallocarborane $2-(\eta-C_5H_5)-2,1,10-CoC_2B_7H_9$ was isolated upon the two electron oxidation of nido-$8-(\eta-C_5H_5)-8,6,7-C_2B_7H_{11}$, (presumably via $6-(\eta-C_5H_5)-6,2,3-CoC_2B_7H_9$, which thermally rearranged to the observed product).

The contraction technique has also been successfully applied to $[3,3'-Co^{III}-(1,2-C_2B_9H_{11})_2]^-$,[28] and a "supercontraction" of the 13-vertex complex $4-(\eta-C_5H_5)-4,1,8-CoC_2B_{10}H_{12}$[31,32] has been observed in which three {BH} units and one {CH} unit were chemically removed in one step.

The degradation of metallocarboranes follows a different course in the presence of excess transition metal salts (with or without excess cyclopentadiene). Upon boron abstraction, the resulting nido-complex combines with the transition metal at the vacant vertex of the polyhedron to form a bimetallic complex with the same number of vertices as the starting material. This reaction, termed "polyhedral subrogation",[25] is actually an extension of the early method of preparation of monometallocarboranes in which a degraded carborane (see Figure 1) was reacted with a transition metal. An example of this synthetic technique is illustrated in Figure 8. In the presence of cyclopentadiene, $FeCl_2$ or $CoCl_2$ and strong alcoholic base, the products from subrogation of $5-(\eta-C_5H_5)-5,1,6-CoC_2B_{10}H_{12}$ were the mixed metallic complex $4,5-(\eta-C_5H_5)_2-4,5,1,6-FeCoC_2B_9H_{11}$[25,33] or the bimetallic complex $4,5-(\eta-C_5H_5)_2-4,5,1,6-Co_2C_2B_9H_{11}$,[9,33] respectively. Note that the products retained the gross "supericosahedral" geometry as dictated by the subrogation sequence. In the presence of $CoCl_2$, subrogation also yielded a trimetallic anion, whose proposed structure (one possible isomer) is shown in the Figure. The polyhedral subrogation reaction has also been applied to $3-(\eta-C_5H_5)-3,1,2-CoC_2B_9H_{11}$[34] and the $[3,3'-Co^{III}(1,2-C_2B_9H_{11})_2]^{1-}$ system.[35]

Electrophilic substitution at boron comprises much of the exo-polyhedral derivative chemistry of the metallocarboranes. As with carboranes,[36,37] the boron atoms prone to electrophilic substitution are those farthest removed from the relatively electropositive carbon

atoms in the polyhedron. These boron atoms possess the highest
ground state electron density.

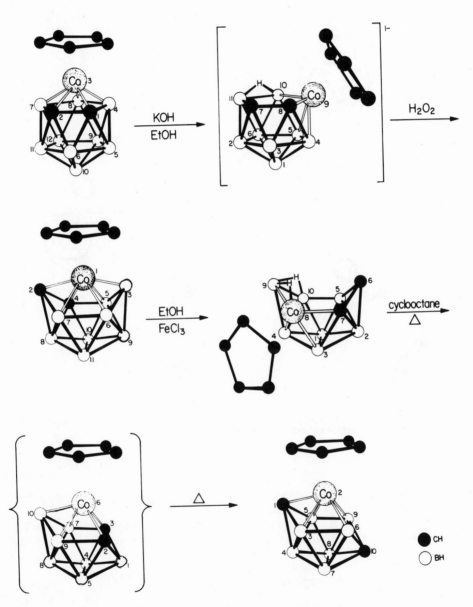

Figure 7. Polyhedral contraction of $3-(\eta-C_5H_5)-3,1,2-CoC_2B_9H_{11}$ and
$1-(\eta-C_5H_5)-1,2,4-CoC_2B_8H_{10}$ (28).

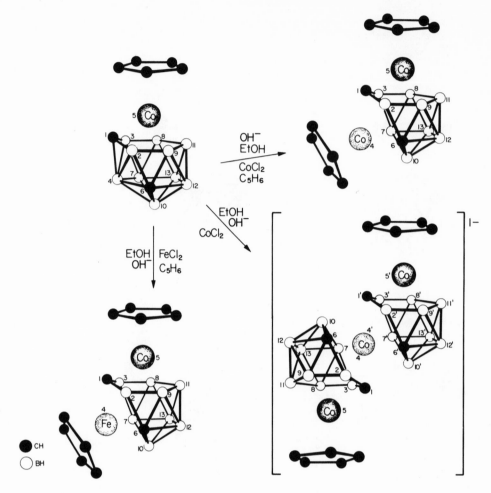

Figure 8.   Polyhedral subrogation of $5-(\eta-C_5H_5)-5,1,6-CoC_2B_{10}H_{12}$ (33) with proposed structures of the products. The <u>trans</u> isomer of the trimetallic complex is depicted.

Treatment of $Rb[3,3'-Co(1,2-C_2B_9H_{11})_2]$ with excess bromine in glacial acetic acid produced a hexabrominated anion.[8] A single crystal x-ray diffraction study[38] confirmed that substitution had occurred at those borons most removed from the carbon atoms, yielding $[3,3'-Co(8,9,12-Br_3-1,2-C_2B_9H_8)_2]^-$.

The zwitterionic complex $\mu-8,8'-S_2CH-3,3'-Co^{III}(1,2-C_2B_9H_{11})_2$ [39,40] in which the dicarbollyl ligands are bridged by the $-SCHS-$ moiety as shown in Figure 9a, was formed in the following reaction:

$$[3,3'-Co(1,2-C_2B_9H_{11})_2]^- + CS_2 + H^+ \xrightarrow{AlCl_3} \mu-8,8'-S_2CH-3,3'-Co-$$

$$(1,2-C_2B_9H_{11})_2 + H_2$$

The electron deficient carbon atoms in the bridging ligand is probably stabilized by the nonbonding electron pairs of the adjacent sulfur atoms. A similar zwitterionic complex was prepared by reaction with acetic acid and acetic anhydride to produce a compound with a briding $-OC^{\oplus}(CH_3)O-$ group.[40]

A final example of electrophilic substitution involved the protonation[41] of $[3,3'-Fe^{II}(1,2-C_2B_9H_{11})_2]^{2-}$ with strong acid followed by treatment with diethyl sulfide to produce the boron substituted complex $8-S(C_2H_5)_2-3,3'-Fe^{III}(1,2-C_2B_9H_{10})(1',2'-C_2B_9H_{11})$, shown in Figure 9b.

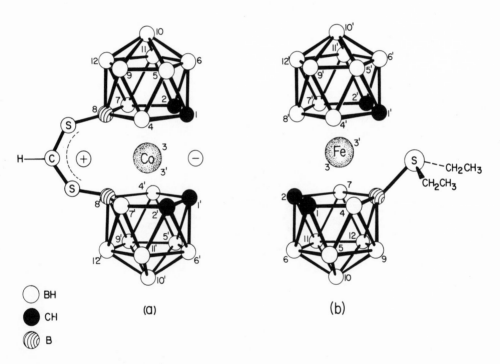

○ BH
● CH
◎ B

(a)   (b)

Figure 9. Products from exopolyhedral electrophilic substitution. (a) $\mu-8,8'-S_2CH-3,3'-Co(1,2-C_2B_9H_{11})_2$ (40); (b) $8-S(C_2H_5)_2-3,3'-Fe(1,2-C_2B_9H_{10})(1',2'-C_2B_9H_{11})$ (41).

Up to this point our concentration had been on the synthesis, characterization, and derivitization of metallocarborane complexes;

little attention was given to their potential applications. The role
of organometallic complexes in homogeneous catalysis has been well
documented, and it was realized that the thermal, oxidative, and
chemical stability of metallocarboranes might make them useful catalysts.

     Prior to the initiation of our exploratory research into the
area of catalysis it was demonstrated that the well known ortho-
metallation[42] of aryl groups in suitable transition metal complexes
could also be effected on a carborane polyhedron. Reaction of [Ir-
$(C_8H_{14})_2Cl]_2$ with $1-P(CH_3)_2-1,2-C_2B_{10}H_{11}$ produced the carboranyl
phosphine analog[43] of $[(C_6H_5)_3P]_3IrCl$; the latter had previously
been shown to undergo facile ortho-metallation.[44] The carboranyl
phosphine species was also found to undergo facile intramolecular
oxidative addition to produce a compound which exhibited a new infrared band, characteristic of an iridium-hydrogen stretch. Deuteration experiments have established that oxidative addition occurs
at a polyhedral {BH} vertex. Figure 10 illustrates this reaction.

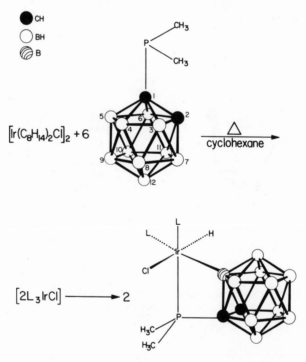

Figure 10. The intramolecular oxidative addition reaction of a B-H
bond (43). One possible isomer of the product is depicted.

The catalytic deuteration of the carboranyl phosphine with deuterium gas and $[(C_6H_5)_3P]_3RuHCl$, a catalyst for the specific ortho-deuteration of triphenylphosphine, incorporated more deuterium into the phosphine than would be possible invoking only an ortho-metallation mechanism. Further studies have shown that a wide variety of homogeneous catalysts can effect hydrogen–deuterium exchange at boron on a wide variety of non–complexing substrates.[45] For example, $[(C_6H_5)_3P]_3RuHCl$ catalyzed the H–D exchange only on boron in $1,2$-$C_2B_{10}H_{12}$. The order of rates of incorporation at available vertices was $B(3,6)>B(4,5,7,11)>B(8,10),B(9,12)$ which is the reverse of that observed for electrophilic substitution.

The preparation[46] of two isomeric hydridometallocarboranes both formulated as $[(C_6H_5)_3P]_2HRhC_2B_9H_{11}$, marked the advent of metallocarboranes as homogeneous catalysts. These complexes were initially synthesized by the oxidative addition reaction of $\{[(C_6H_5)_3P]_3Rh^I\}^+$[47] with $7,8$- or $7,9$-$C_2B_9H_{12}^-$, the source of the metal hydride probably being the bridging proton of the carborane moiety. Shortly thereafter, the iridium congeners were successfully prepared by treatment of $[C_8H_{12}IrCl]_2$ with triphenylphosphine, $C_2B_9H_{12}^-$, and hydrogen.

$$(C_8H_{12}IrCl)_2 + 4(C_6H_5)_3P \xrightarrow[\text{2. } H_2, \Delta]{\text{1. } C_2B_9H_{12}^-} [(C_6H_5)_3P]_2HIrC_2B_9H_{11}$$

The proposed structures of $2,2$-$[(C_6H_5)_3P]_2$-2-H-$2,1,7$-RhC$_2$B$_9$H$_{11}$ and $3,3$-$[(C_6H_5)_3P]_2$-3-H-$3,1,2$-IrC$_2$B$_9$H$_{11}$ with their respective proton nmr spectra in the metal hydride region are depicted in Figure 11.

Benzene solutions of the rhodium complexes at $10^{-3}M$ concentrations were found to readily catalyze the isomerization[46] of terminal olefins to internal olefins. In the presence of a hydrogen atmosphere, both the iridium and rhodium complexes effected the hydrogenation[46] of various alkene substrates as shown in Table Ia.

Both the iridium and rhodium complexes are among the most efficient catalysts we have found to date for the H–D exchange reactions[45] of boron. Table II outlines the preliminary results of catalytic deuterations at boron utilizing various substrates and catalysts.

Preliminary work has also demonstrated that the rhodium complexes are effective catalysts for the hydrosilylation[46] of ketones. The results we have obtained for this catalytic system are presented in Table 1b.

The presence of the carborane moiety bonded to a catalytically active metal may affect the course of catalysis in several ways. The steric effects may be substantial as the carborane cage occupies

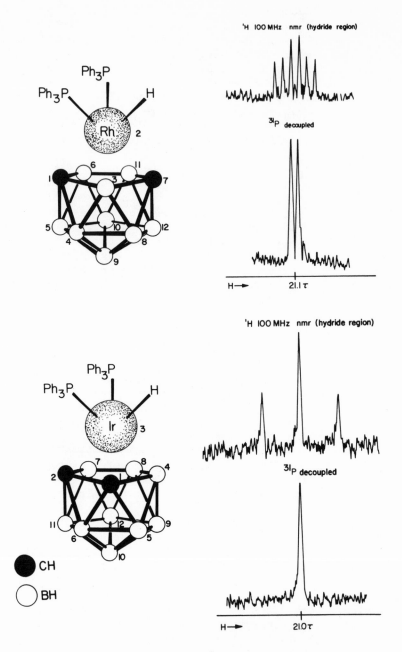

Figure 11. Proposed structures and ^{1}H nmr of the hydride region of $2,2-[(C_6H_5)_3P]_2-2-H-2,1,7-RhC_2B_9H_{11}$ and $3,3-[(C_6H_5)_3P]_2-3-H-3,1,2-IrC_2B_9H_{11}$.

TABLE I

Preliminary Results of Catalytic
Hydrogenations (A) and Hydrosilylations (B)

A

Substrate	Catalyst	Conditions	%reduction
$H_2C=CHOAc$	$2,2-(PPh_3)_2-2-H-2,1,7-RhC_2B_9H_{11}$	a,C_6H_6	~5
	$3,3-(PPh_3)_2-3-H-3,1,2-RhC_2B_9H_{11}$	a,THF	~10
	$1-CH_3-3,3-(PPh_3)_2-3-H-3,1,2-RhC_2B_9H_{10}$	a,C_6H_6	~19
	$1-CH_2OEt-3,3-(PPh_3)_2-3-H-3,1,2-RhC_2B_9H_{10}$	a,C_6H_6	~95
	$1-Ph-3,3-(PPh_3)_2-3-H-3,1,2-RhC_2B_9H_{10}$	a,C_6H_6	100
$H_2C=CHCO_2Et$	$3,3-(PPh_3)_2-3-H-3,1,2-RhC_2B_9H_{11}$	a,C_6H_6	100
	$3,3-(PPh_3)_2-3-H-3,1,2-IrC_2B_9H_{11}$	a,C_6H_6	25
	$2,2-(PPh_3)_2-2-H-2,1,7-IrC_2B_9H_{11}$	a,C_6H_6	16

B

Substrate	Catalyst	Conditions	%reduction
$\overset{O}{\overset{\|}{PhCCH_3}}$	$2,2-(PPh_3)_2-2-H-2,1,7-RhC_2B_9H_{11}$	b	78
	$3,3-(PPh_3)_2-3-H-3,1,2-RhC_2B_9H_{11}$	b	100
	$1-CH_3-3,3-(PPh_3)_2-3-H-3,1,2-RhC_2B_9H_{10}$	b	100
	$1-Ph-3,3-(PPh_3)_2-3-H-3,1,2-RhC_2B_9H_{10}$	b	100

a. Conditions: 40°, 1 atm H_2, 17 hr, 0.6M in substrate, 0.002M
    in catalyst.

b. Conditions: 55°, 1 atm N_2, 17 hr, 1MPh$_2$SiH$_2$, 0.6M in substrate,
    0.002M in catalyst.

a large portion of the catalytic surface of the metal. Electronic
influences exerted by the cage may also be considerable. The electron deficient carborane group may act as an effective <u>trans</u> labil-
izer, thus affecting the bonding interactions of phosphine, olefin,
and hydride ligands on the opposite side of the metal surface.

TABLE II

Preliminary Results of Selected Catalytic Deuterations

Substrate	Catalyst	Conditions[a]	Ave No. D in Product[b]
$1,2-C_2B_{10}H_{12}$	10% Pd on C (1 g)	3 days, 100°	2
	$(PPh_3)_3RhCl$[c]	18 hr, 80°	2
	$(PPh_3)_3RuHCl$	18 hr, 80°	5
	$3,3-(PPh_3)_2-3-H-3,1,2-RhC_2B_9H_{11}$	3 days, 100°	8
	$2,2-(PPh_3)_2-2-H-2,1,7-RhC_2B_9H_{11}$	1 day, 65°	10
$1,7-C_2B_{10}H_{12}$	$(PPh_3)_3RuHCl$	1 day, 65°	10
	$2,2-(PPh_3)_2-2-H-2,1,7-IrC_2B_9H_{11}$	3 days, 100°	8
	$2,2-(PPh_3)_2-2-H-2,1,7-RhC_2B_9H_{11}$	1 day, 100°	10
$1,12-C_2B_{10}H_{12}$	$(PPh_3)_3RuHCl$	1 day, 100°	10
$3-(\eta-C_5H_5)-3,1,2-CoC_2B_8H_{10}$	$(PPh_3)_3RuHCl$	3 days, 100°	5
$B_{10}H_{14}$	$2,2-(PPh_3)_2-2-H-2,1,7-RhC_2B_9H_{11}$	1 day, 100°	6
	$2,2-(PPh_3)_2-2-H-2,1,7-RhC_2B_9H_{11}$[c]	3 days, 100°	3
	$2,2-(PPh_3)_2-2-H-2,1,7-RhC_2B_9H_{11}$[c]	1 day, 100°	6
$(Me_3NH)_2^+[B_{10}H_{10}]^=$	$(PPh_3)_3(CH_3CN)RuHCl$	3 days, 80° (in CH_3CN)	2
Me_3NBH_3	$(PPh_3)_3RuHCl$[c]	1 day, 100°	3

a. Except where noted, reactions were with 1 mmol substrate and 0.05 mmol catalyst in 20 ml of toluene with D_2 bubbling at ~3 ml/min.
b. Estimated from ir, ^{11}B nmr, and mass spectra.
c. Reaction with catalyst was noted.

Although the catalytically active species in these cycles has yet to be identified, we have embarked on a thorough mechanistic study to discover the intriguing role played by the carborane ligand. The complexes described here are the first metallocarboranes which have been shown to exhibit significant catalytic activity, and as this chemistry is still in its infancy, no doubt more efficient metallocarborane catalysts will be tailored in the future.

In the relatively short period of ten years, metallocarborane chemistry has matured into a very active and important specialty. The polyhedral expansion, contraction, and subrogation reactions have allowed the preparation of an enormous number of metallocarboranes. The stability of these products has allowed extensive derivative chemistry and physical studies, thus greatly enriching our knowledge of chemical reactivity, bonding, and structure of the electron deficient boron hydrides. Our successful application of metallocarboranes to homogeneous catalysis has opened another field in boron chemistry, and active researchers will no doubt discover further applications of these compounds.

ACKNOWLEDGMENTS

The authors wish to thank the National Science Foundation, the Army Research Office (Durham), and the Office of Naval Research for their generous support of the research cited in this manuscript.

REFERENCES

1. M. F. Hawthorne, D. C. Young, and P. A. Wegner, J. Amer. Chem. Soc., 87, 1818 (1965).
2. T. J. Kealy and P. L. Pauson, Nature, 168, 1039 (1951).
3. W. Moffitt, J. Amer. Chem. Soc., 76, 3386 (1954).
4. R. A. Wiesboeck and M. F. Hawthorne, J. Amer. Chem. Soc., 86, 1642 (1964).
5. A closo-boron hydride ($B_nH_n^{2-}$, x=n), carborane ($C_2B_nH_{n+2}$, x=n+2), or metallocarborane ($L_yM_mC_2B_nH_{n+2}$, x=m+n+2 where M is a metal and L a ligand), containing x skeletal atoms, assumes a structure in which all x vertices of an exclusively triangular-faced polyhedron (Figure 3) are occupied. A nido-compound containing x skeletal atoms occupies all but one vertex of the polyhedron containing x+1 vertices.
6. M. F. Hawthorne, D. C. Young, P. M. Garrett, D. A. Owen, S. G. Schwerin, F. N. Tebbe, and P. A. Wegner, J. Amer. Chem. Soc., 90, 862 (1968).
7. L. F. Warren, Jr., and M. F. Hawthorne, J. Amer. Chem. Soc., 92, 1157 (1970).
8. M. F. Hawthorne, D. C. Young, T. D. Andrews, D. V. Howe, R. L. Pilling, A. D. Pitts, M. Reintjes, L. F. Warren, Jr., and P. A.

Wegner, J. Amer. Chem. Soc., 90, 879 (1968).

9. a. M. F. Hawthorne and T. D. Andrews, J. Amer. Chem. Soc., 87, 2496 (1965).

    b. M. F. Hawthorne and H. W. Ruhle, Inorg. Chem., 8, 176 (1969).

10. M. F. Hawthorne and R. L. Pilling, J. Amer. Chem. Soc., 87, 3987 (1965).

11. H. W. Ruhle and M. F. Hawthorne, Inorg. Chem., 7, 2279 (1968).

12. M. F. Hawthorne and T. D. Andrews, Chem. Comm., 443 (1965).

13. L. F. Warren, Jr., and M. F. Hawthorne, J. Amer. Chem. Soc., 89, 470 (1967).

14. L. F. Warren, Jr., and M. F. Hawthorne, J. Amer. Chem. Soc., 92, 1157 (1970).

15. L. F. Warren, Jr., and M. F. Hawthorne, J. Amer. Chem. Soc., 90, 4823 (1968).

16. M. F. Hawthorne and T. A. George, J. Amer. Chem. Soc., 89, 7114 (1967).

17. M. F. Hawthorne and A. D. Pitts, J. Amer. Chem. Soc., 89, 7115 (1967).

18. A. D. George and M. F. Hawthorne, Inorg. Chem., 8, 1801 (1969).

19. K. P. Callahan and M. F. Hawthorne, Pure Appl. Chem., in press.

20. M. F. Hawthorne, K. P. Callahan, and R. J. Wiersema, Tetrahedron, in press.

21. K. P. Callahan, W. J. Evans, and M. F. Hawthorne, Ann. N.Y. Acad. Sci., in press.

22. W. J. Evans, G. B. Dunks, and M. F. Hawthorne, J. Amer. Chem. Soc., 95, 4565 (1973).

23. G. Evrard, J. A. Ricci, Jr., I. Bernal, W. J. Evans, D. F. Dustin, and M. F. Hawthorne, J. Chem. Soc., Chem. Comm., 234 (1974).

24. W. J. Evans and M. F. Hawthorne, Inorg. Chem., 13,869 (1974).

25. D. F. Dustin, W. J. Evans, and M. F. Hawthorne, J. Chem. Soc., Chem. Comm., 805 (1973).

26. C. G. Salentine and M. F. Hawthorne, J. Chem. Soc., Chem. Comm., 560 (1973).

27. C. J. Jones, J. N. Francis, and M. F. Hawthorne, J. Chem. Soc., Chem. Comm., 900 (1972).

28. C. J. Jones, J. N. Francis, and M. F. Hawthorne, J. Amer. Chem. Soc., 94, 8391 (1972).

29. C. J. Jones, J. N. Francis, and M. F. Hawthorne, J. Amer. Chem. Soc., 95, 7633 (1973).

30. K. P. Callahan, F. Y. Lo, C. E. Strouse, A. L. Sims, and M. F. Hawthorne, submitted for publication in Inorg. Chem.

31. D. F. Dustin and M. F. Hawthorne, J. Chem. Soc., Chem. Comm., 1329 (1972).

32. D. F. Dustin and M. F. Hawthorne, Inorg Chem., 12, 1380 (1973).

33. D. F. Dustin and M. F. Hawthorne, J. Amer. Chem. Soc., 94, 3462 (1974).

34. C. J. Jones and M. F. Hawthorne, Inorg. Chem., 12, 608 (1973).

35. J. N. Francis and M. F. Hawthorne, Inorg. Chem., 10, 863 (1971).

36. W. N. Lipscomb, "Boron Hydrides", Benjamin, New York, 1966, p. 68.

37. G. B. Dunks and M. F. Hawthorne, Inorg. Chem., 9, 893 (1970).
38. B. G. DeBoer, A. Zalkin, and D. H. Templeton, Inorg. Chem., 7, 2288 (1968).
39. M. R. Churchill, K. Gold, J. N. Francis, and M. F. Hawthorne, J. Amer. Chem. Soc., 91, 1222 (1969).
40. J. N. Francis and M. F. Hawthorne, Inorg. Chem., 10, 594 (1971).
41. M. F. Hawthorne, L. F. Warren, Jr., K. P. Callahan, and N. F. Travers, J. Amer. Chem. Soc., 93, 2407 (1971).
42. G. W. Parshall, Accounts Chem. Res., 3, 139 (1970).
43. E. L. Hoel and M. F. Hawthorne, J. Amer. Chem. Soc., 95, 2712 (1973).
44. M. A. Bennett and D. L. Milner, J. Amer. Chem. Soc., 91, 6983 (1969).
45. E. L. Hoel and M. F. Hawthorne, J. Amer. Chem. Soc., 96, 0000 (1974).
46. T. E. Paxson and M. F. Hawthorne, J. Amer. Chem. Soc., 96, 0000 (1974).
47. P. Legzdins, R. W. Mitchell, G. L. Rempel, J. D. Ruddick, and G. Wilkinson, J. Chem. Soc. A, 3224 (1971).

DISCUSSION

DR. COLLMAN: Are there bimetallic systems in which you have introduced two oxidation states?

DR. HAWTHORNE: Yes.

DR. COLLMAN: Do you know whether the electrons communicate?

DR. HAWTHORNE: Yes.

DR. COLLMAN: Do you see the electronic absorption band in the near infrared?

DR. HAWTHORNE: I don't think we see them out at that low a frequency. We see some weird things closer in to the visible. Most of the work we have done here has been NMR and some EPR.

DR. HALPERN: Do you have any idea how your hydrogenation catalyst might work? Most of the rhodium catalysts are thought to go a Rh^{III}-Rh^{I} cycle.

DR. HAWTHORNE: Yes, we think there is a scheme you can write in which this thing tautaumerizes to form a RH^{I} which is bonded to one side of the open face. The hydrogen jumps back and forth from one side to another.

DR. HALPERN: It was not clear how your ligand permits this.

DR. HAWTHORNE: We are doing some kinetic studies and designing some model systems which might illustrate this mechanism.

DR. HALPERN: It is of course not actually essential that you go through that Rh^{III}-Rh^{I} scheme, although that is the common one. You may remain in a Rh^{III} cycle using heterolytic splitting of hydrogen. We know of such a catalyst, not in a Rh system but in a Ru(II) system. It would be interesting if this one works that way.

be interesting if this one works that way.

DR. HAWTHORNE: We should mention also analogous Ru and iridium systems. This also, as you might expect, isomerizes olefins readily.

DR. COLLMAN: There might be a test for that idea. That is whether a simple innocent olefin will react - if this thing can go into a RH^I state it might be able to react with simple olefins, but if it stayed in the RH^{III} state maybe only alpha-beta unsaturated esters and acids will react.

DR. HALPERN: There are other tests that work, in favorable cases at least. If you go through a heterolytic splitting cycle you always get deuterium incorporation if you work in a deuterated medium whereas frequently in oxidative addition reactions, the hydrogen stays on the metal and is incorporated.

DR. HAWTHORNE: We have a iridium system which corresponds to this which we think may actually capture olefins to give us a stable intermediate.

DR. WHITESIDES: Do you know that stuff really is the catalyst?

DR. HAWTHORNE: Well, I assume it is. What were you driving at?

DR. WHITESIDES: You say that you can recover it. How do you know that you are not getting a few percent of something going away and giving a small amount of active catalyst?

DR. HAWTHORNE: Well, we don't know for sure but we can recover the stuff quantitavely.

DR. PAXON: Well, you're probably thinking of dissociation back to the (-1) anion in the formation of bis-triphenylphosphine rhodium cation which would not be all that neat but we have not proved it conclusively. If it did occur, you would think that the excess monoanion would stop the catalytic reaction, but this has no apparent effect on it.

DR. WHITESIDES: What I was thinking of is the fact that toluenechromium tricaryonyl and things like this act as hydrogenation catalysts by dissociating the arene part of things rather than by dissociating the carbonyls and you can imagine a similar type mechanism taking place here dissociating the carborane part of things.

DR. HAWTHORNE: Then we really would not see any optical activity in the product, would we? We will soon know because we will have some better systems that have both carbon atoms substituted.

DR. KUMADA: Can the cyclopentadienyl ring of the metal carborane undergo aromatic substitution?

DR. HAWTHORNE: When we attempt electrophylic substitution it normally goes to the boron portion of the molecule - the BH group. No CH or CP substitution.

DR. YAMAMOTO: Can you replace the upper portion of the carborane complexes,with other ligands - other hydrocarbon ligands than cyclopentadiene, such as π-allyls?

DR. HAWTHORNE: We are just getting into that at the moment. Carbonyls of course, we have many examples. We now have phosphine derivatives and some metal hydrides. As far as akyl substituents on metals, we have no evidence for those at the moment. When we get back to the lab, we might be able to do so, but it is still anybodys guess.

DR. OSBORN: It would be nice if on those bimetallic systems you could have other ligands apart from cyclopentadienyl which are more labile.

DR. HAWTHORNE: What we want are some catalytic systems in which we have more than one metal center - we might get "push-pull" effects or specific coordination sites due to the presence of a second metal nucleus.

DR. OSBORN: Is the absence of the complexes of the other earlier transition metal elements in your talk significant, or is it that you have concentrated on the latter elements? For example, can you make an analog of titanocene?

DR. HAWTHORNE: Yes, we have an analog of titanocene. We are doing the structure right now. We have 4,000 reflections at the moment. It is very stable.

DR. OSBORN: This is a di anion?

DR. HAWTHORNE: This is using the ligand $B_{10}C_2H_{12}^{-2}$, wo we have a 13 vertex system and that works very nicely with the earlier transition metals.

DR. OSBORN: Does that have any peculiar properties?

DR. HAWTHORNE: Well, we have not seen a reaction yet of nitrogen. We are looking at carbonyl derivatives and perhaps some hydrides. That is not very far long and I did not really wish to talk about it today - but it does exist. In a similar fashion vanadium forms as good a complex as does zirconium. We can work down in that region by changing the ligand system.

DR. IBERS: Does the last rhodium hydride react with oxygen at all?

DR. HAWTHORNE: It is air stable - it will last a day, anyway.

DR. IBERS: Can you substitute that hydrogen by halogen - chloride for example?

DR. HAWTHORNE: Yes.

DR. KUMADA: Did you make group 4-5 or 6 metal analog to the cyclopentadienide?

DR. HAWTHORNE: We have not.

DR. KUMADA: You have not tried it?

DR. HAWTHORNE: We may now, because we know we can make complexes in the same groups. Now it is logical to see if we can put cyclopentadienide on as well.

DR. KUMADA: Is the C_2B_9 system more stable than the C_2B_{10} system?

DR. HAWTHORNE: ·Yes, it appears to be.

NOVEL RHODIUM AND PALLADIUM COMPLEXES WITH BENZOYL AND THIOBENZOYL

ISOCYANATES AS LIGANDS

Seiji Hasegawa, Kenji Itoh[*] and Yoshio Ishii

Department of Synthetic Chemistry, Faculty of Engineering, Nagoya University. Furo-cho, Chikusa-ku, Nagoya 464,
Japan

ABSTRACT

The reaction between benzoyl isocyanate and $RhCl(Ph_3P)_3$ afforded
a complex $[RhCl(Ph_3P)_2(PhCO\cdot NCO)]_2$ (1), which was converted to
$[(bipy)Rh(Ph_3P)_2(PhCO\cdot NCO)]^+BPh_4^-$ (2a) by bipyridine and $NaBPh_4$. The
latter complex was also obtained by the reaction of $[(bipy)Rh(Ph_3P)-_2]^+$ with benzoyl isocyanate. An analogous palladium complex (bipy)-
$Pd(PhCO\cdot NCO)$ (3) was prepared in the reaction of Pd_2(dibenzylideneace-
tone)$_3$ with bipyridine and benzoyl isocyanate. 2-Phenylthiazoline-
4,5-dione(4), a precursor of thiobenzoyl isocyanate, gave a complex
$RhCl(CO)(Ph_3P)_2(PhCS\cdot NCO)$(5) in the reaction either with $RhCl(Ph_3P)_3$
or with $RhCl(CO)(Ph_3P)_2$. A cationic complex $[(bipy)Rh(Ph_3P)_2(PhCS\cdot NCO)]^+BPh_4^-$ (6a) was also obtained. The similar reactions of Pd_2-
$(DBA)_3$ with 4 and bipyridine or o-phenanthroline afforded $(N-N)Pd$-
$(PhCS\cdot NCO)$ complexes (7). Most of these novel complexes have a metallocyclic structure composed of M-O-C-N-C or M-S-C-N-C skeletons.

INTRODUCTION

No systematic investigation of the coordinating properties of
benzoyl and thiobenzoyl isocyanates with low-valent transition metals
have been reported, although some transition metal complexes of simple heterocumulene compounds were known.[1-5] Manuel originally reported $[(CO)_3Fe(PhNCO)]_2$ in the reaction of phenyl isocyanate with
$Fe_3(CO)_{12}$,[1a] however, later structural investigations[1b-1d] concluded
that the original complex was a diphenylurea complex of diironhexa-
carbonyl. Similar type of urea complexes were isolated recently in
the case of arylsulfonyl isocyanates.[1e] Baird and Wilkinson[2] prepared π- and S-bonded phenyl isothiocyanate complexes of nickel,

palladium, platinum, rhodium, and iridium. The preparation and X-ray study of keteneimine complexes of iron were achieved by Otsuka.[3-4] Bycroft and Cotton[5] obtained carbodiimide palladium(II) complexes with the formulae $(R-N=C=N-R')_2PdX_2$. The representative structures of some transition metal complexes induced from simple heterocumulene compounds are shown below.

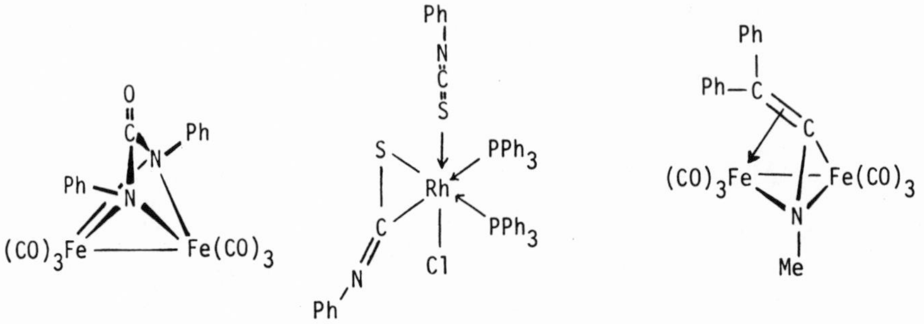

On the other hand, the five-membered metallocyclic complexes containing heteroatoms as the ring member have attracted much attention in recent years.[6-9] In particular, N-acylhydrazine or diazenes were found to be an excellent precursor to generate various five-membered metallocyclic systems (M-O-C-N-N), for instance, $[(PhCO-N=N)_2Cu]^+$[6], $MoCl_2(N-Ar)(ArCO-N=N-Ar)(PhMe_2P)$[7], $[(diphos)_2WCl(N=N-COR)]$[8] and $Pt(Ph_3P)_2(PhCO-N=N-COPh)$.[9] The five-membered structure was definitely concluded by Ittel and Ibers[9] for the final complex by means of X-ray structural determination.

The following canonical structures (X=O or S) can be represented for (thio)benzoyl isocyanate, and are corresponding to three possible modes of coordination.

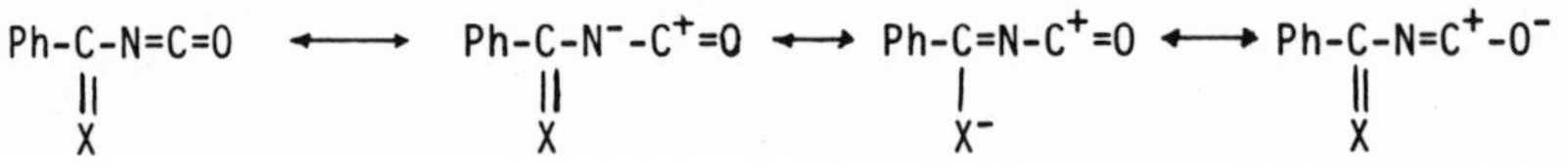

Our studies on the addition-elimination reactions of benzoyl and thiobenzoyl isocyanates with group IVb organometallic compounds disclosed a lot of interesting results indicative of the occurrence of the 1,4-addition of these heterocumulene system(mode b)[10], which had stimulated us to investigate the interaction of these heterocumulene systems with low-valent transition metals.

In this context, Collman et al.[11] obtained $IrCl(Ph_3P)_2(ArCO \cdot NCO)$ in the reaction between Vaska's complex and aroyl azide, and proposed the following structure in which nitrogen and carbonyl carbon atoms of the isocyanate group coordinated to iridium atom (mode a). Another probability, the 1,4-addition of (thio)benzoyl isocyanate by mode b, should induce a new five-membered metallocyclic complex containing two heteroatoms (N and O or S).

In this paper, the results of reactions between benzoyl or thiobenzoyl isocyanate and various rhodium(I) or palladium(0) complexes are described.

RESULTS AND DISCUSSION

Reactions of Benzoyl Isocyanate with Rhodium(I) Complexes

The reaction of tris(triphenylphosphine)chlororhodium; RhCl-$(Ph_3P)_3$ with excess of benzoyl isocyanate[15] in benzene at room temperature under nitrogen gave yellow needle-like crystals of $[RhCl-(Ph_3P)_2(PhCO\cdot NCO)]_2$ (1) in 89% yield. From large lower frequency shifts of infrared absorptions: $\nu(NCO)$ 1710 and $\nu(CO)$ of benzoyl group 1400 cm^{-1} (absorptions of the free ligand were 2240 and 1690 cm^{-1}, respectively), both of the isocyanate and the benzoyl carbonyl groups of benzoyl isocyanate may coordinate to rhodium atom, just like the keteneimine group coordination to cobalt in the case of $\pi C_5H_5Co(Ph_2C=C=N-Me)$[3] (with a change of $\nu(CCN)$ from 1998 to 1565 cm^{-1} on coordination). The complex 1 is very stable in air and in solvent. When 1 was heated up to 200° under reduced pressure, benzoyl isocyanate was recovered quantitatively.

Evidence for the structure of 1 was supplemented by the following reaction. When excess of sodium tetraphenylborate and bipyridine(bipy) were added to 1 in CH_2Cl_2-CH_3OH at room temperature, pale yellow plate-like crystals of $[(bipy)Rh(Ph_3P)_2(PhCO\cdot NCO)]^+BPh_4^-$ (2a) were isolated in 72% yield. The $\nu(CO)$ absorption of 2a (1623 cm^{-1}) implies the acyl-type bonding with rhodium metal as a five-membered ring structure; Rh-O-C=N-C as depicted in scheme 1. The conversion from 1 to 2a could be followed by recording the infrared spectrum of the reaction mixture at different times (Figure 1). As shown in Figure 1, $\nu(CO)$ of 1 at 1710 cm^{-1} began to disappear gradually, in accordance with the increase of $\nu(CO)$ of 2a at 1623 cm^{-1}, which means that bipyridine displaces the bridging oxygen atom and the

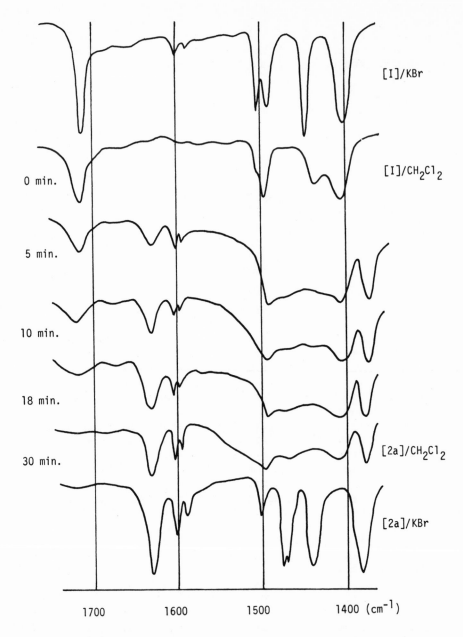

Figure 1. Variation of infrared spectrum with time for the conversion from [RhCl(Ph$_3$P)$_2$(PhCO·NCO)]$_2$ (1) to [(bipy)Rh(Ph$_3$P)$_2$(PhCO·NCO)]$^+$BPh$_4$$^-$ (2a) by bipyridine and sodium tetraphenylborate in CH$_2$Cl$_2$-CH$_3$OH.

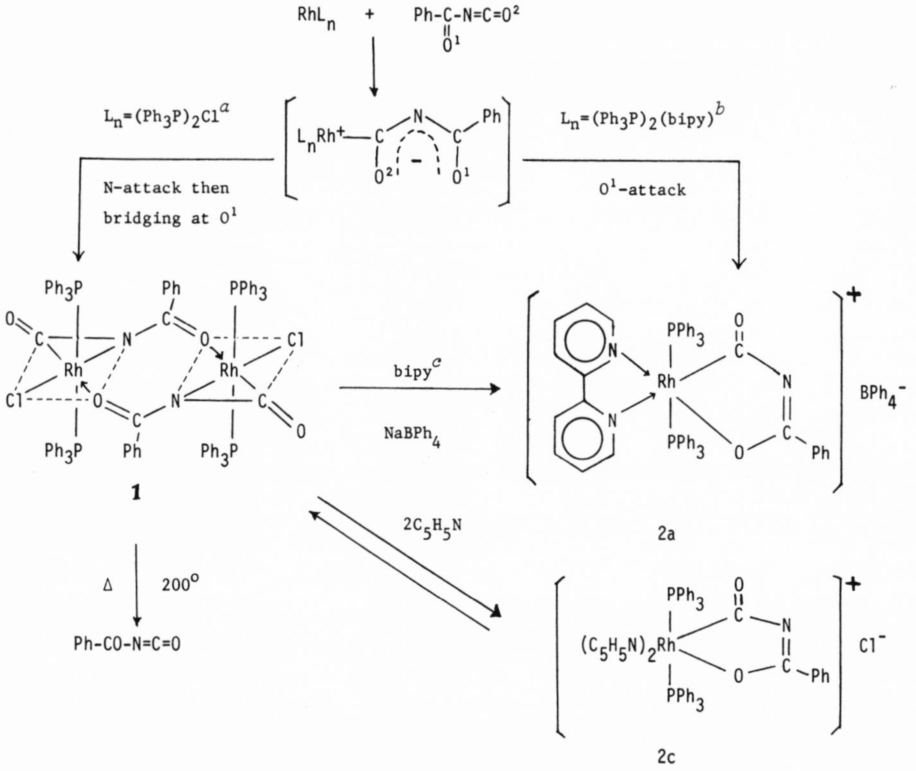

Scheme 1.   [a]$Rh(Ph_3P)_3Cl-Ph_3P$.   [b]$[(bipy)Rh(cyclooctadiene)]^+ + 2Ph_3P-$
cyclooctadiene. [c]When o-phenanthroline was used instead of bipyridine, $[(o-phen)Rh(PhCO \cdot NCO)]^+BPh_4^-$ (2b) was obtained.

nitrogen atom of 1, followed by the bond formation between Rh metal
and benzoyl oxygen with simultaneous liberation of chloride anion
which was trapped by sodium cation, to result the more stable 1,4-
metallocyclic complex(2a).

The cationic complex(2a) could be prepared independently by the
reaction of $[(bipy)Rh(Cyclooctadiene)]^+BPh_4^{-16}$ with benzoyl isocyanate under excess of triphenylphosphine in refluxing CH_2Cl_2 in lower
yield(34%).

An analogous cationic complex (2b) which involves o-phenanthro-
line(o-phen) as a stabilizing ligand instead of bipy was obtained
in 73% yield from 1, o-phen and sodium tetraphenylborate($\nu(CO)$ 1628
and 1379 cm^{-1}). In contrast to the reaction with bipy or o-phen,
the reaction between 1 and pyridine in CH_2Cl_2 produced a reversible
change to $[(py)_2Rh(PPh_3)_2(PhCO \cdot NCO)]^+Cl^-(\nu(CO)$ at 1625 cm$^{-1})$(2c) in-

stantaneously, which returned to 1 on recrystallization from CH_2Cl_2-$(C_2H_5)_2O$.

These reactions were summarized in Scheme 1. The nucleophilic attack of rhodium(I) complexes on the carbon atom of the isocyanate group may be the primary step, followed by N,O^1-attack (more precisely, at first N-attack and the following O^1-bridging to dimerization) in the case of $RhCl(PPh_3)_3$, or by O^1-attack on the $[(bipy)$-$Rh(PPh_3)_2]^+$ species. Due to dissociation of one molecule of triphenylphosphine from $RhCl(PPh_3)_3$, the resulting coordinatively unsaturated $RhCl(PPh_3)_2$ may accept benzoyl isocyanate as a tridentate ligand $(C,N$ and $O^1)$[mode a and b] to complete six coordination around Rh(III), while $[(bipy)Rh(PPh_3)_2]$ has a coordination number of four and benzoyl isocyanate behaves as a bidentate ligand $(C$ and $O^1)$ [mode b]. Because the metal basicity of $Rh(CO)(Ph_3P)_2Cl$ or Rh^+-$(Ph_3P)_2$(cyclooctadiene) are lower than that of $RhCl(Ph_3P)_3$ or Rh^+-$(Ph_3P)_2$(bipy), the former did not react with benzoyl isocyanate. The X-ray structure of 1 and 2a is in progress by Professor J. A. Ibers at Northwestern University.

Reactions of Benzoyl Isocyanate with
Tris(dibenzylideneacetone)dipalladium

Tris(dibenzylideneacetone)dipalladium; $Pd_2(DBA)_3$ was found as useful complex in oxidative addition reactions, ligand exchange reactions and preparation of (p-quinone)PdL_2 and palladiacyclopentadiene complexes[12-13]. When an excess of bipyridine was added to an ether solution of $Pd_2(DBA)_3$ and benzoyl isocyanate under nitrogen at room temperature, orange-colored crystals of (bipy)$Pd(PhCO \cdot NCO)$-$1/2$ CH_2Cl_2 (3) was obtained in 75% yield after recrystallization from methylenechloride. In this case, benzoyl isocyanate acts as a bidentate ligand(C and O^1) to form stable 1,4-coordination of benzoyl isocyanate to square planar palladium (mode b; $\nu(CO)$ 1630 cm^{-1}).

No stable complex was obtained when an excess of triphenylphosphine was used instead of bipyridine in the above-mentioned reaction.

Reactions of 2-Phenylthiazoline-4,5-dione with Rh(I) Complexes

2-Phenylthiazoline-4,5-dione(4) is a precursor of thiobenzoyl isocyanate which was generated by heating 4 in refluxing methylcyclohexane.[14]

In the reaction between $RhCl(Ph_3P)_3$ and an excess of 4 in benzene at room temperature under nitrogen, a precipitate of yellow crystals of $RhCl(CO)(Ph_3P)_2(PhCS \cdot NCO)$ (5) was obtained in 50% yield. The infrared spectrum of 5 shows the acyl carbonyl absorption at 1623 and Rh-CO band at 2070 cm^{-1}.

The same complex 5 could also be prepared in 64% yield in the reaction of 4 with $RhCl(CO)(Ph_3P)_2$ in refluxing toluene with the evolution of carbon monoxide. On heating 5 in refluxing chloroform, decomposition to thiobenzoyl isocyanate and $RhCl(CO)(Ph_3P)_2$ was observed.

On the other hand, the reaction of 4 with $[(bipy)Rh(COD)]^+BPh_4^-$ or $[(o-phen)Rh(COD)]^+BPh_4^-$ in the presence of an excess triphenylphosphine in refluxing mixed solvent $(CH_2Cl_2:$ toluene = 1:2) for 6 hr gave yellow needle-like crystals of $[(bipy)Rh(Ph_3P)_2(PhCS \cdot NCO)]^+$-$BPh_4^-$ (6a ; 69% yield) or yellowish prism-like crystals of $[(o-phen)Rh(Ph_3P)_2(PhCS \cdot NCO)]^+BPh_4^-$ (6b ; 85% yield), respectively. Their structures were identified by analysis and the $\nu(CO)$ absorption bands (6a 1620 ; 6b 1615 cm^{-1}) indicative of a five-membered metallocyclic structure(mode b).

These results were summarized in Scheme 2.

Reactions of 2-Phenylthiazoline-4,5-dione
with Tris(dibenzylideneacetone)dipalladium

When the benzene solution of $Pd_2(DBA)_3$ with an excess of bipyridine and 4 was heated to 50°, evolution of carbon monoxide occurred violently. The product was reddish needle-like crystals of (bipy)Pd(PhCS·NCO) (7a) in 92% yield. The same complex (7a) was also obtained in 63% yield in the reaction of $Pd_2(DBA)_3$ with thiobenzoyl isocyanate, generated in situ from 4, in the presence of bipyridine. When o-phenanthroline was used instead of bipyridine by the reaction of 4 with $Pd_2(DBA)_3$, (o-phen)Pd(PhCS·NCO) (7b) was obtained in 84% yield.

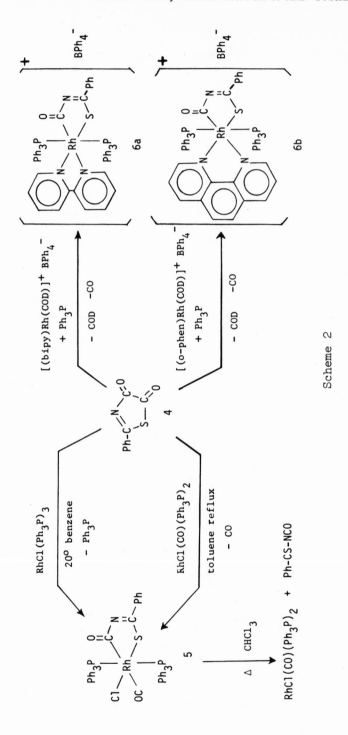

Scheme 2

Table I. Analytical Results for Benzoyl or Thiobenzoyl Isocyanate Complexes of Rhodium and Palladium

Complex	Yield (%)	M. p. (°C)	Color	Analysis: Found(Calcd.)			IR(KBr) ν(C=O)(cm^{-1})
				C	H	N	
[RhCl(Ph$_3$P)$_2$(PhCO·NCO)]$_2$ (1)	89	206-208(dec.)	Yellow	65.18 (65.24)	4.24 (4.36)	1.76 (1.73)	1710
[(bipy)Rh(Ph$_3$P)$_2$(PhCO·NCO)]$^+$ BPh$_4$$^-$(2a)	72	187-188	Pale Yellow	74.64 (74.95)	5.25 (5.08)	3.50 (3.36)	1623
[(o-phen)Rh(Ph$_3$P)$_2$(PhCO·NCO)]$^+$ BPh$_4$$^-$(2b)	71	212-214(dec.)	Pale Yellow	75.55 (75.45)	5.25 (4.99)	3.18 (3.38)	1628
(bipy)Pd(PhCO·NCO)· 1/2 CH$_2$Cl$_2$ (3)	75	202-207(dec.)	Orange	48.36 (49.14)	3.12 (3.24)	9.34 (9.29)	1630
RhCl(CO)(Ph$_3$P)$_2$(PhCS·NCO) (5)	50	210-212(dec.)	Yellow	63.41 (63.28)	4.11 (4.13)	1.62 (1.64)	1623
[(bipy)Rh(Ph$_3$P)$_2$(PhCS·NCO)]$^+$ BPh$_4$$^-$ (6a)	69	197-199(dec.)	Yellow	73.82 (74.00)	5.08 (5.02)	3.21 (3.32)	1620
[(o-phen)Rh(Ph$_3$P)$_2$(PhCS·NCO)]$^+$ BPh$_4$$^-$(6b)	85	217-219(dec.)	Yellow	74.24 (74.48)	4.88 (4.92)	3.32 (3.26)	1615
(bipy)Pd(PhCS·NCO) (7a)	63	185-190(dec.)	Red	50.44 (50.78)	3.20 (3.08)	9.79 (9.87)	1620
(o-phen)Pd(PhCS·NCO) (7b)	84	200-205(dec.)	Red	53.59 (53.41)	2.81 (2.91)	9.31 (9.34)	1630
(PhN=CMe-MeC=NPh)Pd⟨$^{CO-CO}_{S-C}$⟩$^N_{Ph}$ (8)	76	159-163(dec.)	Brown	55.77 (56.24)	4.19 (3.96)	8.05 (7.87)	1660

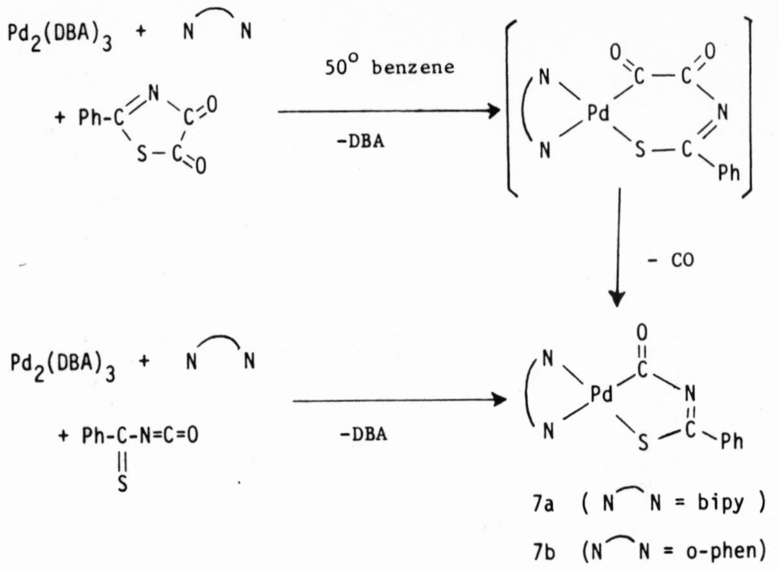

Scheme 3

However, when biacetyldianil was used in the analogous reaction, no evolution of carbon monoxide was observed. The sole product was brownish needle-like crystals of the complex with the composition of [(PhN=CMe-MeC=NPh)Pd($C_9H_5NO_2S$)] (8) in 76% yield. In the infrared spectrum, a broad absorption band at 1660 cm^{-1} is quite different from those of either 7a or 7b. Consequently, 8 may correspond to the unidentified reaction intermediate shown in brackets in Scheme 3. The complex 8 did not evolve carbon monoxide upon heating, only caused decomposition to metallic palladium.

REFERENCES

1. (a) T. A. Manuel, Inorg. Chem., 3, 1703 (1964).
   (b) J. A. J. Jarvis, B. E. Job, B. T. Kilbourn, R. H. B. Mais, P. G. Owston, and P. F. Todd, Chem. Comm., 1149 (1967).
   (c) J. Piron, P. Piret, and M. Van Meerssche, Bull. Soc. Chim. Berges., 76, 505 (1967).
   (d) W. T. Flannigan, G. R. Knox, and P. L. Pauson, Chem. & Ind., 1094 (1967).
   (e) W. Beck, W. Hieber, S. Cenini, F. Porta, and G. LaMonica, J. Chem. Soc. Dalton Trans., 298 (1974).
2. M. C. Baird, G. Hartwell, and G. Wilkinson, J. Chem. Soc (A)., 865, 2037 (1967).
3. S. Otsuka, A. Nakamura, and T. Yoshida, J. Organometal. Chem., 7, 339 (1967).

4. K. Ogawa, A. Torii, H. Kobayashi-Tamura, T. Watanabe, T. Yoshida, and S. Otsuka, J. Chem. Soc (D), Chem. Comm., 991 (1971).
5. B. M. Bycroft and J. D. Cotton, J. Chem. Soc. Dalton Trans., 1867 (1973).
6. R. J. Baker, S. C. Nyburg, and J. T. Szymanski, Inorg. Chem., 10, 138 (1971).
7. J. Chatt and J. D. Dilworth, J. Chem. Soc. Chem. Comm., 549 (1972).
8. J. Chatt, G. A. Heath, and G. J. Leigh, ibid., 444 (1972).
9. S. D. Ittel and J. A. Ibers, Inorg. Chem., 12,2290 (1973).
10. I. Matsuda, K. Itoh, and Y. Ishii, J. Chem. Soc (C)., 701 (1969) and 1870 (1971); Tetrahedron Letts., 2675 (1969); J. Organometal. Chem., 19, 339 and 347 (1969); J. Chem. Soc. Perkin Trans., 1. 1678 (1972).
11. J. P. Collman, M. Kubota, F. D. Vastine, J. Y. Sun, and J. W. Kang, J. Amer. Chem. Soc., 90, 5430 (1968).
12. (a) Y. Takahashi, Ts. Ito, S. Sakai, and Y. Ishii, J. Chem. Soc (D). Chem. Comm., 1065 (1970).
    (b) Ts. Ito, S. Hasegawa, Y. Takahashi, and Y. Ishii, ibid., 629 (1972).
    (c) T. Ukai, H. Kawazura, Y. Ishii, J. J. Bonnet, and J. A. Ibers, J. Organometal. Chem., 65, 253 (1974).
13. K. Moseley and P. M. Maitlis, J. Chem. Soc (D). Chem. Comm., 1604 (1971) and J. Chem. Soc. Dalton Trans., 169 (1974).
14. J. Goerdeler and H. Schenk, Chem. Ber., 98, 2954 (1965).
15. A. J. Speziale and L. R. Smith, J. Org. Chem., 27, 3742 (1963).
16. C. Cocevar, G. Mestroni, and A. Camus, J. Organometal. Chem., 35, 389 (1972).

DISCUSSION

DR. PARSHALL: Your results with the benzoyl isocyanate are very interesting in helping to explain what is happening with the carbon dioxide complexes that Professor Volipin has reported. There is quite a spectrocopic similarity between some of your products and I wonder if perhaps there is not a structural similarity?

DR. ITOH: I think, in this case where the carbon dioxide is coordinating to a transition metal, the pi-type bonding complexes of isocyanates are similar. However, I think the pi-bonding type of CO_2 is close to our system, in particular for Wilkinson's rhodium complex. But in the case of 5 membered metal systems, we cannot guess, and I am not sure.

DR. PARSHALL: Isn't the first intermediate in the CO_2 system very much like the first intermediate which you write in which the nucleophilic metal has attacked the carbon atom? Then a secondary product is formed from that?

DR. ITOH: Yes. We did some work on the addition reactions of

heterocumulenes in group IV organometallics. Benzoyl and thio-ben-
zoyl isocyanates are very electrophylic agents, and the most reactive center is the carbon atom at that position; I am sure that this
is the first step. Of course, in the case of generation of pi-bonding isocyanate complexes simultaneous formation of carbon rhodium
and nitrogen rhodium bonds occurs and this is quite different in the
case of five membered metallocyclic formation.

DR. IBERS: With respect to your compound 8 might you expect

easy expulsion of carbon dioxide from that compound to form a five
membered ring?

DR. ITOH: No, not really.

DR. COLLMAN: You would need substantial rearrangement to eli-
minate CO_2. It does not look likely (general murmur of agreement.)

DR. IBERS: You expect the two adjacent carbonyls to be stable?

DR. COLLMAN: Well, alpha dicarbonyls are certainly stable e-
nough.

DR. IBERS: Yes, but they are not next to a metal, either.

DR. WHITESIDES: Are there any conditions in which you can displace the isothiocyanates or related species from these? Are there
exchange reactions?

DR. ITOH: They are quite stable ligands. In particular the
formation of a five membered ring is extremely stable, and pyrolysis
can be achieved only at 200° in vacuum. We have treated it with
phosphines in solution, and saw only bridge cleavage - no exchange
of the coordinated heterocumulenes.

POLYCYANOVINYL TRANSITION METAL DERIVATIVES

R. B. King

Department of Chemistry, University of Georgia, Athens,
Georgia 30602

An important principle in cyanocarbon chemistry[1] is the analogy between the oxygen atom and the dicyanomethylene moiety in many
compounds.[2,3] On the basis of this analogy polycyanovinyl halides
(I) correspond to acyl halides and thus should be reactive towards
nucleophiles in contrast to ordinary vinyl halides, which are relatively unreactive towards nucleophiles. Such considerations suggested reactions of polycyanovinyl halides with metal carbonyl anions[4] as a method for the preparation of novel polycyanovinyl transition metal derivatives.

I	II	III

In accord with such expectations reactions of the metal carbonyl anions $Mn(CO)_5^-$ and $C_5H_5M(CO)_3^-$ (M=Mo and W) with the polycyanovinyl halides I (X=Cl, H, and CN) led to the isolation of an extensive series of polycyanovinyl transition metal derivatives of
structural types II and III summarized in Table 1.[5,6] More recently some pentacyanobutadienyl transition metal derivatives (Table 1)
were analogously prepared from the same metal carbonyl anions and
1-pentacyanobutadienyl halides.[7] All of the compounds listed in
Table 1 were crystalline somewhat light-sensitive solids which could

be handled in air and which were stable at room temperature for extended periods. They thus were attractive substances for more extensive study.

These polycyanovinyl transition metal compounds can be regarded
as derivatives of tetracyanoethylene and related polycyanoolefins in
which one of the cyano groups is replaced by a transition metal moiety. One characteristic of tetracyanoethylene[1] is the extreme electron deficiency of the carbon-carbon double bond because of the
strong electron withdrawing properties of the multiple cyano groups.
The polycyanovinyl ligands in the compounds listed in Table 1 should
have a similar electron deficient carbon-carbon double bond. If a
polycyanovinyl group is bonded directly to a transition metal such
as in the polycyanovinyl derivatives II and III, the empty π^*-anti-
bonding orbitals of the polycyanovinyl carbon-carbon double bond
will have a strong tendency to remove electron density from the
filled d orbitals of the transition metal. Polycyanovinyl groups
should thus function as extremely strong π-acceptor ligands like
carbonyl groups. Strong spectroscopic evidence for such π-acceptor
properties of polycyanovinyl groups is provided by a $\sim$100 cm.$^{-1}$ reduction of the $\nu(C=C)$ frequency from 1540-1570 cm.$^{-1}$ to 1450-1480
cm.$^{-1}$ when the chlorine in a polycyanovinyl chloride I is replaced
by an $Mn(CO)_5$ or $C_5H_5M(CO)_3$ (M=Mo and W) group. A similar but much
smaller reduction of 15-20 cm.$^{-1}$ in the $\nu(CN)$ frequency of the polycyanovinyl chlorides upon replacement of the chlorine with a transition metal moiety suggests that some electron density from the
transition metal can even reach the π^* antibonding orbitals of the
cyano carbon-nitrogen triple bonds.

Further consideration of the oxygen-dicyanomethylene analogy
mentioned above[2,3] makes the dicyanovinylidene ligand (IV) an analogue of carbon monoxide, the basic building block of metal carbonyls. An interesting synthetic objective, therefore, was the preparation of metal dicyanovinylidene complexes analogous to the long-
known metal carbonyls. A major difficulty, however, in the synthesis of such metal dicyanovinylidene complexes was the absolute lack
of any known methods for preparing the necessary dicyanovinylidene
(IV), even as an unstable and presumably reactive intermediate.

$$C = C \begin{array}{c} CN \\ CN \end{array}$$

IV

Further studies to extend the range of polycyanovinyl transition metal derivatives led fortuitously to the first transition metal dicyanovinylidene complex. The syntheses of the polycyanovinyl
derivatives II and III mentioned above used the metal carbonyl anions
$Mn(CO)_5^-$ and $C_5H_5M(CO)_3^-$ (M=Mo and W), which are only moderately nu-

Table 1

Some Polycyanovinyl Transition Metal Derivatives

Compound	Color	M.P.
A) 2,2-Dicyanovinyl Derivatives		
$(NC)_2C=CHMo(CO)_3C_5H_5$	yellow	dec. 127°
$(NC)_2C=CHW(CO)_3C_5H_5$	yellow	170-171°
$(NC)_2C=CHMn(CO)_5$	light yellow	dec. 155°
$(NC)_2C=CHFe(CO)_2C_5H_5$	yellow	115-116°
B) Tricyanovinyl Derivatives		
$(NC)_2C=C(CN)Mo(CO)_3C_5H_5$	yellow-brown	133° (dec.)
$(NC)_2C=C(CN)W(CO)_3C_5H_5$	yellow	>167° (dec.)
$(NC)_2C=C(CN)Mn(CO)_5$	light brown	141-142° (dec.)
C) 1-Chloro-2,2-dicyanovinyl Derivatives		
$(NC)_2C=C(Cl)Mo(CO)_3C_5H_5$	yellow	134-136° (dec.)
$(NC)_2C=C(Cl)W(CO)_3C_5H_5$	yellow	dec. ~132°
$(NC)_2C=C(Cl)Mn(CO)_5$	yellow	133-135° (dec.)
D) 1-Pentacyanobutadienyl Derivatives		
$C_4(CN)_5W(CO)_3C_5H_5$	yellow	185-190° (dec.)
$C_4(CN)_5Mn(CO)_5$	yellow	140° (dec.)

cleophilic (i.e. relative nucleophilicities[5] of 60 to 500). Similar reactions of polycyanovinyl halides with the extremely nucleophilic metal carbonyl anion $C_5H_5Fe(CO)_2^-$ (70,000,000 on the same nucleophilicity scale) gave different results. Although the reaction of 2,2-dicyanovinyl chloride (I: X=H) with $NaFe(CO)_2C_5H_5$ gave the normal product $(NC)_2C=C(H)Fe(CO)_2C_5H_5$ (V), the corresponding reaction of 1,1-dichloro-2,2-dicyanoethylene with $NaFe(CO)_2C_5H_5$ led mainly to extensive coupling to give $[C_5H_5Fe(CO)_2]_2$. However, also produced from this reaction in low yields were two isomers of the stoichiometry $(C_5H_5)_2Fe_2(CO)_3[C=C(CN)_2]$ in which both chlorine atoms of the 1,1-dichloro-2,2-dicyanoethylene were replaced by cyclopentadienyliron carbonyl residues, with, however, a subsequent loss of one carbonyl group.[5,6] The originally proposed formulation of these $(C_5H_5)_2Fe_2(CO)_3[C=C(CN)_2]$ products as the cis and trans bridging di-

cyanovinylidene derivatives VI and VII, respectively, which was
based on spectroscopic data, has been recently confirmed by X-ray
crystallography.[8]

V   VI   VII

The $(C_5H_5)_2Fe_2(CO)_3[C=C(CN)_2]$ complexes VI and VII may be regarded as dicyanomethylene analogues of the cis and trans isomers,
respectively, of the well-known $[C_5H_5Fe(CO)_2]_2$.[9] The bridging carbonyl in the bridging dicyanovinylidene derivatives VI and VII exhibits a $\nu(CO)$ frequency at 1828 ± 3 cm.$^{-1}$ which is appreciably
higher than the bridging carbonyl $\nu(CO)$ frequencies of 1779 ± 2 cm.$^{-1}$
in the carbonyl analogue $[C_5H_5Fe(CO)_2]_2$. Interpretation of this upwards $\nu(CO)$ frequency shift upon replacing a bridging carbonyl group
by a bridging dicyanovinylidene group in otherwise identical compounds by the usual relationships between $\nu(CO)$ frequencies and π-
acceptor abilities of the other ligands[10] leads to the interesting
conclusion that the dicyanovinylidene ligand is even a stronger π-
acceptor than carbon monoxide, which is generally regarded as among
the strongest π-acceptors in transition metal chemistry.

The discovery of the two $(C_5H_5)_2Fe_2(CO)_3[C=C(CN)_2]$ isomers VI
and VII with bridging dicyanovinylidene ligands demonstrates that
the dicyanovinylidene ligand can exist in transition metal complexes.
At that time the preparation of a complex with a terminal dicyanovinylidene ligand remained an unsolved synthetic problem. Attempts
to prepare $C_5H_5Fe(CO)[C=C(CN)_2]$ I with a terminal dicyanovinylidene
ligand by iodine cleavage of the iron-iron bond in the more abun-
dant cis isomer VI of $(C_5H_5)_2Fe_2(CO)_3[C=C(CN)_2]$ were completely unsuccessful in contrast to the facile iodine cleavage of the iron-
iron bond in $[C_5H_5Fe(CO)_2]_2$ to give $C_5H_5Fe(CO)_2I$ with resulting conversion of the bridging carbonyl groups to terminal carbonyl groups.
This suggests that the dicyanovinylidene ligand is more reluctant
than the carbonyl group to occupy a terminal position in otherwise
similar systems. Further evidence for the reluctance of dicyanovinylidene to occupy terminal positions is suggested by the slow
interconversion at room temperature of the cis and trans isomers of
$(C_5H_5)_2Fe_2(CO)_3[C=C(CN)_2]$, as demonstrated by their chromatographic
separation at room temperature, in contrast to the ease of interconverting cis and trans isomers of $[C_5H_5Fe(CO)_2]_2$ at room temperature[11] through an intermediate containing only terminal carbonyl

groups.[12]

Before we had a chance to contemplate extensively some possible alternative synthetic approaches to metal complexes containing terminal dicyanovinylidene ligands, a chance observation led to the serendipitous discovery of the first terminal dicyanovinylidene complex. We were interested in evaluating the reactivity of the remaining chlorine atom in $(NC)_2C=C(Cl)Mo(CO)_3C_5H_5$ towards nucleophilic reagents. In this connection, the reaction of this molybdenum complex with triphenylphosphine even only slightly above room temperature was found to proceed by replacement of all three carbonyl groups with two triphenylphosphine ligands according to the following equation:[13,14]

$$(NC)_2C=C(Cl)Mo(CO)_3C_5H_5 + 2(C_6H_5)_3P \longrightarrow C_5H_5Mo[P(C_6H_5)_3]_2CCl=C(CN)_2$$
$$+ 3CO$$

This reaction had the following two unusual features: (1) The replacement of all three carbonyl ligands in a $RMo(CO)_3C_5H_5$ derivative by a Lewis base is very rare since such reactions normally lead to the loss of one or, at most, two carbonyl groups; (2) The replacement of three carbonyl groups by only two triphenylphosphine ligands is very unusual since both ligands are always two-electron donors[15] in transition metal chemistry. This latter feature suggested that something was happening to the $CCl=C(CN)_2$ moiety to convert it from a one-electron donor in $(NC)_2C=C(Cl)Mo(CO)_3C_5H_5$ into a three-electron donor. A clue to this puzzle lay in the observation of two sharp cyclopentadienyl proton n.m.r. resonances of approximate 3:1 relative intensities on an analytically pure sample. Since the compound $C_5H_5Mo[P(C_6H_5)_3]_2CCl=C(CN)_2$ has only one molybdenum atom, this n.m.r. observation can only be explained by postulating the presence of two isomers. Isomerism is possible in a cyclopentadienylmolybdenum derivative of the type $C_5H_5MoA_2BC$ but not in one of the type $C_5H_5MoA_2B$. This leads to the conclusion that the $CCl=C(CN)_2$ moiety has split into two monodentate units. The most likely such possibility is for the $CCl=C(CN)_2$ moiety to become a terminal chlorine ligand and a terminal dicyanovinylidene ligand. This led to our postulation of the two structures VIII and IX ($R=C_6H_5$), both with terminal dicyanovinylidene ligands, for the two isomers of $C_5H_5Mo[P(C_6H_5)_3]_2CCl=C(CN)_2$. Further support for this postulate was provided later by the observation that a similar reaction of $(NC)_2C=C(Cl)Mo(CO)_3C_5H_5$ with the tritertiary phosphine $C_6H_5P[CH_2CH_2P(C_6H_5)_2]_2$ followed by metathesis with ammonium hexafluorophosphate gave the hexafluorophosphate salt X containing a similar terminal dicyanovinylidene ligand in the absence of accompanying coordinated chlorine.

More detailed studies on the corresponding trimethyl phosphite derivative $C_5H_5Mo[P(OCH_3)_3]_2[C=C(CN)_2]$ Cl (VIII: $R=OCH_3$) provided unambiguous proof of the terminal dicyanovinylidene ligand in these

VIII   IX   X

complexes. The carbon-13 n.m.r. spectrum of the trimethyl phosphite
complex exhibited a very low field resonance (-158.5 p.p.m. relative
to CS_2) which can be assigned to the coordinating carbene carbon in
the dicyanovinylidene ligand.[16] A similar extremely low field car-
bon-13 resonance from a carbene carbon coordinated to a transition
metal was previously[17] found in the methoxymethylcarbene complex
$CH_3C(OCH_3)Cr(CO)_5$. The final proof of the terminal dicyanovinylidene ligand in $C_5H_5Mo[P(OCH_3)_3]_2[C=C(CN)_2]Cl$ came from its structure
determined by X-ray crystallography.[18] Characteristic features of
this structure are the bent dicyanovinylidene ligand with a Mo-C-C
angle of 166.6° and a relatively long molybdenum-chlorine distance
of 2.467 Å indicating a strong trans effect of the terminal dicyanovinylidene ligand.

XI

The ability of the 1-chloro-2,2-dicyanovinyl derivative $(NC)_2C=$
$C(Cl)Mo(CO)_3C_5H_5$ (III: M=Mo, X=Cl) to form terminal dicyanovinylidene complexes upon reactions with tertiary phosphines and related
ligands[13,14] suggested that this molybdenum complex might function
as a source of free dicyanovinylidene when heated alone possibly
according to the following general scheme:

$$(NC)_2C=C(Cl)Mo(CO)_3C_5H_5 \longrightarrow C=C(CN)_2 + C_5H_5Mo(CO)_3Cl$$

Support for this possibility was provided by the observation that the
thermal reaction of $(NC)_2C=C(Cl)Mo(CO)_3C_5H_5$ with diphenylacetylene
gives the novel green 6,6-dicyano-1,2,3,4-tetraphenylfulvene(XI).[19]
The yield of the fulvene XI was increased from ∿15% to ∿50% by adding cerium (IV) to the reaction mixture before product isolation
thus suggesting the intermediacy of a metal complex of the fulvene
XI.

The rather strange preparation of terminal dicyanovinylidene complexes of the type VIII and IX has not been the only unusual feature of the chemistry of polycyanovinyl molybdenum and tungsten derivatives which we have found. Another unusual type of complex arose from our efforts to prepare tricyanovinyltungsten derivatives from the reaction of $NaW(CO)_3C_5H_5$ with tricyanovinyl chloride. Reaction of $NaW(CO)_3C_5H_5$ with tricyanovinyl chloride followed by chromatography on Florisil gives the usual yellow tricyanovinyltungsten derivative III (M=W, X=CN). However, if the same reaction mixture is chromatographed on alumina rather than Florisil a different orange compound is produced.[20] The orange compound was formulated as the novel chelate $C_2(CN)_2(CHNH)W(CO)_2C_5H_5$ (XII) on the basis of elemental analyses, the observation of appropriate infrared $\nu(NH)$, $\nu(CH)$, $\nu(CN)$, and $\nu(CO)$ frequencies, and the observation of apparent methine and imine n.m.r. resonances in addition to the cyclopentadienyl resonance. Furthermore, reaction of $C_2(CN)_2(CHNH)W(CO)_2C_5H_5$ (II) with triphenylphosphine in boiling methylcyclohexane resulted in the replacement of one carbonyl group with a triphenylphosphine ligand without destruction of the unusual chelate ring.

The conversion of yellow $(NC)_2C=C(CN)W(CO)_3C_5H_5$ to orange $C_2(CN)_2(CHNH)W(CO)_2C_5H_5$ upon chromatography on alumina can be visualized as a hydrolysis with decarboxylation according to the following equation:

$$(NC)_2C=C(CN)W(CO)_3C_5H_5 + H_2O \longrightarrow C_2(CN)_2(CHNH)W(CO)_2C_5H_5 + CO_2$$

Similar alumina treatment of the 1-chloro-2,2-dicyanovinyl derivatives $(NC)_2C=C(Cl)M(CO)_3C_5H_5$ III (M=Mo or W, X=Cl) results not only in analogous hydrolytic conversions to dicarbonyl derivatives but also in hydrolysis of the chlorine atom to give the red-orange $C_2(CN)(OH)(CHNH)M(CO)_2C_5H_5$ formulated as XIII (M=Mo and W) on the basis of their infrared and n.m.r. spectra.[20]

XII   XIII   XIV

The formation of the unusual chelates XII and XIII upon alumina treatment of appropriate polycyanovinyl transition metal derivatives suggested that reactions of other basic reagents with various polycyanovinyl transition metal derivatives might give other types of novel complexes. Solutions of sodium hydroxide or sodium alkoxides

in the corresponding alcohols (CH_3OH or C_2H_5OH) were found to react
with $(NC)_2C=C(Cl)M(CO)_3C_5H_5$ (M=Mo and W) at room temperature within
minutes to give not the compounds $C_2(CN)(OH)(CHNH)M(CO)_2C_5H_5$ (XIII)
obtained from the alumina treatment but instead the red esters C_2-
$(CN)(OR)[C(CO_2R)NH]M(CO)_2C_5H_5$ (XIV: M=Mo and W, R=CH_3 and C_2H_5).[20]
The infrared spectra of these esters exhibited a strong ester $\nu(CO)$
frequency at 1698 $\pm$ 4 cm.$^{-1}$ in addition to two metal carbonyl $\nu(CO)$
frequencies, a single $\nu(C\equiv N)$ frequency, a single $\nu(C=N)$ frequency,
and a $\nu(NH)$ frequency. The proton n.m.r. spectra of the esters XIV
demonstrate the non-equivalence of the two alkoxy groups in accord
with the proposed structure.

The formation of the esters XIV upon treatment of $(NC)_2C=C(Cl)-$
$M(CO)_3C_5H_5$ with alcohols in the presence of base can be considered
as an alcoholysis according to the following equation:

$$(NC)_2C=C(Cl)M(CO)_3C_5H_5 + 2ROH \longrightarrow C_2(CN)(OR)[C(CO_2R)NH]M(CO)_2C_5H_5 +$$

$$HCL$$

The failure of the alcoholysis of the 1-chloro-2,2-dicyanovinyl derivatives to result in carbon dioxide elimination in contrast to the
alumina-catalyzed hydrolyses discussed above suggests that the carbon dioxide elimination comes from decarboxylation of a carboxylic
acid intermediate, which can be stabilized if esterified. The demonstrated tendency of metal-bonded carboxyl groups to undergo facile
decarboxylation[21] unless esterified suggests intermediates with metal-bonded carboxyl groups in these reactions. However, at the present time insufficient experimental evidence is available to postulate definitive mechanisms.

The formations of the esters XIV upon alcoholysis of $(NC)_2C=C-$
$(Cl)M(CO)_3C_5H_5$ (III: X=Cl, M=Mo and W) suggested that the 1-chloro-
2,2-dicyanovinyl derivatives might form novel amides upon amino-
lysis with amines containing at least one N-H bond. Reactions of
$(NC)_2C=C(Cl)M(CO)_3C_5H_5$ with the secondary amines dimethylamine and
piperidine gave complex mixtures which were separated by chromatography on Florisil into bright yellow products of general stoichio-
metries $C_5H_5M(CO)_2[R_2N=C=C(CN)_2]$ (M=Mo and W; R=CH_3 or $R_2=-(CH_2)_5-$)
and mixtures of more strongly absorbed reddish products, which have
not yet been separated into pure components.[20] The infrared spectra
of the yellow $C_5H_5M(CO)_2[R_2N=C=C(CN)_2]$ derivatives exhibit a medium
$\nu(C\equiv N)$ frequency at 2215 $\pm$ 3 cm.$^{-1}$, two strong metal $\nu(CO)$ frequencies at 1991 $\pm$ and 1918 $\pm$ 2 cm.$^{-1}$, and a single strong band at 1584
$\pm$ 7 cm.$^{-1}$ which can arise from the $\nu(C=N)$ of an uncomplexed carbon-
nitrogen double bond. On the basis of this type of spectroscopic
data, the yellow $C_5H_5M(CO)_2[R_2N=C=C(CN)_2]$ are formulated as the dicyanoketeneimmonium complexes XV (M=Mo and W; R=CH_3 or $R_2=-(CH_2)_5-$).
The proton and carbon -13 n.m.r. spectra of the complexes XV(R=CH_3)
indicate that both cyano groups are equivalent but that the two

N-methyl groups are non-equivalent demonstrating that in these complexes the dicyanoketeneimmonium ligand is bonded to the metal
through the carbon-carbon rather than the carbon-nitrogen double
bond. The organic ligand in the complexes XV is closely related to
that in the complex XVI, which we prepared almost concurrently[22]
from $NaMo(CO)_3C_5H_5$ and the α-chloroenamine $(CH_3)_2C=C[N(CH_3)_2]Cl$ in
another research program.

XV   XVI

    The work described in this paper shows that polycyanovinyl
transition metal derivatives are a new and interesting class of stable organometallic derivatives which are readily accessible in good
yields from metal carbonyl anions and polycyanovinyl halides. The
chemistry of these polycyanovinyl derivatives is very unusual and
has led to dicyanovinylidene and dicyanoketeneimmonium complexes
as well as novel types of metal chelates. Our work has only barely
scratched the surface of this extremely interesting and exciting
area of chemistry. Systems yet to be investigated which are almost
certain to be of considerable interest and significance include
these from reactions of polycyanovinyl derivatives other than those
of the type III with bases including bases other than water, alcohols, and amines as well as all reactions of the pentacyanobutadienyl
derivatives. Furthermore, a more intensive study of the decomposition of 1-chloro-2,2-dicyanovinyl derivatives such as $(NC)_2C=C(Cl)$-
$Mo(CO)_3C_5H_5$ could lead to the isolation of free dicyanovinylidene
(IV), possibly as a species only stable at low temperatures. Dicyanovinylidene (IV) would certainly be a significant and unusual
intermediate for organic chemists with applications in the synthesis
of novel organic compounds.

ACKNOWLEDGMENT

    The painstaking key experimental contributions of Dr. Mohan
Saran to the development of this new field of chemistry are gratefully acknowledged as well as extensive discussions with him since
the beginning of this project. We thank the Office of Naval Research and the National Science Foundation for financial support of
this research during 1970-1972 and 1972-1973, respectively.

REFERENCES

1. E. Ciganek, W. J. Linn, and O. W. Webster, "The Chemistry of the Cyano Group", (Z. Rappoport, Ed.), Interscience, New York, 1970: pp. 423-638.
2. K. Wallenfels, Chimia, 20, 303 (1966).
3. H. Kohler, B. Eichler, and R. Salewski, Z. anorg. allgem. Chem., 379, 183 (1970).
4. R. B. King, Accts. Chem. Research, 3, 417 (1970).
5. R. B. King and M. S. Saran, J. Am. Chem. Soc., 94, 1784 (1972).
6. R. B. King and M. S. Saran, J. Am. Chem. Soc., 95, 1811 (1973).
7. R. B. King, J. W. Howard, Jr., and M. S. Saran, to be published.
8. J. A. Ibers et al., unpublished results.
9. F. A. Cotton, T. S. Piper, and G. Wilkinson, J. Inorg. Nucl. Chem., 1, 165 (1955); O. S. Mills, Acta Cryst., 11, 620 (1958).
10. F. A. Cotton, Inorg. Chem., 3, 702 (1964).
11. R. F. Bryan, P. T. Greene, D. S. Field, and M. J. Newlands, Chem. Comm., 1477 (1969).
12. J. G. Bullitt, F. A. Cotton, and T. J. Marks, Inorg. Chem., 11, 671 (1972).
13. R. B. King and M. S. Saran, Chem. Comm., 1053 (1972).
14. R. B. King and M. S. Saran, J. Am. Chem. Soc., 95, 1817 (1973).
15. R. B. King, Advan. Chem. Ser., 62, 203 (1967).
16. O. A. Gansow, A. R. Burke, R. B. King, and M. S. Saran, Inorg. Nucl. Chem. Lett., 10, 291 (1974).
17. L. F. Farnell, E. W. Randall, and E. Rosenberg, Chem. Comm., 1078 (1971).
18. R. M. Kirchner, J. A. Ibers, M. S. Saran, and R. B. King, J. Am. Chem. Soc., 95, 5775 (1973).
19. R. B. King and M. S. Saran, to be published.
20. R. B. King and M. S. Saran, paper presented at the 167th National Meeting of the American Chemical Society, Los Angeles, California April, 1974: paper INOR240 in abstracts.
21. E. O. Fischer, K. Fichtel, and K. Ofele, Ber., 95, 249 (1962); T. Kruck and M. Noack, Ber., 97, 1693 (1964).
22. R. B. King and K. C. Hodges, J. Am. Chem. Soc., 96, 1263 (1974).

DISCUSSION

DR. COLLMAN: Some of your heterocycles should react with disubstituted acetylenes under conditions in which a carbonyl is lost to form a pyridine. This might be an attractive route to some bizarrely substituted pyridynes?

DR. KING: That is an idea; there are so many possibilities in the cyanocarbon field. I tried to pick out the highlights. We have not tried such reactions mainly because I have only one man part time on this project, furthermore he is occupied with some other things.

DR. COLLMAN: But this is in general true, of course, for all carbon substituted metallocycles of this kind in which there are two double bonds; it might well apply to a quite general synthesis of heterocycles.

DR. KING: Yes, that is a very interesting idea on which we unfortunately have no information.

A NEW PREPARATION OF ORGANOCOPPER(I)-ISONITRILE COMPLEXES AND THEIR REACTIONS

Yoshihiko Ito and Takeo Saegusa[*]

Department of Synthetic Chemistry, Faculty of Engineering, Kyoto University, Kyoto, Japan

In the course of our continuing studies[1] on the synthetic reactions catalyzed by copper isonitrile complexes, it has been found that metallic copper and Cu_2O are dissolved in excess of isonitrile under nitrogen. The mixture of metallic copper or Cu_2O with isonitrile has been characterized spectroscopically. Esr spectrum of the mixture of metallic copper and isonitrile showed an unresolved absorption band (g value = 2.0041), suggesting a species of Cu(0) isonitrile complex. Moreover, esr spectra[3] of the mixture of metallic copper or Cu_2O and π-substrates such as p- and m-dinitrobenzene, nitrobenzene, benzoquinone, fluorene and tetracyanoethylene in isonitrile exhibited the resolved signals of the corresponding organic radical anions.

The metallic copper/isonitrile and Cu_2O/isonitrile systems reacted with the so-called active methylene compounds, accompanying the formation of H_2 gas and H_2O, respectively.[1] The mixture thus obtained reacted[4] with ketone and aldehyde in the manner of aldol condensation and with α,β-unsaturated carbonyl and nitrile compounds in the manner of the Michael reaction,[5] as shown in Scheme I. On the basis of these observations, we have proposed that organocopper-(I)/isonitrile complexes (1) are involved in the reaction of the active hydrogen compound with the metallic copper/isonitrile and Cu_2O/isonitrile systems. Here, we wish to describe a new preparation of organocopper(I)/isonitrile complexes by the reaction of the so-called active methylene compounds such as cyclopentadiene,[4] indene,[4] acetylacetone,[6] acetoacetate,[6] malonate[6] and cyanoacetate[6] with the Cu_2O/isonitrile and metallic copper/isonitrile systems. As was expected, these organocopper(I) isonitrile complexes (1) underwent the insertion reaction of isonitrile into their copper-carbon bonds.

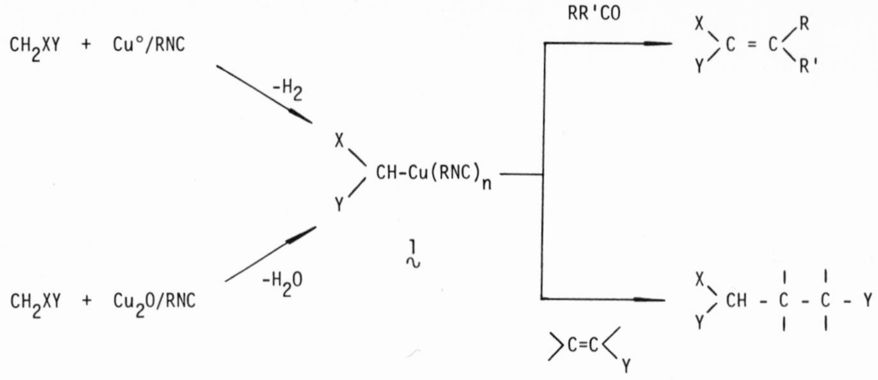

$$CH_2XY + Cu^{\circ}/RNC \xrightarrow{-H_2}$$

$$\underset{Y}{\overset{X}{>}}CH\text{-}Cu(RNC)_n \underset{\underset{\sim}{1}}{}$$

$$CH_2XY + Cu_2O/RNC \xrightarrow{-H_2O}$$

$$\xrightarrow{RR'CO} \underset{Y}{\overset{X}{>}}C = C\underset{R'}{\overset{R}{<}}$$

$$\xrightarrow{>C=C<_Y} \underset{Y}{\overset{X}{>}}CH - \overset{|}{\underset{|}{C}} - \overset{|}{\underset{|}{C}} - Y$$

Scheme I

PREPARATION OF ORGANOCOPPER(I)/ISONITRILE COMPLEX

When a mixture of cyclopentadiene, tert-butyl isocyanide and Cu_2O was warmed up to 50°, an exothermic reaction occurred, and Cu_2O went into solution. From the reaction mixture, pentahapto-cyclopentadienyl (tert-butyl isocyanide) copper (I) ($\underset{\sim}{2}$) was isolated by recrystallization as a white crystalline solid. The structure of $\underset{\sim}{2}$ was established by ir and nmr spectra and elemental analysis. The pentahapto structure of $\underset{\sim}{2}$ is supported by a singlet ir absorption ν_{C-H} of the cyclopentadienyl ring at 3084 cm^{-1} [7] and by singlet nmr absorption at τ 3.76 of the cyclopentadienyl ring. As to the cyclopentadienyl copper-isonitrile complex, Cotton[7] prepared pentahapto-cyclopentadienyl(methyl isocyanide)copper(I) ($\underset{\sim}{3}$) for the first time by reaction of $Tl(C_5H_5)$ with $CuI(CH_3NC)_n$. It is interesting that $\underset{\sim}{2}$ was much more stable than $\underset{\sim}{3}$. $\underset{\sim}{3}$ is reported to decompose in several minutes at room temperature. But $\underset{\sim}{2}$ remained unchanged for at least 1 month at room temperature under nitrogen.

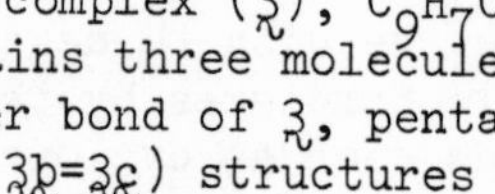

$$\text{[cyclopentadienyl]} - Cu(\underline{t}\text{-}C_4H_9NC)$$

$$\underset{\sim}{2}$$

From a mixture of indene, tert-butyl isocyanide and Cu_2O , an indenylcopper(I)isonitrile complex ($\underset{\sim}{3}$), $C_9H_7Cu(\underline{tert}\text{-}C_4H_9NC)_3^2$, was isolated. Unlike $\underset{\sim}{2}$, $\underset{\sim}{3}$ contains three molecules of isonitrile ligand. As to the indene ring-copper bond of $\underset{\sim}{3}$, pentahapto ($\underset{\sim}{3a}$) and rapid-interconverting monohapto ($\underset{\sim}{3b}=\underset{\sim}{3c}$) structures were possible from nmr spectral evidence. The nmr spectrum of $\underset{\sim}{3}$ (in C_6D_6) is shown in Figure 1.

In the nmr spectrum of $\underset{\sim}{3}$ in C_6D_6, a broad singlet at τ 9.10

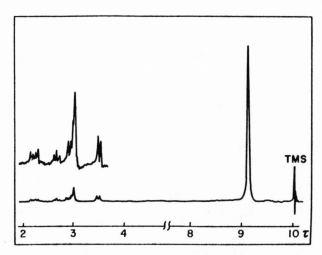

Figure 1. Nmr spectrum of 3 in C_6D_6 at room temperature.

(27 H) is assigned to the <u>tert</u>-butyl protons. The doublet at 3.50
(2H) is assigned to the C-1 and C-3 protons of the indene ring, and
the triplet at τ 2.65 (1H) to the C-2 proton. The coupling constant
of these two absorptions is identical (J=3 Hz). These absorptions
are well explained by an A_2B pattern. On the other hand, four proton on the benzene ring of the indenyl group may be analyzed by an
A_2B_2 pattern. The quartet centered at τ 2.22 (2H) is assigned to
the C-4 and C-7 protons. However, another quartet near τ 3.0 (2H)
which is to be assigned to the C-5, and the C-6 protons is partially
superimposed by the residual protons of the C_6D_6 solvent. Full analysis of the nmr spectrum of 3 has not been made, because 3 is not
stable and decomposes in most organic solvents.

The nmr spectrum of 3 suggests two possible structures. One
is the pentahapto structure 3a. The other is the rapid interconverting monohapto structure (3b⇌3c). Although the structure of 3 has not
been conclusively established, the monohapto structure may be preferred on the basis of the inert gas rule.

The present preparative method of organocopper(I)/isonitrile complex is applicable to the so-called active methylene compound such as acetylacetone, acetoacetate, malonate. On stirring a mixture of acetylacetone ($4a$), tert-butyl isocyanide and Cu_2O for a half hour at room temperature under nitrogen, copper(I) acetylacetonate bis-(tert-butyl isocyanide) ($5a$) was produced in an almost quantitative yield. The complex $5a$ was reprecipitated from the benzene solution with ether or with petroleum ether. In a similar way, the corresponding organocopper(I)/isonitrile complexes ($5b$-$5e$) were prepared by the reaction of acetoacetate, malonate or cyanoacetate with Cu_2O and tert-butyl isocyanide.

$$CHXYZ \quad + \quad Cu_2O/tert\text{-}C_4H_9NC \quad \longrightarrow \quad \begin{pmatrix} X \\ Y \text{---} C \\ Z \end{pmatrix} Cu(\underline{tert}\text{-}C_4H_9NC)_2$$

$$\underset{4}{} \qquad\qquad\qquad\qquad\qquad\qquad\qquad\qquad\qquad\qquad\qquad \underset{5}{}$$

$\quad$ a : $X = Y = COCH_3$, $Z = H$, $n = 2$

$\quad$ b : $X = Y = CO_2C_2H_5$, $Z = H$, $n = 2$

$\quad$ c : $X = COCH_3$, $Y = CO_2C_2H_5$, $Z = H$, $n = 2$

$\quad$ d : $X = CN$, $Y = CO_2CH_3$, $Z = H$, $n = 2$

$\quad$ e : $X = Y = CO_2C_2H_5$, $Z = C_2H_5$, $n = 2$

All of these complexes were soluble in most organic solvents, and thermally stable under nitrogen but rapidly oxidized in air. A freshly prepared metallic copper was also reactive toward active methylene compounds to produce organocopper(I)/isonitrile complexes, but its reactivity was less than that of Cu_2O. Ir and nmr data are summarized in Table I.

The infrared spectrum of copper(I)acetylacetylacetonate bis-(tert-butyl isocyanide) complex $5a$ showed a strong band at 1610 cm^{-1} which may be taken to indicate a structure of oxygen-bonded copper-(I)acetylacetonate.[8] It may be pertinent to note that a carbon-bonded gold(I)acetylacetonate phosphine complex exhibits two sharp ir bands at 1660 and 1645 cm^{-1}.[9] As to the structure of other organocopper(I)/isonitrile complexes ($5b$-$5e$), no information is now at hand. The structure determination of $5b$-$5e$ is the subject of the future.

INSERTION REACTION OF ORGANOCOPPER(I)/ISONITRILE COMPLEXES

Insertion of carbon monoxide and isonitrile into carbon-metal bond is a reaction characteristic of the transition metal alkyls, which constitutes a key step of many synthetic reactions by transi-

Table I
Organocopper(I) Isonitrile Complexes (5)

(XYZC)Cu(tert-$C_4H_9NC)_2$	Ir [a]	Nmr	
X=Y=$COCH_3$, Z=H, n=2 (5a)	2166, 2142 1610, 1525	5.18 (s, 1H)[b] 1.42 (s, 18H)	1.87 (s, 6H)
X=Y=$CO_2C_2H_5$, Z=H, n=2 (5b)	2172, 1690 1580	4.19 (q, 4H)[c] 1.25 (t, 6H)	1.47 (s, 18H)
X=$COCH_3$, Y=$CO_2C_2H_5$, Z=H, n=2 (5c)	2167, 2144 1690, 1655 1580, 1520	4.06 (q, 2H)[b] 1.43 (s, 18H)	1.87 (s, 3H) 1.21 (t, 3H)
X=CN, Y=CO_2CH_3, Z=H, n=2 (5d)	2175, 1640 1620	3.80 (s, 3H)	1.02 (s, 18H)
X=Y=$CO_2C_2H_5$, Z=C_2H_5, n=2 (5e)	2172, 1690 1585	4.30 (q, 4H)[c] 1.40 (t, 3H) 0.94 (s, 18H)	1.80-1.60 (2H) 1.21 (t, 6H)

a) nujol   b) $CDCl_3$ solution with TMS   c) C_6D_6 solution with TMS

tion metal catalysts. The possibility of the insertion reaction of isonitrile with the organocopper(I)/isonitrile complexes 2, 3 and 5 was examined, and it has been found that refluxing complexes 2 and 3 with excess of isonitrile in benzene under nitrogen yielded enamine derivatives 6 and 7, respectively, in good yields after hydrolysis, which are assumed to have been derived via the expected insertion products 8 and 9. Attempts to isolate 8 and 9 are now being carried out. The products 6 and 7 were also obtained simply by heating cyclopentadiene and indene with excess of isonitrile in benzene in the presence of Cu_2O or metallic copper as the catalyst. Similarly, the reaction of organocopper(I)/isonitrile complexes 5 with excess of i-sonitrile in benzene afforded the corresponding enamine derivatives (10), after hydrolysis. Concerning the mechanism of the insertion reaction of the organocopper(I)/isonitrile complexes (5), it may be plausibly assumed that the oxygen-bonded copper(I) acetylacetonate bis(tert-butyl isocyanide) 5a is converted in the presence of excess isonitrile to the carbon-bonded copper(I) acetylacetonate isonitrile complex (12) prior to the insertion reaction, as illustrated in Scheme II.

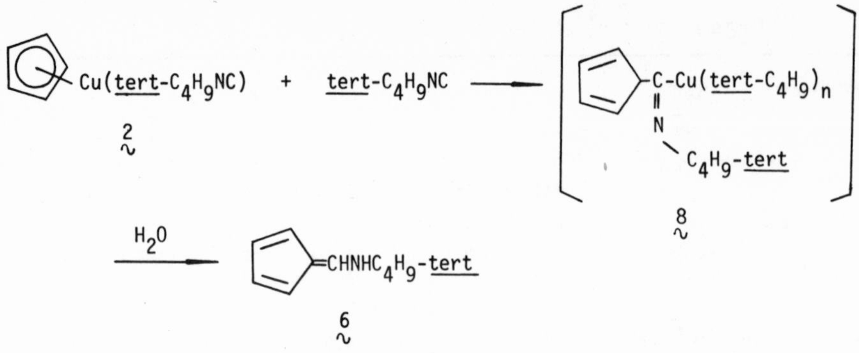
Cu(tert-C4H9NC)
+
tert-C4H9NC
C-Cu(tert-C4H9)n
N
C4H9-tert
2
8
H2O
=CHNHC4H9-tert
6

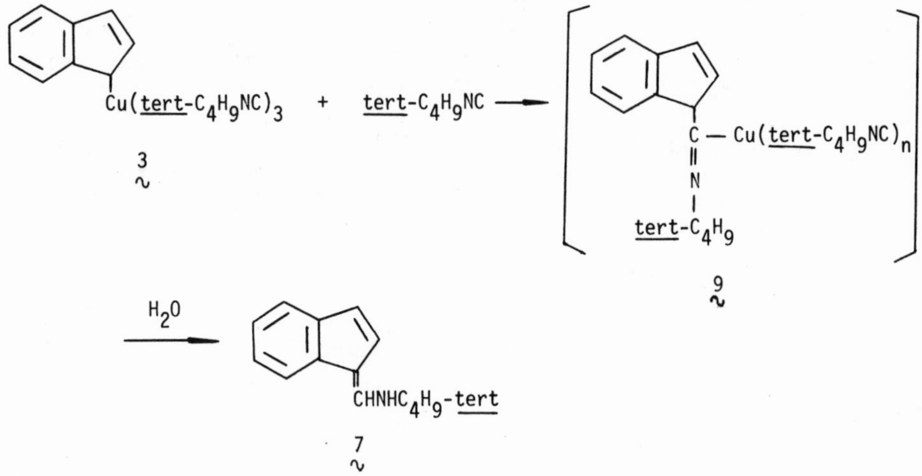
Cu(tert-C4H9NC)3
+
tert-C4H9NC
C — Cu(tert-C4H9NC)n
N
tert-C4H9
3
9
H2O
CHNHC4H9-tert
7

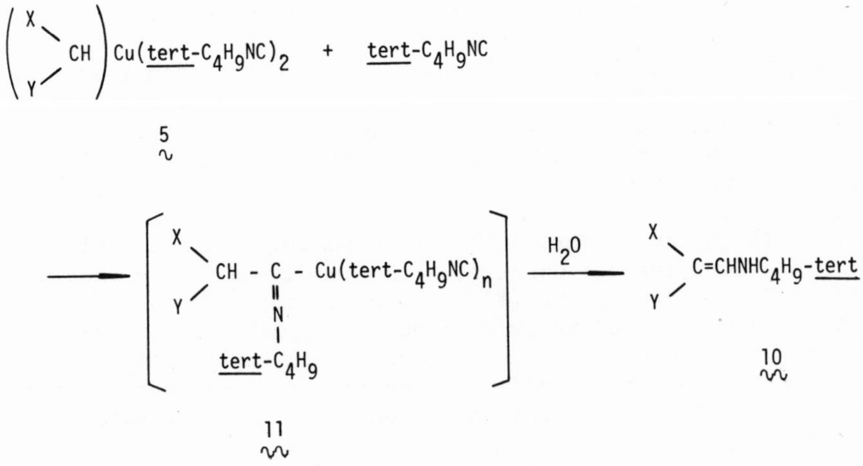
X
CH
Y
Cu(tert-C4H9NC)2
+
tert-C4H9NC
5
X
CH - C - Cu(tert-C4H9NC)n
Y
N
tert-C4H9
H2O
X
C=CHNHC4H9-tert
Y
10
11

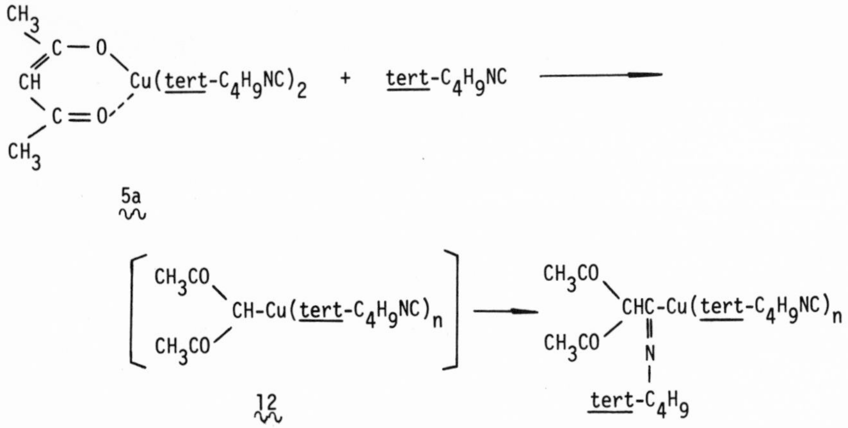

5a

12

Scheme II

REFERENCES

1. T. Saegusa, K. Yonezawa, I. Murase, T. Konoike, S. Tomita and Y. Ito, J. Org. Chem., 38, 2319 (1973).
2. T. Saegusa, K. Yonezawa and Y. Ito, Synth. Commun., 2, 431 (1972).
3. Y. Ito, T. Konoike and T. Saegusa, Tetrahedron Lett., 1974, 1287.
4. T. Saegusa, Y. Ito and S. Tomita, J. Amer. Chem. Soc., 93, 5656 (1971).
5. T. Saegusa, Y. Ito and S. Tomita, Bull. Chem. Soc. Japan, 45, 496 (1972).
6. Y. Ito, T. Konoike and T. Saegusa, submitted to J. Organometal. Chem.
7. F. A. Cotton and T. J. Marks, J. Amer. Chem. Soc., 92, 5114 (1970).
8. R. Nast and W. H. Lepel, Chem. Ber., 102, 3224 (1969).
9. D. Gibson, B. F. G. Johnson, J. Lewis and C. Oldham, Chem. & Ind., 1966, 342.

AN UNUSUAL BEHAVIOR OF π-VINYL ALCOHOL COMPLEXES OF TRANSITION METALS

J. Frances*, M. Ishaq, and M. Tsutsui

Chemistry Department, Texas A&M University, College
Station, Texas 77843

Although the usefulness of polyvinyl alcohol [PVA] polymer has
long been recognized, no work has been reported on the vinyl alcohol
moiety itself. This is due to the fact that the existence of free
vinyl alcohol has not yet been shown. It has been postulated that
a small percentage of this unstable specie, vinyl alcohol, exists in
situ in an equilibrium with acetaldehyde.[1] However, it has been
claimed that vinyl alcohol forms stable organometallic complexes
with transition metal ions. These vinyl alcohol complexes of transition metals have been proposed as being π-bonded to the metal.

One of the major interests in chemical research is the stabilization of a variety of hitherto unisolatable organic moieties such
as cyclobutadiene,[2] benzyne,[3] carbene[4] and ketenimine[5] by coordination with transition metal ions. Interest in trapping of vinyl alcohol has been stimulated by the proposed π-vinyl alcohol intermediate[6] in the catalytic oxidation of ethylene to acetaldehyde by palladium ions which has been known as the Wacker's process.

So far, four transition metal complexes containing the vinyl
alcohol moiety π-bonded to the metal have been reported in the literature. These complexes have either been prepared by the rearrangement of β-oxoalkyl complexes or by the hydrolysis of vinyl silylether
derivative of transition metals. The first stable vinyl alcohol
transition metal complex was reported by Green and Ariyaratne.[8]
During the study of oxoalkyl iron complexes, the authors prepared
the dicarbonylcyclopentadienyl (β-oxoalkyl) iron complex [A]. The
treatment of a petroleum ether solution of the complex [A] with dry
hydrogen bromide yielded the dicarbonylcyclopentadienyl (π-enol)
iron cation [B]. The authors assigned the structure [B] for the
complex on the basis of infrared and elemental analysis. However,

J. FRANCES, M. ISHAQ AND M. TSUTSUI

$$[CpFe(CO)_2]^- Na^+ + CH_2Cl\ CHO \longrightarrow Fe-CH_2-\overset{O}{\overset{\|}{C}}-H + NaCl \quad [A]$$

$$[A] + HB \longrightarrow [B] \quad \text{(Br}^-\text{)} \qquad [C] \quad \text{(Br}^-\text{)}$$

the complex exhibits an A_2X pattern in the nmr spectrum which was not expected for the structure [B]. The alternate structure [C] proposed by the authors seems to satisfy the nmr data [A_2X pattern], but the ir and nmr spectra of the complex did not show Fe---H bond. [See Table I for ir and nmr spectra] The elucidation of the structure of the complex needs further work and probably x-ray diffraction analysis would provide a conclusive answer.

Very recently, Rosenblum and co-workers have reported other methods for the preparation of the complexes [A] and [B] which gave better yield than that reported by Green and co-workers previously. The authors have shown that treatment of sodium salt of dicarbonyl-cyclopentadienyl iron and chloroacetaldehyde dimethyl or diethyl gave a yellow complex [D]. The complex [D] on passing through an alumina column gave the complex [A] in 90% yield. The complex was protonated to give π-vinyl alcohol complex [B] on treatment with hexafluorodiethyl etherate.

$$[CpFe(CO)_2]^- Na^+ + CH_2Cl\ CH(OC_2H_5)_2 \longrightarrow Cp(CO)_2Fe-CH_2CH(OC_2H_5)_2$$
$$[D]$$

Murahashi and co-workers[10] have claimed the preparation of a π-vinyl alcohol complex of platinum(II). The authors have used a different technique than that of Green and Ariyaratne. They used Zeise's dimer $(Cl_2PtC_2H_4)_2$ which upon treatment with vinyl trimethyl-silyl ether gave a yellow solid of 1,3-bis(vinyl trimethylsilyl ether) 2,4-dichloro-μ-dichloroplatinum [E]. The complex [E] was hy-

drolyzed with moist benzene to give π-vinyl alcohol complex [F],
1,3-bis(π-vinyl alcohol) 2,4-dichloro-μ-dichloroplatinum(II). These

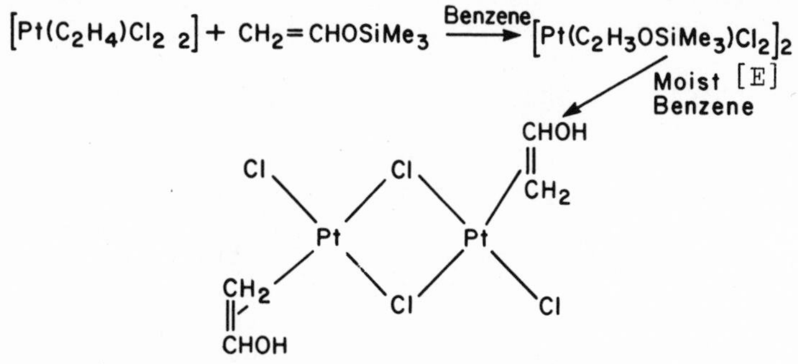

[F]

Table I
Complex [B]

I.R. Spectrum (the typical stretching frequencies are shown) as Nujol Mull	NMR Spectrum in (τ) (in SO_2) with Reference to TMS
3095	1.43, s (OH)
2910	
2810	1.83, t (CHO) J, 8.0
2580	
2500 g	4.62, s (C_5H_5)
2350	
2060] vs	7.15, d (=CH_2) J, 8.0
2010] vs	
1550 k	
1433 m	
1425 m	s = singlet
1375 s	d = doublet
1280 s	t = triplet
1017 ml	J = complex constant
935 m	cycle/sec.
890 w	
875 w	
852 ml	
836 w	
826 m	
816 m	

g = O-H stretch
k = Coordinated C=C stretch
l = π-C_5H_5 frequencies

authors have assigned the structure [F] of the complex on the basis
of infra-red studies alone. The ir spectrum measured in KBr showed
band at 3200 cm^{-1} (vs) due to (OH), 3020 cm^{-1}(w) (C-H) and at 1548
cm^{-1} (vs) due to C=C stretches. The nmr of the complex was not studied because of the poor solubility of the complex.

The authenticity of the work reported by Murahashi has been
questioned by Thyret who reported that he was unable to reproduce
Murahashi's results. Instead, he obtained a mixture of products
comprised of acetaldehyde paraldehyde, and hexamethyldisiloxane.

Thyret has reported the preparation of tetracarbonyl (π-vinyl
alcohol) iron complex [H][11] by using essentially the same experimental techniques as used by Murahashi, i.e. the use of vinyl trimethylsilyl ether, which upon treatment with diiron nonacarbonyl
gave tetracarbonyl (π-vinyl trimethyl silyl) iron [G] which on hydrolysis in acetone at $-90°$ gave the tetracarbonyl (π-vinyl alcohol)
iron complex [H].

$$Fe_2(CO)_9 + CH_2=CHOSiMe_3 \xrightarrow{-80°C} \underset{\underset{Me_3SiO}{}{\overset{CH_2}{\underset{\diagdown}{\overset{\|}{C}}}}\diagup H}{} \longrightarrow Fe(CO)_4 + Fe(CO)_5$$

[G]

$$\underset{\underset{Me_3SiO}{}{\overset{CH_2}{\overset{\|}{C}}}\diagup H}{} \longrightarrow Fe(CO)_4 \xrightarrow{H_2O} \underset{\underset{HO}{}{C}}{\overset{H_b \diagdown C \diagup H_a}{\overset{\|}{C}} Hc} \longrightarrow Fe(CO)_4$$
(dec. : $-70°C$)

[G]   [H]

The characterization of the π-vinyl alcohol complex [H] was based
only on its nmr spectra. The complex [H] is said to be unstable a-
bove $-70°C$. The nmr spectra of the complex [H] at $-75°C$ as reported
showed the expected ABX pattern for the three vinylic protons [τ:
3.1(H_1, $J_{1,2}$=10.0H$_z$); 7.2(H_2, $J_{1,3}$=4.9H$_z$); 7.65(H_3, $J_{2,3}$=3.0)]. No
further characterization such as ir and x-ray structure data for the
complex were given.

Recently, Tsutsui and co-workers[12] have reported the preparation
of an air stable π-vinyl alcohol complex of platinum(II), i.e. chlo-
ro(acetylacetone) π-vinyl alcohol platinum(II) [J]. This complex
[J] was prepared by using the general procedure employed by Murahashi, i.e. the treatment of chloro(acetylaceto) π-ethylene platinum
(II) with vinyl trimethylsilyl ether in benzene gave the complex [I]
which on subsequent hydrolysis in hexane yielded yellow crystals of
π-vinyl alcohol complex [J]. The stability, solubility, and the
ease with which the complex [J] can be handled made it possible for

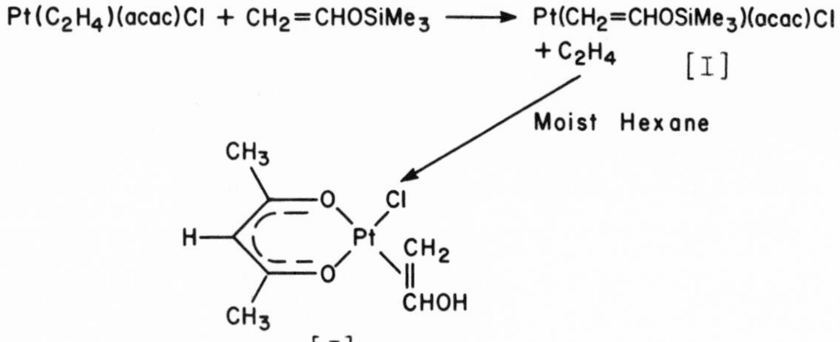
Pt(C2H4)(acac)Cl + CH2=CHOSiMe3 ⟶ Pt(CH2=CHOSiMe3)(acac)Cl
+ C2H4
[I]
Moist Hexane
CH3
H
O
Cl
Pt
CH2
O
CHOH
CH3

[J]

well characterization by ir, nmr, and x-ray diffraction[13] and exchange studies[14] (for ir and nmr see Table II).

The nmr spectrum of the π-vinyl alcohol complex [J] showed A_2X pattern like that of Green's complex [B]. Moreover, the complex [J] is a strong acid and can be titrated with sodium hydroxide in 50% aqueous acetone which gives anionic β-oxoethyl complex [K]. The ir spectrum of the complex [K] showed typical aldehydic [C=O] stretch

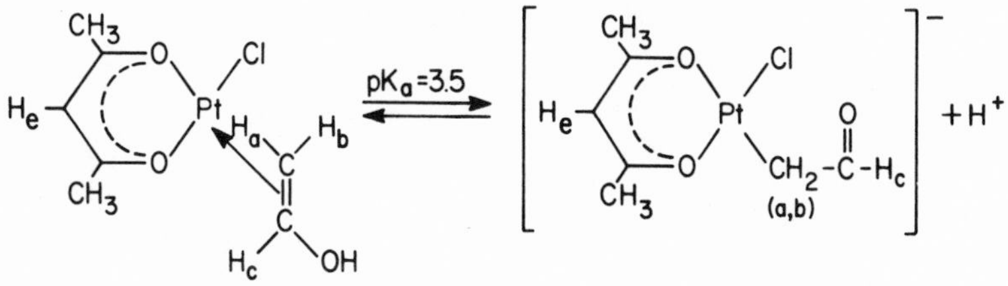
CH3
He
O
Cl
Pt
O
CH3
Ha
Hb
C
C
Hc
OH
pKa=3.5
CH3
He
O
Cl
Pt
O
CH3
O
CH2-C-Hc
(a,b)
+H+

[J] [K]

at 1650 cm^{-1} and the nmr spectrum at 0.7 τ (H_1,t) and 6.7 τ (2 H,d) again A_2X pattern. (See Table II) The nmr spectrum observed for the mixture showed a rapid equilibrium between the complexes [J] and [K], Table II.

Viewing the acidic behaviors of the complex [J], it is not surprising to see the equivalence of protons (H_a) and (H_b) as there is some degree of ionization (as expected) in polar solvents.

The x-ray studies reported for the complex [J] by the authors have shown that metal to vinyl alcohol linkage is intermediate between a conventional π-olefin and a sigma bonded aldehyde complex. It is shown (see Figure I) that the principal coordination plane of the platinum atom does not bisect the olefin bond as is usually found in platinum(II) olefin complexes. The midpoint of the olefin bond is 0.59 Å above the plane and C(1) lies 0.09 Å below the plane. Thus, the platinum-carbon bond lengths differ significantly though

Table II

Infrared spectrum of [J] chloro (acetylacetonate) π-vinyl alcohol platinum(II) in [KBr]

3260 s	OH stretch
3080 w	CH stretch
3020 vw	CH_2 asym
1550 s	C=C stretch, CH bonding, CH_2 bonding
1405 w	CH_2 bonding, CH bonding, OH bonding
1330 s	OH bonding, CO stretch
1270 vs	CH bonding, CO stretch, OH bonding
1145 w	C=C stretch, CO stretch
1013 vw	
968 vs	out of plane bonding
934 m	
850 w	CH_2 rocking
497 s	C=CO bonding

NMR data for the complex [J] and [K] and the mixture in the ratio [1:2] of the complex [J] & [K] respectively

	[J]	[K]	J:K (1:2)	% stuff
Hc ppm	2.63	1.84	1.52	64
JHa, b-Hc H_2	2.00	4.5	3.5	60
JPt-Hc H_2	71	20	38	65
Ha, b ppm	6.11	6.72	6.51	66
JPt-ab H_2	76	113	99	64
CH_3 ppm	8.01	8.32	8.20	61
CH_3 ppm	8.11	8.36	8.27	64
He ppm	4.36	4.68	4.59	72

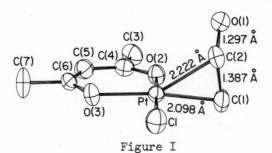

Figure I

both carbon atoms are within bonding distance of the metal. Tsutsui and co-workers have also reported that a very small percentage [0.46%] of free labelled ^{14}C acetaldehyde exchanges with the complex [J]. They carried out the reaction between free labelled ^{14}C acetaldehyde and the complex [J] in dioxane for 45 hours at room temperature. At the end of the period, they noticed an activity corresponding to 0.46% exchange which shows a very small amount of exchange. It has been suggested that if the exchange proceeds via attack of free vinyl alcohol then the limited exchange is attributable to the small amount of vinyl alcohol in equilibrium with acetaldehyde. e.g.

$$(acac)ClPt{\leftarrow}||_{CHOH}^{CH_2} + CH_3{*}CHO \rightleftharpoons (acac)ClPt{\leftarrow}||_{{*}CHOH}^{CH_2} + CH_3CHO$$

$$< 0.46\ \%$$

A more simple and convenient procedure for the preparation of the complex [J] has been reported by Tsutsui and co-workers.[15] The complex [J] was prepared upon treatment of a THF solution of complex chloro(acetylacetonate) ethylene platinum(II) and excess acetaldehyde in presence of aqueous KOH gave an anionic complex [K] which on protonation with mineral acid gave the complex [J].

$$Pt(C_2H_4)(acac)Cl + CH_3\overset{O}{\overset{||}{C}}R \xrightarrow{KOH} K\left[Pt(CH_2\overset{O}{\overset{||}{C}}R)(acac)Cl\right] + C_2H_4 + H_2O$$

$$[K]$$

$$K\left[Pt(CH_2\overset{O}{\overset{||}{C}}R)(acac)Cl\right] + HCl \longrightarrow Pt(CH_2{=}\overset{OH}{\overset{|}{C}}{-}R)(acac)Cl + KCl$$

$$[K] \qquad\qquad [J]$$

R = H or CH$_3$

Thus, the studies made on the complexes [J] and [B] unambiguously show that the π−vinyl alcohol group has exhibited a strong acidity in association with σ−bonded β−oxoalkyl complexes. This possibility can not be ruled out for the complexes [F] and [H]. The complex [F] was characterized in the solid phase due to its poor solubility. The complex [H] however, showed ABX behavior in nmr which is unlike the A$_2$X pattern for the complexes [J] and [B].

This change in the nmr studies of these complexes could be at-

tributed to the experimental conditions. Since π-vinyl alcohol complexes of only two metals are known so far, it remains yet to be determined whether this unusual behavior is common to π-vinyl alcohol
complexes of other transition metal in general.

REFERENCES

1. A. Gero, J. Org. Chem., 19, 469 (1954).
2. R. Criegee and G. Schroder, Angew Chem., 71, 70 (1959);
   M. Avram, E. Mrica and C. D. Nenitzescu, Chem. Ber., 92, 1088
   (1959);
   A. Brown, E. H. Braye, A. Clauss, E. Weiss, U. Kruerke, W. Hubel,
   G. D. I. King and C. Hoogzand, J. Inorg. Nucl. Chem., 9, 204
   (1959).
3. E. W. Gowling, S. F. A. Kettle, and G. M. Sharples, Chem. Comm.,
   21 (1968).
4. E. O. Fischer and A. Massbol, Chem. Ber., 100, 2945 (1967).
5. J. K. P. Ariyaratne and M. L. G. Green, J. Chem. Soc., 2976
   (1963).
6. R. Jira, J. Sedlmeier and J. Smidt, Liebigs., Ann. Chem. Internat. Edn., 1, 80 (1962).
7. J. Smith, W. Hafner, R. Jira, L. Sedlmeier and A. Sabel, Angew.
   Chem. Internat. Ed., 1, 80 (1962).
8. J. K. P. Ariyaratne and M. L. H. Green, J. Chem. Soc., 1 (1964).
9. M. Rosemblum, A. Cutler and S. Raghu, J. Organometal. Chem. in
   press and personal comm. to Prof. M. Tsutsui.
10. Y. Wakatsuki, S. Nozakura and S. Murahashi, Bull. Chem. Soc.
    Japan, 42, 273 (1969).
11. H. Thyret, Angew. Chem. Internat. Ed., 11, 520 (1972).
12. M. Tsutsui, M. Ori and J. Francis, J. Chem. Soc., 94, 1414
    (1972).
13. F. A. Cotton, J. N. Francis, B. A. Frenz and M. Tsutsui, J. Am.
    Chem. Soc., 95, 2483 (1973).
14. M. Tsutsui and J. N. Francis, Chem. Letters, 663 (1973).
15. J. Hillis and M. Tsutsui, J. Am. Chem. Soc., 95, 7907 (1973).

THE MODE OF METAL TO CARBON BOND FORMATION BY OXIDATIVE ADDITION

John A. Osborn

Department of Chemistry, Harvard University, Cambridge,
Massachusetts 02138

At the outset of this work, two mechanisms for the oxidative addition[1] of alkyl halides to low valent metal complexes had been postulated. One process involved the metal functioning as a nucleophile,[2] using two essentially non-bonding electrons, as depicted below:

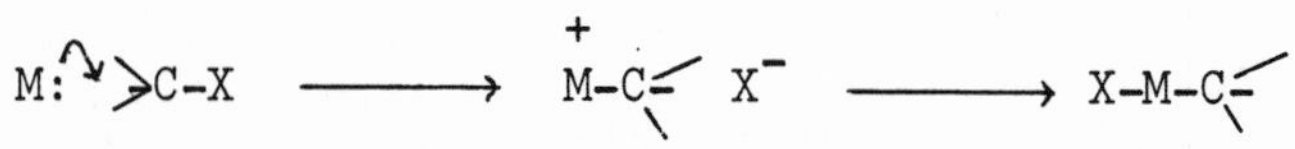

$$M: \overset{\curvearrowright}{} >C-X \longrightarrow M-C \overset{+}{\underset{\diagdown}{<}} \; X^- \longrightarrow X-M-C \overset{\diagup}{\underset{\diagdown}{}}$$

The second suggested mechanism involved a concerted three-center addition,[3] in which metal and alkyl halide orbitals of the appropriate symmetry interact so that smooth cleavage of the RX bond and formation of the MR and MX bonds occur synchronously.

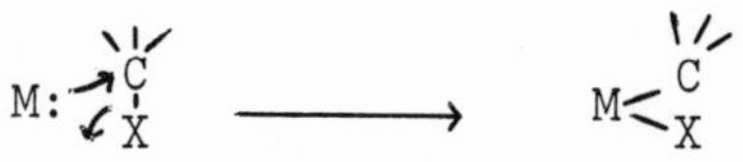

$$M: \overset{\diagup\!\diagdown}{\underset{\diagdown}{\rightharpoonup}} \overset{\displaystyle C}{\underset{X}{|}} \longrightarrow M \overset{\diagup\!\diagdown}{\underset{\diagdown}{<}} \overset{\displaystyle C}{\underset{X}{}}$$

During the last few years we have systematically studied the reactions of two classes of complexes of the general formulas 1 and 2 with alkyl, aryl, vinyl and related halides. For alkyl halides, an experimental distinction between the above two processes is clear: the observation of inversion or retention of configuration at carbon in the adduct.

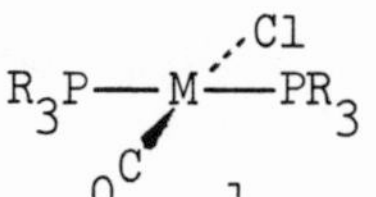

$$R_3P \overset{}{\underset{\underset{O}{\diagup}C}{\overset{\diagdown Cl}{\underset{}{M}}}} PR_3$$

1

M = Ir[1] 1a
  = Rh[1] 1b

$$R_3P \overset{\diagdown}{\underset{\underset{PR_3}{|}}{\overset{M}{}}} \overset{\diagup}{PR_3}$$

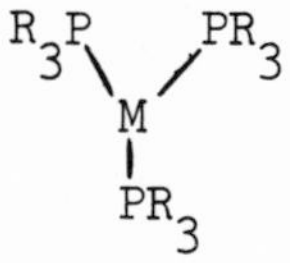

2

M = Pt[0] 2a
  = Pd[0] 2b

Additions to Ir^I and Rh^I

Initial experiments involving $\underset{\sim}{1a}$ ($PR_3 = P(CH_3)_3$) and using appropriately designed substrates indicated that neither mechanism was operative. Figure I summarizes results of addition of $\underset{\sim}{1a}$ to erythro and threo forms of $C_6H_5CHFCHDBr$, which can be conveniently studied by ^1H and ^{19}F nmr. Racemisation has occurred at the reacting carbon center.[4] Similarly an α-bromoester analogue, $C_6H_5CHFCH(COO-C_2H_5)Br$, reacts also with racemisation.[5] In this case, it is possible to separate out one diastereomeric pair and show that this does not epimerise to the other pair in the presence of 1a. Thus, racemisation after metal-carbon bond formation by rapid nucleophilic displacement on the metal-carbon bond by iridium(I) is not occurring.

During these investigations we made a further important observation. Many of these reactions (but not all) were greatly inhibited by efficient radical scavengers such as duroquinone or galvinoxyl. Five mole percent of scavenger routinely caused a 20-fold decrease in overall rate (see below). Certain other substrates (CH_3I, $C_6H_5CH_2Cl$, CH_3OCH_2Cl, $CH_2=CHCH_2X$) were, however, totally unaffected. At least two categories of alkyl halide reaction were therefore evident, one involving a radical chain path and the other some non-chain process.

Competitive experiments established an ordering typical of radical reactions for those alkyl halides reacting by the chain process, i.e., $n\text{-}C_4H_9Br < s\text{-}C_4H_9Br < t\text{-}C_4H_9Br$ and $C_6H_5CH_2CH_2Br \ll C_6H_5CHFCH_2Br$. Further evidence for a radical chain mechanism includes rate acceleration by radical initiators (e.g., AIBN) and formation of polyacrylonitrile when the reactions were carried out in the presence of $CH_2=CHCN$[6].

Based on earlier observations of the reaction of Co^{II} with alkyl halides, a reasonable propagation mechanism for the oxidative addition of RX to Ir^I can be postulated:

$$R^{\cdot} + Ir^I \longrightarrow Ir^{II}\text{-}R$$

$$Ir^{II}\text{-}R + RX \longrightarrow X\text{-}Ir^{III}\text{-}R + R^{\cdot}$$

which overall is

$$Ir^I + RX \longrightarrow X\text{-}Ir^{III}\text{-}R$$

The nature of the initiation step is unclear except that in some cases, small quantities of molecular oxygen accelerate addition.

For the analogous rhodium complexes ($\underset{\sim}{1b}$, $R = C_2H_5$), only the reactive halides (e.g., CH_3I, $C_6H_5CH_2X$, $CH_2=CHCH_2X$) readily added and the addition was unaffected by scavengers. Those halides which

undergo addition to Ir^I by a radical chain process appear very unreactive toward Rh^I. One-electron processes may not be as available
in rhodium(I) as in iridium(I) chemistry.

The nature of the process by which CH_3I, $C_6H_5CH_2X$, and $CH_2=$
$CHCH_2X$ add to 1a and 1b is unclear. The relative reactivities of
benzyl halides towards 1a ($R=CH_3$) ($C_6H_5CH_2Br/C_6H_5CH_2Cl \sim 50$) is that
expected for a nucleophilic displacement process. However, for allyl
halides the ratio $CH_2=CHCH_2Br/CH_2=CHCH_2Cl$ is about 2 and this may
indicate that prior coordination of the double bond precedes addition (as in an S_N2' process) (vide infra).[6]

Additions to Pt^o and Pd^o

For reactions of 2a ($R=C_2H_5$) the situation is more complex.[7]
1-Bromobutane in benzene (25°) reacts with 2a to yield initially
trans-$PtBuBr(PEt_3)_2$ (95%), trans-$PtHBr(PEt_3)_2$ ($\sim4\%$) and trans-$PtBr_2$-
$(PEt_3)_2$ ($\sim1\%$) along with $BuPEt_3^+Br^-$ and small quantities of butane
and but-1-ene ($< 5\%$). Further reaction causes growth of the hydride
at the expense of the butyl complex, followed by an even slower reaction in which the hydride is converted into the dibromide.Eventu-
ally the reaction products consist of only ca. equimolar quantities
of dibromide, alkene and alkane (and phosphonium salt). For secondary alkyls the reaction generally occurs more slowly, and produces
significantly larger initial quantities of hydride: i.e., maximum
yields of alkyl complexes do not generally exceed 30-40%.

Again many lines of evidence strongly suggest a chain reaction
for a large number of substrates studied (for exceptions, vide infra). The reactions are greatly inhibited by duroquinone or galvinoxyl. Neopentyl bromide (and many secondary alkyl bromides) when
allowed to react with 2a in toluene give significant quantities of
trans-$Pt(benzyl)Br(PEt_3)_2$. The large isotope effect ($k_H/k_D \sim 6$) for
the formation of the benzyl adduct in toluene-d_8 as solvent suggests
that benzyl radicals are generated by direct hydrogen atom abstraction from toluene by neopentyl radicals. No platinum benzyl products
are observed however for most unhindered primary halide reactions
in toluene, indicating that radicals in these cases are captured too
rapidly to permit hydrogen atom abstraction. For neopentyl bromide
(assuming an approximate value for the rate of abstraction of a hydrogen atom from toluene by a neopentyl radical), a rate constant
for the capture of neopentyl radical by Pt^o can be estimated as ca.
10 $M^{-1}sec^{-1}$. Correspondingly, for most unhindered primary halides,
the rate constant must be $>10^4$ $M^{-1}sec^{-1}$. In support of this, addition of 6-bromo-hex-1-ene to 2a gives ca. 3:1 product ratio of cy-
clized to linear alkyl metal adducts. Since the rate constant for
the closure of the 5-hexenyl radical is known ($\sim10^5$ sec^{-1}), the rate
constant for capture by Pt^o in this case can be estimated at ca. 10^6
$M^{-1}sec^{-1}$. Further, only racemic products have been observed for ad-

dition of α-haloesters to Pt^{o}. Many of these observations recur in
the reactions of the palladium complex $Pd(PEt_3)_3$ (2b).

All these data point to a radical chain process for oxidative
addition to Pt^o and Pd^o by many alkyl halides, similar to the iri-
dium(I) process already described:

$$Pt^o + R^. \longrightarrow Pt^I-R$$

$$Pt^I-R + RX \longrightarrow X-Pt^{II}-R + R^.$$

One puzzling aspect of the problem is that the primary halides
react more rapidly with 2a and 2b than secondary halides, which, at
first, is unexpected for a radical reaction. However, we observe
that addition of free phosphine to the reactions of 2a with primary
halides has only slight effect (due to formation of inactive Pt-
$(PEt_3)_4$ whereas with secondary halides a very pronounced effect is
observed. We suggest that the active metal capturing species for
secondary alkyl radicals is $Pt(PEt_3)_2$, which is present in low concentration in solutions of 2a. 2a appears to be much less reactive
toward secondary radicals (steric effects?) whereas 2a is clearly
involved in the predominant reaction pathway of primary halides.
The reversed ordering, primary > secondary, thus results from [2a]
>> $[Pt(PEt_3)_2]$ under normal experimental conditions.

The initial formation of trans-$PtHBr(PEt_3)_2$ is too fast to ori-
ginate from decomposition of trans-$PtRBr(PEt_3)_2$. This latter rate
can be measured separately under a variety of conditions. We believe that the hydrido-complex is produced by rapid olefin elimination in the intermediate species $RPt^1(PEt_3)_2$, followed by abstraction
of halide from RX by $HPt^1(PEt_3)_2$. Note that if the usual mode of
elimination via a complex of the type $HPt^1(olefin)(PEt_3)_n$ (n=2,3),
for n=2 this intermediate has 17 electrons, whereas for n=3, the
electron count is 19--which should disfavor this latter route.

For the less reactive bromides, the formation of trans-$PtBr_2$-
$(PEt_3)_2$ probably results from reaction of trans-$PtHBr(PEt_3)_2$ with
RBr in a radical chain process. Certainly this reaction is accelerated by radical initiators such as AIBN.

These observations are brought together in Scheme I and II.

Certain aspects of these reactions still remain obscure.
Firstly, some halides are not affected by scavengers and must follow a different route; as for iridium(I), CH_3I, $C_6H_5CH_2Cl$, $CH_2=$
$CHCH_2X$ are examples. But in contrast to iridium(I) chemistry, there
are, in addition, certain fuctionalized halides which also react by
a non-chain route, e.g., C_6H_5X (X=I, Br, Cl, CN), vinyl halides,
$CH_2=CH(CH_2)_2X$, and $BrCH_2CH_2COOEt$ (but not $CH_3CH(Br)COOEt$). Secondly, very reactive halides such as $C_6H_5CH_2Br$, $CH_3CH(Br)COOEt$ or

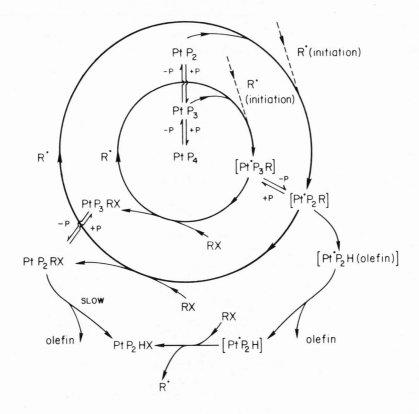

Scheme I

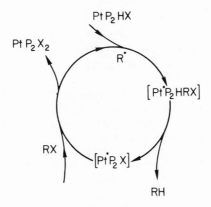

Scheme II

$(CH_3)_2CHI$ produce $Pt(PEt_3)_2X_2$ too rapidly for the reaction to occur via the chain processes described above. Moreover, these reactions are also largely unaffected by radical scavengers.

Let us discuss these latter reactions first. Lappert,[8] on the basis of trapping experiments using Bu^tNO, has suggested that CH_3I, $C_6H_5CH_2Br$ and C_2H_5I react with $Pt(P(C_6H_5)_3)_3$ to produce $PtRX(P(C_6H_5)_3)_2$ complexes via a non-chain mechanism, viz.

$$Pt(PPh_3)_3 \longrightarrow Pt(PPh_3)_2 + PPh_3$$

$$Pt(PPh_3)_2 + RX \longrightarrow [PtX(PPh_3)_2]\cdot + R\cdot$$

$$[PtX(PPh_3)_2]\cdot + R\cdot \longrightarrow trans\text{-}PtXR(PPh_3)_2$$

Several points must be made concerning this mechanism. Observation of the trapped alkyl radical, RBu^tNO, does not, in the absence of further experiments, discount either a chain mechanism or non-radical routes in total or in competition with other pathways. Indeed, we have shown that C_2H_5I (but not CH_3I or $C_6H_5CH_2Br$) reacts with $Pt(PPh_3)_3$ by a predominately chain route. Moreover, in order that good second order kinetics be followed, the fast step must result from a radical pair, $\overline{PtX(PPh_3)_3\cdot \ R\cdot}$. It was suggested that also the formation of both $PtHBr(PPh_3)_2$ and $PtBr_2(PPh_3)_2$ in these systems would result from reactions of $Pt^IX(PPh_3)_3$. Certainly we have observed the formation of $PtHBr(PEt_3)_2$ only in chain reactions or in the subsequent decomposition of $PtBuBr(PEt_3)_3$ and only in reactions of alkyl halides with a β-hydrogen. It seems that this non-chain radical route for hydride formation is not a major one. However, it does appear an attractive path for $Pt(PEt_3)_2X_2$ formation for very reactive bromides and iodides. Indeed, the reaction of $Pd(PEt_3)_3$ with $(CH_3)_2CHI$ yields $Pd(PEt_3)_2I_2$, propene, propane and $(CH_3)_2CHCH(CH_3)_2$. We now find that if this reaction is followed by ^1H nmr spectroscopy, CIDNP effects are observed in the propene and propane products and in the reactant $(CH_3)_2CHI$. The polarization effects are characteristic of a diffusive encounter of isopropyl radicals to form a radical pair. Similar observations are found in the reaction of 2a with $(CH_3)_2CHI$ and $C_6H_5CH_2Br$. However, we have observed these effects only when PtP_2X_2 complexes are rapidly produced, and, as yet not in cases when only the regular $PtRBr(PEt_3)_2$ complexes are formed. For example, $C_6H_5CH_2Cl$ reacts cleanly with $Pt(PEt_3)_3$ to yield the regular adduct $Pt(C_6H_5CH_2)Cl(PEt_3)_2$ in 100% yield. No CIDNP effects have been observed. On the other hand, $C_6H_5CH_2Br$ reacts to give a mixture of $Pt(C_6H_5CH_2)Br(PEt_3)_2$ and $PtBr_2(PEt_3)_2$ as immediate products. CIDNP is observed in the aromatic resonances. However, the rate of formation of both products is unaffected by radical scavengers.

We have written a generalized scheme for these reactions (Scheme)

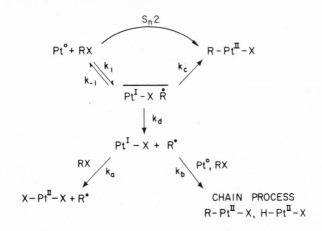

Scheme III   Generalized Scheme for Alkyl Halide
             Reaction with Pt(o)

III). Platinum(0) may react initially with an organic halide by two
paths (but vide infra): (i) an S_N2 displacement on carbon to give an
incipient cationic complex (possibly strongly ion-paired) followed
by reentry of the anion to give the regular oxidative addition adduct; (ii) halide abstraction (possibly preceded by electron transfer) to form a radical pair. This pair can either collapse (Lappert
route) to the regular adduct or can diffusively separate to form
Pt^IX and R'. Here depending on the reactivity of the reactant alkyl
halide two pathways can develop. For very reactive halides (containing weak C-X bonds), the Pt^IX can rapidly abstract further halide to yield dihalide and organic radicals. Here CIDNP effects are
seen in the diffusive encounter of these product radicals. Alternatively, for less reactive halides, the generated organic radical
will set up a chain process (as described earlier) to produce the
regular adduct and hydrido-complex. Accordingly, in cases where the
chain reaction dominates, the halide abstraction step and subsequent
separation of the radical pair can provide the initiating radical.
Like mechanisms, of course, could also provide initiation for the
iridium(I) reactions.

    Returning to the benzyl bromide reaction with platinum(0) we
see (since the chain mechanism is not competitive) two ways in which
regular adduct can be formed: (i) via the S_N2 process or (ii) by
radical pair collapse. However, PtP_2Br_2 is also formed and should
its formation be competitive with the collapse mechanism (and not
the S_N2), then the product ratio should be dependent on solvent viscosity (k_d will decrease with respect to k_c as viscosity increases).
No viscosity effects on product ratio is found, however, This observation points to the conclusion that for benzyl bromide addition

to platinum(0), PtP_2RX formation is by S_N2 (and not pair collapse) whereas PtR_2X_2 formation is via a radical abstraction mechanism. In benzyl chloride presumably only the S_N2 is operative because in comparing RCl vs. RBr reactivities, $Cl/Br \sim 10^{-2}$ for nucleophilic displacements, whereas for radical abstraction processes $Cl/Br \sim 10^{-4}$. The radical process would then become non-competitive for $C_6H_5CH_2Cl$.

Another example concerns α-haloester reactions with platinum (0). Using ethyl α-chloropropionate, a regular adduct to 2a is cleanly produced--and here the reaction can be shown to occur by a radical chain process. In the corresponding α-bromoester reaction, two products are formed--the dibromide as well as the regular adduct. In this case only the regular adduct formation is inhibited by the presence of 5% duroquinone. Here it appears that the radical chain and abstraction processes are competitive, but the S_N2 process is not.

In summary then, several possible routes exist for these reactions. The choice of a particular pathway will depend on the nature of the carbon-halogen linkage, the nucleophilicity of the metal complex, the ability of the metal complex to undergo one electron processes (i.e., relative positioning of oxidation levels, availability of suitable mechanisms for the donation of one electron), steric effects, and ligand exchange processes. A subtle balance of these factors will direct which path (if any) dominates the oxidative addition process.

Reactions of Alkenyl Halides and Related Substrates

A noticeable difference between the reactivity of 1 and 2 is that certain substrates (aryl halides, vinyl halides) appear to undergo addition to iridium(I) via a chain process, whereas with platinum(0) an alternative path must be accessible.

We have systematically studied the interaction of alkenyl halides with 2a.[7] At low temperatures (-55°) in toluene-benzene (4:1) ^{31}P data shows that rapid π-complexation occurs and the following equilibrium is set up:

$$Pt(PEt_3)_3 + CH_2=CH(CH_2)_nX \rightleftharpoons Pt(PEt_3)_2(\pi-CH_2=CH(CH_2)_nX)$$

$$(n = 0,1...4) \qquad\qquad + Pt(PEt_3)_4$$

For allyl halides (n=1), no π-complex could be detected, since oxidative addition occurred too rapidly even at -55°. For all other values of n, π-complexes could be detected and approximate values of K at -55° determined. Relative to hex-1-ene, the stability of the π-complexes follows the order:

hex-1-ene(1) < 5-hexenyl bromide(3) < 4-pentenyl bromide(10) << 3-butenyl bromide(200) << vinyl bromide(10^4).

Three further observations can be made:
(a) Duroquinone has no effect on the reactions of 2a with alkenyl halides when n=0, 1 and 2. A moderate inhibition is observed for n=3, but very strong inhibition for n=4.
(b) The rates of oxidative addition to 2a for alkenyl bromides when n=0, 1 or 2 are considerably greater than in the reactions of 2a with the corresponding alkyl bromides. Little difference in rate is observed, however, between 5-hexenyl bromide and n-hexyl bromide; 4-pentenyl bromide adds somewhat faster to 2a than pentyl bromide.
(c) The rearrangement of $Pt(PEt_3)_2(CH_2=CHBr)$ to the oxidative addition product $Pt(PEt_3)_2(CH_2=CH)Br$ follows a first order rate law. Further, for cis- and trans-styryl halides, addition to 2a proceeds with total retention of configuration.

It appears that a further mechanism for oxidative addition is operative, in which prior coordination of a donor function on the reacting substrate occurs. If the carbon-halogen bond is strategically positioned as shown below, a facile intramolecular addition can occur. (See Scheme IV which illustrates the reaction of $CH_2=CHCH_2CH_2Br$ with 2a).

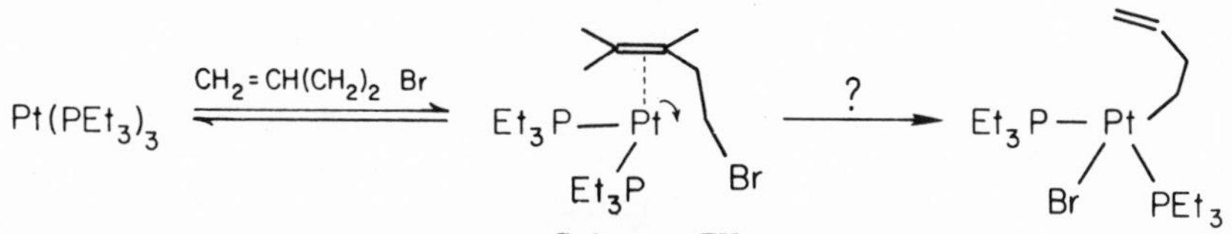

Scheme IV

Two points may be made concerning this process. The intramolecular addition could involve an S_N2, concerted or radical process, although the former would intuitively be favored. Secondly, it is clear that such a rearrangement can only occur on a complex with an electron count of 16. Hence it is the accessibility of the 14-electron moiety, $Pt(PEt_3)_2$, which allows both π-complexation and further rearrangement, that allows this template process to take place. Here is a possible reason why iridium(I) with such substrates still reacts via a chain path. trans-$Ir(PMe_3)_2COCl$ forms stable, 18-electron, complexes with olefins. Even if a reactive carbon-halogen bond is available on the olefin substrate, facile rearrangement cannot occur, since 20-electron configurations are very much higher in energy. Hence alkenyl halides (except allyl) react by a non-template route for 1a. As we have previously indicated, allyl halides may also react via prior coordination, but in this case, an S_N2' process is possible, which can occur on a complex with an 18-electron configuration. (Scheme V) This process, however, has yet to be experimentally established.

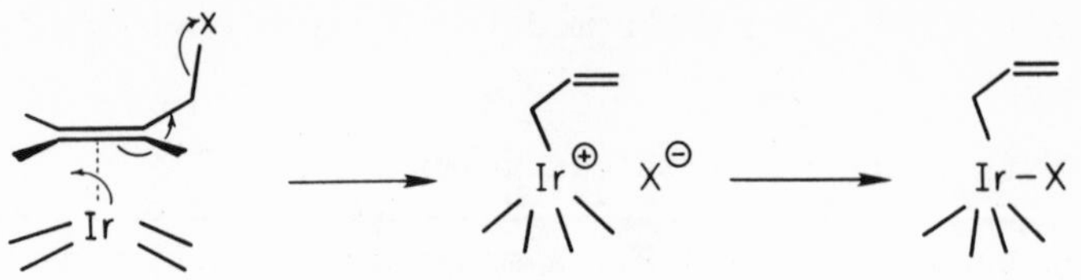

Scheme V

The operation of this template process may also account for the fact that, whereas $CH_3CH(Br)COOC_2H_5$ adds to 2a by a radical chain process, the isomeric substrate, $BrCH_2CH_2COOC_2H_5$, is unaffected by radical scavengers. Clearly, as results on the alkenyl bromides (n=2,3, and 4) illustrate, for the template mechanism to operate, the size of the ring developed in the transition state for rearrangement is crucial.

Finally, it is tempting to speculate that aryl halides, which add very readily to complexes of the type 2, but only sluggishly to type 1 complexes, react with 2 via a template mechanism similar to that suggested here, i.e., via prior coordination (via a 14-electron species?) followed by rearrangement.

CONCLUSIONS

Oxidative addition is a term used to describe a transformation relating reactants and products. It contains no mechanistic information; it is essentially a thermodynamic statement. Transition metals have a variety of properties (some of which we have described above) and it is to be expected that subtle changes in either metal complex or substrate may have a profound effect on the mechanistic pathway followed. It is, of course, these subleties which both stimulate our interests and provide impetus for both synthetic and mechanistic transition metal chemistry. The implications of these studies in designing complexes to perform specific tasks in organic and catalytic chemistry are evident.

ACKNOWLEDGEMENTS

The author is greatly indebted to coworkers J. A. Labinger and A. V. Kramer who carried out most of the work described here, and to J. S. Bradley for some of the earlier material. Grants from the Alfred P. Sloan and Henry and Camille Dreyfus Foundations made much of this work possible.

APPENDAGE

A summary of oxidative addition mechanisms, some established, some still speculative, is collected below. Suggestions of functional groups which may react according to the various paths are also made.

1. Two Step-One Electron Processes
    a) Radical chain (e.g., $Bu^{n}Br + Ir^{I}$ (Pt^{0}, Pd^{0}))

$$M + R^{\cdot} \longrightarrow M-R$$

$$M-R + R-X \longrightarrow X-M-R + R^{\cdot}$$

   b) Radical pair ($CH_3I + Pt^{0}$??)

$$M + R-X \longrightarrow \overline{M-X\ R^{\cdot}}$$

$$\overline{M-X\ R^{\cdot}} \longrightarrow R-M-X$$

   (Other possibilities: metal-halide, boron-halide additions)

2. Metal as a Nucleophile
    a) S_N2 process (e.g., $RX + Rh^{I}(DO)DOH)pn$; $CH_3I + Ir^{I}$; $C_6H_5CH_2Cl + Pt^{0}$)

$$M:\overset{\curvearrowright}{} R-X \longrightarrow M-R^{+}X^{-} \longrightarrow X-M-R$$

   b) S_N2' process (α-haloacetylene + Ir^{I} ?)

   (Possibility: $CH_2=CHCH_2X$ additions to Ir^{I}, Pt^{0})

3. Metal as a Template
    (e.g., $CH_2=CHX + Pt^{0}$ (Pd^{0}), $CH_2=CH(CH_2)_2X + Pt^{0}$)

   (Possibilities: $R_3SiH + Pt^{0}$, $C_6H_5X + Ni^{0}$ (or Pt^{0}))

4. Concerted Three-Center Addition

$$(\text{e.g., } H_2 + Ir^I, Pt^0)$$

(Possibilities: $>$C-H, $>$C-C$<$, $>$B-H, $>$Si-H additions)

REFERENCES

1. For a review see J. P. Collman, Accounts Chem. Res.
2. P. B. Chock and J. Halpern, J. Amer. Chem. Soc., 88, 3511 (1966).
3. R. G. Pearson and W. R. Muir, J. Amer. Chem. Soc., 92, 5519 (1970).
4. J. S. Bradley, D. E. Connor, D. Dolphin, J. A. Labinger and J. A. Osborn, J. Amer. Chem. Soc., 94, 4043 (1972).
5. J. A. Labinger, A. V. Kramer, and J. A. Osborn, J. Amer. Chem. Soc., 95, 7908 (1973).
6. J. A. Labinger, Ph.D. Thesis (Harvard) 1973.
7. A. V. Kramer, Ph.D. Thesis (Harvard) 1974.
8. M. F. Lappert and P. W. Lednor, Chem. Comm., 948 (1973).

DISCUSSION

DR. IBERS: May I ask a naive question on this olefin mechanism you talked about last, with bromoethylene on platinum. Wouldn't you expect in terms of the vinylization to get a cis product rather than a trans product?

DR. OSBORN: The rate of isomerization of cis to trans is very fast in the presence of phosphines in solution, and the liberated triethylphosphine will catalyze this isomerization.

DR. IBERS: You don't think you will see that in the NMR?

DR. OSBORN: We have not seen it. Maybe we should look harder over shorter reaction times.

DR. IBERS: Yes, it seems to me that the initial step might be expected to be cis.

DR. OSBORN: Very possibly - I think this problem is worth looking at carefully.

DR. OTSUKA: In our experience, in the reaction of vinyl halides you always get trans.

DR. IBERS: In the final product?

DR. OTSUKA: Yes.

DR. OSBORN: If you noticed on the slide of the template reaction, I put a question mark on the second step. It is not clear whether that is going by a nucleophilic displacement, by a concerted cleavage, or by a radical pathway. If it goes by a nucleophilic displacement you would end up with the trans product if it is thermodynamically controlled. If it is concerted you might anticipate prior formation of cis adduct.

DR. HALPERN: On this point of the template reactions the scheme you have considered is one in which you bind at the olefins sites and react by a non radical path at the other site, i.e. at the carbon halogen bond. You could just as well invoke a radical reaction at the carbon halogen bond and this might explain the results on the homoallyl system because if you had halogen abstraction by platinum in that case, the radical which you form and which is initially bound through the olefin would have its unpaired electron rather close to the metal, and it might be trapped so that you could not detect any spin.

DR. OSBORN: That is why I left the question mark on that step. That is a possibility. If one believes the kind of mechanism Lappert was suggesting then that is a strong possibility. I think it is important to do the stereochemical experiments.

DR. BERGMAN: Because benzyl bromide is in your uninhibited category, it would be really nice to know the stereochemistry of the addition. I wonder whether it might be reasonable to try using an optically active phenyl CHDBr and at least see if there is any optical activity in the oxidative addition product.

DR. OSBORN: I hear that a closely related experiment has been done, although to my knowledge it has never appeared in print. I think Stille has studied the addition of optically active $\phi CH(CF_3)$-Br to Pd(0).

DR. WHITESIDE: That has turned out to be more complicated.

DR. OSBORN: Yes.

DR. BERGMAN: That is the secondary bromide, though.

DR. OSBORN: Yes.

DR. BERGMAN: Was it on iridium?

DR. OSBORN: No, I believe it was on a Pd(0) complex.

DR. BERGMAN: But is is a secondary system, and you don't know whether that is on the inhibited or uninhibited category, right?

DR. OSBORN: We have not studied that substrate.

DR. BERGMAN: Thats why I'd like to know the stereochemistry of the primary benzylic addition, since it has been well studied. It would be nicest of course to get the absolute configuration of the adduct but that is going to be difficult; one could at least find out if it has any optical rotation. The other question I would like to ask is whether it might be reasonable to compare the products formed from homoallyl and cyclopropylcarbonyl halides? Is there any possibility you are getting any interconversion between homoallyl and cyclopropyl?

DR. OSBORN: No, we get the normal adduct; we don't get any cyclopropyl carbonyl adduct.

DR. BERGMAN: If you carry out the reaction with cyclopropyl carbonyl halide, do you get the cyclopropyl carbonyl adduct or do you get the homoallyl analog?

DR. OSBORN: I think we have done that experiment but I am not certain. We have thought of this obvious arrangement and written it down to do, but I am not sure whether or not we have done it yet.

DR. COLLMAN: I have three questions. In the addition of aryl halides to these low valent transition metal systems, the iodides are invariably faster than the bromides, and presumably chloride. I think some recent work by W. F. Little at Chapel Hill has shown that chloride is very unreactive. This is the opposite of the order for traditional nucleophilic displacement behaving in a classic fashion. It is hard to understand, although there may be some subtle factors in the metal reaction. You had a case of a square planar system d8 system reacting with allyl bromide or chloride and the bromide is only slightly faster. Did you in fact do those particular comparisons in the presence of galvinoxyl because there is always the possibility of picking up the other mechanistic pathway?

DR. OSBORN: The allyl halide reactions have been carried out in the presence of galvinoxyl, which has no effect, but we have not studied a direct competitive reaction of allyl halides.

DR. COLLMAN: The criteria of studying the stereochemistry of the carbon being attacked is not always valid either. Self displacements are common and their mechanisms are not well understood although a one electron possibility has been pointed to as the case Espenson has pointed out. You said there was no self displacement in one of your early cases in your vicinal system. Were the products at equilibrium? If they were equilibrium the comment is not valid.

DR. OSBORN: The iridium adducts do not readily equilibrate.

DR. BERGMAN: Don't you take one isomer out and subject it to the reaction conditions?

DR. OSBORN: Yes. We have separated out one of the diasteriomeric pairs because they have different physical properties, and reacted it with some more iridium(I) - the other diasteriomeric pair is not formed under mild conditions.

DR. WHITESIDES: In the reactions with the additions to iridium you write the species that is trapping the radical as iridium III. Why isn't it some other iridium species? For example, iridium II or for that matter iridium III.

DR. OSBORN: Mainly we don't see any iridium(II) in solution. I agree the collapse with radicals would be extremely fast but if you estimate the concentration of free radical, and an upper limit for the concentration of Ir(II) one might expect, the overall rate would be very small.

DR. HALPERN: That is a termination step in your process?

DR. OSBORN: Yes, that would be a termination step.

DR. WHITESIDES: What is the rate of disproportionation of iridium II?

DR. OSBORN: I do not know, although there are some stable iridium II species.

DR. HALPERN: What is known about the reactivity of radicals toward platinum(0)?

DR. OSBORN: Nothing, I think, on platinum(0). For platinum(II) Lappert has looked at the t-butoxide radical attacking platinum(II) alkyls and gets displacement of the alkyl radical and production of the corresponding platinum(II) alkoxide. That is extremely fast but is a displacement process. There is a brief note in Chem. Comm. about this a year ago. The t-butoxide radical does appear to associate with platinum(II) fairly readily.

DR. WHITESIDES: In principle, one might "change" some of these schemes in such a fashion that you would have the radical being trapped by other metal valence states and still come out with the same sort of answers.

DR. OSBORN: Certainly, that might be correct. At the present stage of the art, one has to put the most reasonable schemes down and test them out to see whether they hold water. At the moment, I don't see any reason to postulate anything more complicated than what we have advocated. It could be more complex, but I would not at present apply the principle of Occams razor.

DR. PARSHALL: Have you accumulated any results yet with nickel?

DR. OSBORN: We have done a few experiments with Ni. It appears as though we can inhibit some nickel (0) reactions with radical scavengers.

DR. PARSHALL: From our results it looks as though you might see a reversal with aryl halides, that is, they may react via the radical pathway. Whereas the Pd and the Pt compounds work very nicely in toluene as a solvent, the reactions become very messy with Ni, and we had to go to an aliphatic solvent.

DR. OSBORN: The trouble with Ni is that it is more difficult to study the reaction.

DR. PARSHALL: The Ni(I) is much more stable, but it looks like a very good pathway.

DR. PETTIT: Is anything known about the stereochemistry of the vinyl chloride when you replace the chlorine by a metal?

DR. OSBORN: The stereochemistry is always retained in Pt and Pd, to my knowledge.

DR. COLLMAN: One other quick point, do other olefins inhibit these vinyl halide reactions, because that would be a test of your template mechanism.

DR. OSBORN: Phosphines do.

DR. COLLMAN: They are not olefins.

DR. OSBORN: Let me get on the phone and I will call my man to do the experiments today.

DR. HALPERN: I do not agree that would be a test of the template reaction because even in the case of conventional oxidative addition, if it required a 2 - coordinate bis-phosphine platinum it would still be inhibited by excess ligand in the same way as the template reaction.

DR. OSBORN: That is right.

DR. HALPERN: And in fact, the oxidative addition of methyl iodide to platinum(0) is inhibited by excess phosphine.

DR. MANGO: I don't really understand your template mechanism as it applies to bromobenzene. Are you proposing that the metal is going to coordinate in a face-on way, and then somehow reach around and bond to the Br?

DR. OSBORN: If you write the aryl halide in sort of a localized bond form, then one can regard the X group either as vinylic or allylic:

so that the metal binds in a dihapto fashion on the face. This structure is reminiscent of a pi-benzyl system, and, in that way, one has a quasi-allylic halide. The problem is, then, how does the rearrangement occur? For model substrates we are looking at - for example - acrylonitrile and allyl cyanide. The former does not readily add oxidatively; the latter we have not yet studied.

ORGANOACTINIDES: COORDINATION PATTERNS AND CHEMICAL REACTIVITY

Tobin J. Marks

Department of Chemistry, Northwestern University,
Evanston, Illinois 60201

The explosive growth of organometallic chemistry over the past two decades, has, until very recently, largely bypassed the actinide elements. The reason for this partly stems from early unsuccessful attempts to synthesize uranium tetraalkyls. However, the actinide metal ions offer unique electronic (5f valence orbitals) and stereochemical (high coordination numbers, unusual coordination geometries) features which suggest intriguing prospects as organometallic reagents and catalysts. This paper summarizes recent developments in two important fields: sigma-bonded organoactinides and actinide ions as templates in cyclooligomerization reactions. In both areas, metal ion coordination geometry plays a central role in directing the path of various chemical transformations.

SIGMA-BONDED ORGANOACTINIDES

It has recently been shown[1] that the appropriate selection of stabilizing ligands allows the synthesis of a large number of uranium(IV) alkyl and aryl compounds. The molecular structure of the R = phenylacetylide compound is shown in Figure 1.[2]

$$(\eta^5\text{-}C_5H_5)_3UCl + RLi \longrightarrow (\eta^5\text{-}C_5H_5)_3U\text{-}R + LiCl \qquad (1)$$

R = variety of alkyl and aryl groups

A striking feature of these compounds is their extraordinary thermal stability in solution (Table I). That such monohapto organometallics are not unique to uranium is demonstrated by the high yield synthesis of thorium(IV) analogs as shown in eq (2).[3] These also have exceptional thermal stability (Table I).[3b]

$$(\eta^5\text{-}C_5H_5)_3ThCl + RLi \longrightarrow (\eta^5\text{-}C_5H_5)_3Th\text{-}R + LiCl \qquad (2)$$

R = various alkyl groups

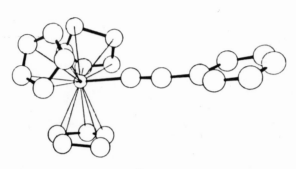

Figure 1. X-ray structure of $(\eta^5\text{-}C_5H_5)_3U(\text{-}C{\equiv}CC_6H_5)$ from reference 2.

TABLE I

Comparative Kinetic Data for Thermolysis of
$(C_5H_5)_3MR$ Compounds in Toluene Solution[a]

	M = U	M = Th
R =	$\Delta G^{\ddagger}$ (kcal/mole)	$\Delta G^{\ddagger}$ (kcal/mole)
$n\text{-}C_4H_9$	33.3 (97°)	37.6 (167°)
neopentyl	32.2 (97°)	41.4 (167°)
$i\text{-}C_3H_7$	29.8 (72°)	34.3 (167°)
allyl	28.7 (72°)	38.6 (167°)

[a]All compounds exhibited good first-order kinetic behavior.

In view of the reported[4] instability of uranium tetraalkyls, we have undertaken an investigation of those factors which stabilize actinide-to-carbon sigma bonds. Thermolysis of both $(C_5H_5)_3UR$ and $(C_5H_5)_3ThR$ compounds in toluene solution does not occur via the commonly observed β-hydrogen elimination sequence, eq (3) sometimes attended by eq (4). Rather, for both thorium and uranium, elimination of R-H takes place. Kinetic data (good first-order plots) and

$$CH_2 = CHR$$
$$\downarrow$$
$$M-CH_2CH_2R \rightleftharpoons M \underline{\quad\quad} H \rightleftharpoons M-H + CH_2 = CHR \qquad (3)$$

$$\underline{A}$$

$$M-H + M-CH_2CH_2R \longrightarrow 2M + CH_3CH_2R \qquad (4)$$

deuterium labelling studies indicate that the H is derived from a
cyclopentadienyl ring in an intramolecular abstraction.[1c,3b] Analysis of pmr spectra together with kinetic data for the rate of dis-
appearance of resonances corresponding to various groups on the same
$(C_5H_5)_3MR$ molecule, demonstrates that facile β-elimination prior to
thermolysis, such as in the isomerization of eq (5), does not occur
to any appreciable extent. For $(C_5H_5)_3U$(n-butyl) the hydrogen abstraction exhibits a kinetic isotope effect (k_H/k_D) of about 8 ± 1.[5]

$$\begin{array}{c} CH_3 \\ | \\ M-CH \\ | \\ CH_3 \end{array} \rightleftharpoons \begin{array}{c} \overset{H}{CH_2 = C-CH_3} \\ \downarrow \\ M-H \end{array} \rightleftharpoons M-CH_2CH_2CH_3 \qquad (5)$$

$$\underline{A}$$

Complete retention of configuration is observed in the 2-butenes
derived from decomposition of 2-cis- and 2-trans-2-butenyl uranium
and thorium tris(cyclopentadienyls).[1c,3b] The mechanism of thermolysis for both the uranium and thorium complexes can be discussed
in terms of either a homolytic bond scission, (B), followed by stereospecific ring hydrogen abstraction within a very tightly constrained
solvent cage, a 4-center concerted elimination, (C), or an oxidative
addition of a C-H bond to the actinide, (D), followed by stereospecific R-H elimination.

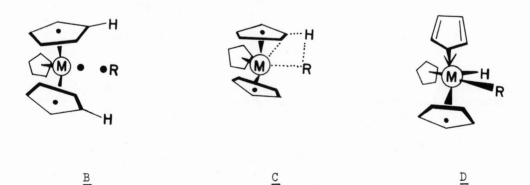

$$\underline{B} \qquad\qquad\qquad \underline{C} \qquad\qquad\qquad \underline{D}$$

Assuming the same mechanism is operative in both systems, then
the enhanced thermal stability of $(C_5H_5)_3ThR$ over $(C_5H_5)_3UR$ may be
consistent with either pathway (B) or (D) since both can involve
(formally) a change in oxidation state for M, and thorium(IV) is
both more difficult to reduce and to oxidize than uranium(IV). This
argument assumes the oxidative addition in (D) proceeds via M(IV)
$\longrightarrow$ M(VI). However, the cyclopentadienylene carbenoid ligand in
(D) actually permits considerable flexibility in oxidation state.

Hence reductive elimination subsequent to formation of (D) does not
require invocation of an unlikely divalent actinide. The uranium-
containing product of the thermolysis has not yet been isolated in
a completely pure state;[1c] analytical and spectroscopic data indicate it still possesses η^5-C_5H_5 groups. We have succeeded in isolating and crystallizing the thorium analog.[3b] Spectroscopic (vibrational, pmr, mass) and analytical data are compatible with structure (E).

One of the most striking features of $(C_5H_5)_3MR$ thermal stability and thermolysis behavior is the reluctance of these systems to
undergo β-hydrogen elimination, eq (3). This apparently reflects
coordinative saturation of the actinide (destabilization of structure (A)) and is likely a major reason for the high thermal stability of the R = i-propyl and n-butyl compounds, and for the absence
of eq (5). Further evidence for the importance of coordinative saturation is provided by our observations on the supposed uranium
tetraalkyls.[6] An example is shown in eq (6). Here coordinative

$$UCl_4 + 4\ n\text{-}C_4H_9Li \xrightarrow[\text{ether} -30°]{\text{hexane or}} [(n\text{-butyl})_4U] \qquad (6)$$

$$\downarrow$$

$$\text{butane} + \text{1-butene}$$

$$49\% \qquad\quad 46\%$$

saturation is unlikely, and the thermally unstable complex undergoes
extensive (essentially quantitative) β-elimination, as do all other
tetraalkyls where β-elimination, as do all other tetraalkyls where
β-hydrogens are present.[6]

Intermediate degrees of coordinative saturation can also be
achieved (eq (7)).[7] The stability and chemistry of these products
are also under investigation.

$$\text{(Cyclopentadienyl ligand)} + UCl_4 \xrightarrow{THF} \quad UCl_2 \cdot THF \xrightarrow{2\,RLi} \quad \textbf{Stable} \quad (7)$$
$$\textbf{dialkyl}$$

ACTINIDE IONS AS TEMPLATES

Template reactions constitute a large class of chemical transformations (both stoichiometric and catalytic) in which a metal ion
serves as a framework for the coordinative cyclization of organic
ligands. A fascinating question related to such reactions is whether increasing the ionic radius of the template (e.g. employing an
actinide ion) will yield an expanded macrocyclic ligand. The pronounced tendency of the uranyl ion to achieve a pentagonal bipyra-
midal ($\underline{F}$) or hexagonal bipyramidal ($\underline{G}$)coordination geometry led us

to investigate the modification of the usual phthalocyanine condensation, eq (8), by employing M = UO_2^{+2}. Our structural results[8]

$$4 \quad \text{(benzene-1,2-dicarbonitrile)} + M^{+2} \longrightarrow \text{(super phthalocyanine complex)} \quad (9)$$

indicate (Figures 2 and 3) that the so-called uranyl phthalocyanine
(known since 1956) is in reality a "super phthalocyanine" complex.
It is seen that the coordinative preferences of the uranyl ion can
dramatically alter the normal course of the cyclization reaction,

eq (9). Further examples of this phenomenon are under investigation.

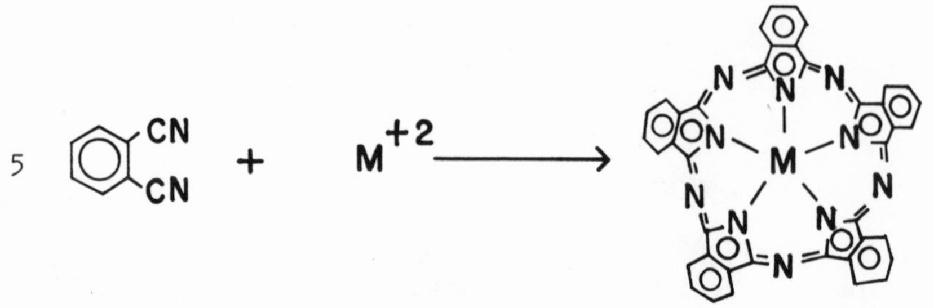

$$5 \quad \overset{CN}{\underset{CN}{\bigcirc}} \quad + \quad M^{+2} \quad \longrightarrow \quad \text{(structure)} \qquad (9)$$

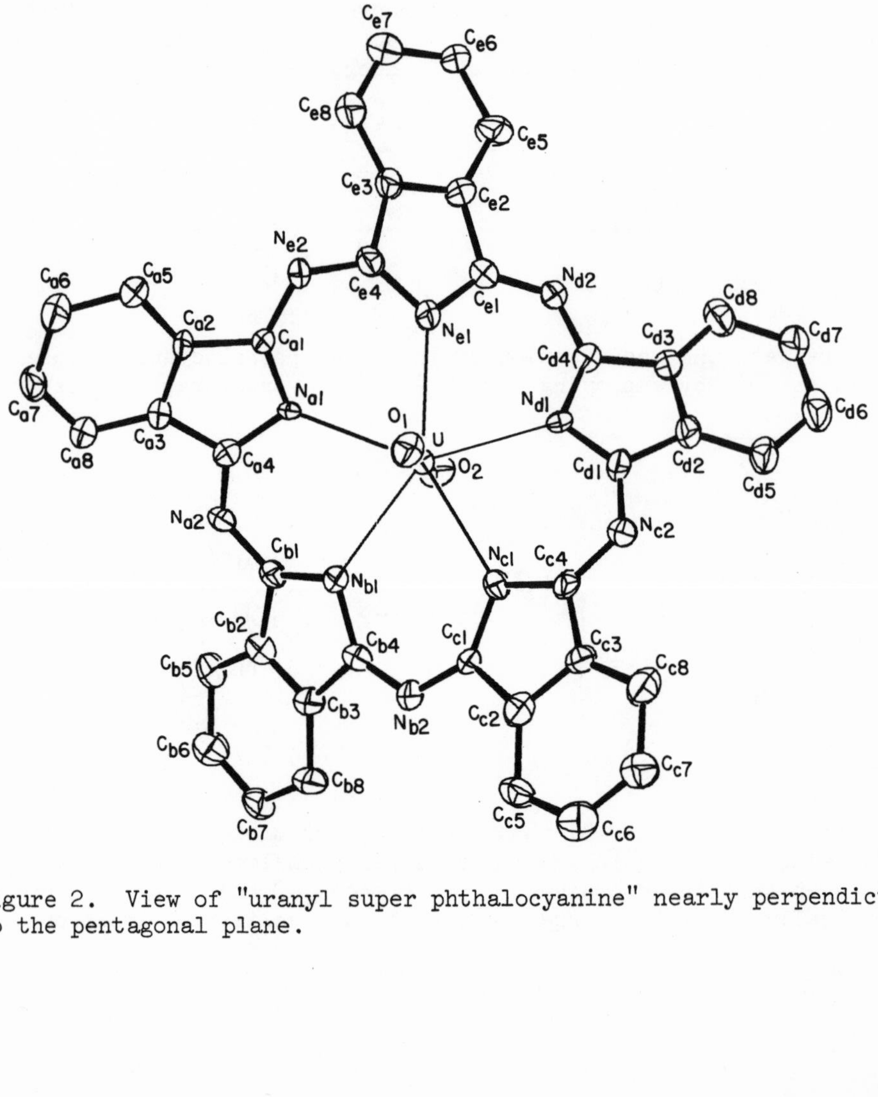

Figure 2. View of "uranyl super phthalocyanine" nearly perpendicular to the pentagonal plane.

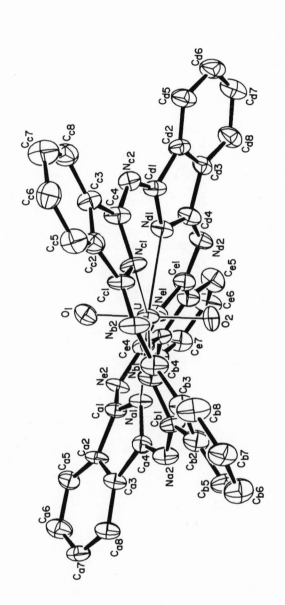

Figure 3. View of "uranyl super phthalocyanine" perpendicular to the O=U=O axis.

ACKNOWLEDGEMENTS

We (my co-workers and I) are grateful to the National Science Foundation and the Research Corporation for generous support of this research.

REFERENCES

1a. T. J. Marks and A. M. Seyam, J. Amer. Chem. Soc., 94, 6545 (1972).
  b. A. E. Gebala and M. Tsutsui, ibid., 95, 91 (1973).
  c. T. J. Marks, A. M. Seyam, and J. R. Kolb, ibid., 95, 5529 (1973).
  d. G. Brandi, M. Brunelli, G. Lugli, and A. Mazzei, Inorg. Chim. Acta, 7, 319 (1973).
2. J. L. Atwood, C. F. Haines, Jr., M. Tsutsui, and A. E. Gebala, J.C.S. Chem. Comm., 452 (1973).
3a. T. J. Marks, J. R. Kolb, A. M. Seyam, and W. A. Wachter, Proc. Sixth Int. Conf. Organometal. Chem., Amherst, Mass., (1973), abstract 114.
  b. T. J. Marks and W. A. Wachter, submitted for publication.
4a. H. Gilman, R. G. Jones, E. Bindschadler, D. Blume, G. Karmas, G. A. Martin, Jr., J. F. Nobis, J. R. Thirtle, H. L. Yale, and F. A. Yoeman, J. Amer. Chem. Soc., 78, 2790 (1956). This paper begins with a summary of unpublished results.
  b. J. J. Katz, private communication.
5. T. J. Marks and A. M. Seyam, unpublished work.
6. T. J. Marks and A. M. Seyam, J. Organometal. Chem., 67, 61 (1974).
7. T. J. Marks and W. J. Kennelly, manuscript in preparation.
8. V. W. Day, T. J. Marks, and W. A. Wachter, J. Amer. Chem. Soc., in press.

DISCUSSION

DR. HALPERN: This refers to your discussion of the thermal decomposition of these compounds. G. Whitesides and I were kicking around something which raises the question of whether you have correctly represented the process of intramolecular oxidative addition of the cyclopentadienyl. One of the arguments against your representation is that it involves going from an oxidation state of +4 to +6.

DR. MARKS: If you can represent this final product as a carbene complex:

DR. HALPERN: Yes, that is the way, I think, many of us would prefer to represent the oxidative adduct.

DR. MARKS: This is something we discussed last fall. In general terms, oxidative addition would simply imply U(IV) going to U(VI), but the carbene is a reasonable representation in the present case How you represent these things is a problem. Whatever the product is after the R-H is split out, however, it seems to have some finite lifetime.

DR. HALPERN: It may be an important point. It may explain why the cyclopentadienyl ligand is so prone to under such intramolecular oxidative additions, because it does not involve, formally, a change in oxidation state of the metal, or at least to a much lesser degree than other oxidative additions - because you can get some of the increased charge back by forming a carbene complex.

DR. MARKS: We do have very large activation energies to do that.

DR. OSBORN: Have you carried this out in the presence of olefins?

DR. MARKS: To try to trap it (the intermedate) that way?

DR. OSBORN: I just wondered if the intermediate carbene might be transferred.

DR. MARKS: No, we have not tried that.

DR. BERGMAN: I am a little bit bothered by this picture of the alkane being eliminated from the complex via a transition state with all those dotted lines. If you follow this through it would leave a species which has one electron on the metal and one electron on the ring, which seems to me would have much higher energy than the species that Jack is calling (correctly) a carbene complex. The product you obtain could be formed equally well from dimerization of the radical or carbene complex. Which way do you think this thing really goes?

DR. MARKS: I like the caged di-radical, because its a lot harder to get from Th(IV) to Th(III) than from U(IV) to U(III), and that could explain the differences in thermal stability. We are hoping to get more evidence for these things. The trouble is, when you push the activation energies up into this almost non organometallic range, many other mechanisms for which there is no precedent may become important. Sometimes you are exciting these systems to such high energies that they are entering onto a new potential energy surface that no one has seen before, so if we can get something that is somewhat less stable, we may have better luck learning about what really are the factors which control the stability of actinide-to-carbon sigma bonds.

DR. BERGMAN: Well, I believe the energetics of biradical formation present a real problem. You are breaking a metal alkyl bond and a cyclopentadienyl hydrogen bond, and all you are getting back is an R-H bond if you don't get anything back from the complex. So even neglecting any activation energy, this is going to cost at least the energy of the Thorium-alkyl bond plus the difference in energy between the two C-H bonds, and that is a lot of energy.

DR: MARKS: Sure, that is why I drew 3 possible mechanisms. In many systems, it may turn out, and I think Lappert is beginning to realize this also, when you put two radicals close enough together it becomes a concerted mechanism and you may be talking about things that are going to be very difficult to differentiate, because you are talking about whether something is a quarter of an Angstrom one way or the other sometimes, that is what it boils down to when you start drawing these kinds of structures. Sometimes it reduces to a purely semantic argument.

DR. TSUTSUI: That template reaction is remarkable. I would like to ask what the oxidation state of uranium is in the final product?

DR. MARKS: +6. The ligand has a dinegative charge. It is possible to write a valence bond structure for the whole ligand.

DR. TSUTSUI: Lux and coworkers reported uranium tetraiodide was used in a template reaction with phthalonitrile.

DR. MARKS: This is Franz Lux at Garching, part of German AEC, they actually looked at the uranyl compound also; they got as far as the mass spectrum. The compound has actually been known since 1955. What I started doing was looking at some crystal structure data for uranyl compounds, and some data for phthalocyanines. It is quite obvious you could never get one uranyl group into that ring, so that is why we started the crystal structure.

DR. TSUTSUI: At what temperature did you determine your NMR's of tetra-alkyl uranium (IV)?

DR. MARKS: We have not looked at them, but I did have some correspondence with S. A. Cotton at East Anglia, and they tried to make things like this:

because you can make stable lanthanide complexes with this ligand. All he wrote me was that the uranium products were very unstable and fell apart. They have not published anything. They are still working on them so we kind of stayed away. This ligand is kind of a last resort. When something like this fails, which worked with some of the early lanthanides, you really do not know where to go. We have also been unable to stabilize uranium tetraalkyls by the addition of ligands such as bipyridyl. We obtained a pmr spectrum of "$U(CH_3)_4$" at $-40°$.

DR. YAMAMOTO:You have compared the activation energy for the thermolysis of copper compounds, and the activation energy for thermolysis of primary alkyl groups, which is higher than that for secondary ones.

DR. MARKS: Which, in turn, is higher than the tertiary?

DR. YAMAMOTO: Yes, have you compared the activation energy of uranium alkyls of different chain lengths? The reason I asked is that we have found recently in the thernal decomposition of copper alkyls of different chain lengths that the butyl copper was most stable, more stable than propyl, and propyl than ethyl.

DR. MARKS: The only two we had on the slide were the methyl and the n-butyl, and there was a big difference there. In the Thorium, we actually have made the n-propyl and we are going to compare it. In fact, we probably have the data for the n-propyl and the n-butyl one. They are very similar, but all I saw was that the student had the kinetic plots, and I was trying to hold them together to see if they were parallel or not. We have not least squared the data yet. They are not too different.

DR. OSBORN: You mentioned deuterium labeling studies. Did you get any deuterium isotope effects?

DR. MARKS: Yes, I forgot to mention that. We have looked at an uranium compound where we have two deuterated rings and one non-deu- terated and at the R-H and R-D formed, such that we have a kinetic isotope effect of about 7 or 8 - very large. This is comparable to what John Stille has reported on beta-elimination reactions, where he saw about 7. This may of course reflect something fundamentally different from the rate-determining, we just do not know exactly what that number means. Except if it were 0 or 1, then I think it would give you some information. But when it is large, all you can say is it is large. It is not easy to form it, and there is some competition.

DR. OSBORN: You talked about oxidative additions, especially from an alpha position. Are hydrides of Uranium of this type sta- ble?

DR. MARKS: We have tried to make them. So far, no luck. I think we will have better luck with Thorium. What happens is it looks like you split out hydrogen, and you just get tris-cyclopenta- dienyl uranium, which is a known compound. We are still trying that, though. The trouble with some of these systems is that you get some- thing that is very insoluble, and you have to soxlet extract for 3 weeks before you convince yourself that you have something that is not going to work.

RECENT DEVELOPMENTS IN CHEMISTRY OF ORGANOLANTHANIDES AND ORGANO-ACTINIDES

Minoru Tsutsui[*], Carol Hyde, Alan Gebala and Neal Ely

Department of Chemistry, Texas A&M University, College Station, Texas 77843

SIGMA-BONDED ORGANOMETALLIC DERIVATIVES
OF THE LANTHANIDES AND ACTINIDES

In contrast with pi-bonded organometallic compounds of the lan-thanides[1] and actinides[2], which date to the mid 1950's, encompass a large variety of ligands[3] and culminate with the synthesis of bis-(cyclooctatetraenyl)uranium ("uranocene"),[4] the preparation of sig-ma-bonded lanthanide and actinide organometallics is relatively recent. Early attempts to prepare simple organometallic derivatives, such as $U(CH_3)_4$[5], $Y(C_2H_5)_3$, and $Sc(C_2H_5)_3$[6], often resulted in poor-ly characterized or irreproducible products. Sigma-bonded deriva-tives of both the lanthanide[7,8] and actinide[9] elements are now known and well characterized.[10]

Considering previous unsuccessful attempts to prepare sigma-bonded organometallic compounds of the lanthanides and actinides, and the passage of nearly fifteen years from the first report of pi-bonded species, the simple syntheses and resulting thermal sta-bilities of these compounds are indeed surprising. With the excep-tion of divalent derivatives of Yb, Sm, and Eu, which are prepared via a Grignard-type reaction between an alkyl or aryl iodide and the appropriate lanthanide metal (equation 1)[7],

$$(1) \quad M^{\circ} + RI \xrightarrow[THF]{} RMI$$

(M = Yb, Sm, Eu; R = CH_3, C_2H_5, o-MeC_6H_4, 2,6-$Me_2C_6H_3$, 2,4,6-$Me_3C_6H_2$), these sigma-bonded species are prepared by the reaction of one or more metal-halide bonds with lithium and sodium derivatives or a Grignard reagent. Generally these are carried out in THF, ethyl ether or mixtures of these solvents:

$$(2a) \quad MCl_3 + 3 \text{ LiR} \longrightarrow MR_3 + LiCl$$

$$(2b) \quad MCl_3 + 4 \text{ LiR} \longrightarrow Li[MR_4] + 3 \text{ LiCl}$$

$$(3) \quad Cp_2\text{-MCl} + NaR \longrightarrow Cp_2\text{-MR} + NaCl$$

Tricyclopentadienyl scandium has also been synthesized by reaction between the anhydrous metal trifluoride and dicyclopentadienylmagnesium:[11,12]

$$(4) \quad ScF_3 + 1.5 \text{ MgCp}_2 \longrightarrow ScCp_3 + 1.5 \text{ MgF}_2$$

The presently known sigma-bonded lanthanide organometallic compounds are given in Table I with the method of preparation and some physical and chemical properties. Tsutsui and Gysling have reported a series of triindenyl lanthanide monotetrahydrofuran complexes[13]: $(Ind)_3Ln \cdot THF$, Ln = La, Sm, Gd, Tb, Dy, and Yb. For $(Ind)_3La \cdot THF$, the nmr spectrum was found to resemble that of sodium indenide (ionic), $(Ind)_2Fe$[14], $(Ind)FeCp$[15], and $(Ind)_2Ru$[14] (pi-bonded species); with an A_2X pattern for the protons of the 5-membered ring. However, an ABX pattern was found for $(Ind)_3Sm \cdot THF$, similar to that reported for the covalently sigma-bonded indenyl compound $(\underline{h}^5\text{-}C_5H_5)Fe$-$(CO)_2(1\text{-indenyl})$[16], and a similar bonding mode was postulated for $(Ind)_3Sm \cdot THF$. The lanthanide contraction, and increased "hardness" of Sm^{3+} relative to La^{3+}, was suggested as a reason for the different binding modes; it has been found for $[M(2,6\text{-}Me_2C_6H_3)_4]^-$ that lighter lanthanides could not be isolated.[10a]

Atwood, et al.[17] have prepared $(Ind)_3Sm$,

$$(5) \quad SmCl_3 + 1.5 \text{ Mg(Ind)}_2 \longrightarrow Sm(Ind)_3 + 1.5 \text{ MgCl}_2$$

and have done the single crystal X-ray structure determination. They have found that the five-membered rings have an approximately trigonal configuration about the samarium atom and the Sm-C distances for carbon atoms and the bridgehead of the five-membered rings show no significant differences and therefore no evidence for enhanced covalent bonding to the 1-carbon position in the indenyl group.

So far, only two lanthanide compounds containing both pi-and sigma-bonds have been prepared: one containing both sigma- and pi-cyclopentadienyl ligands and one with a sigma-bonded ligand and two pi-cyclopentadienyl groups. In 1970, Coutts and Wailes isolated dicyclopentadienylscandiumphenylacetylide from the reaction of dicyclopentadienylscandium chloride and sodium phenylacetylide in THF in 67% yield.[8c] It is a yellow, air sensitive solid which decomposes at 254° and does not sublime. The molecular weight determined in THF is somewhat greater than that for a monomer which leads the authors to conclude that the $[PhC \equiv C]^-$ groups are bridging and that the compound is at least a dimer. Further evidence for this is seen in the ir where the $C \equiv C$ stretch has been reduced to 2045 cm^{-1}; some 40 cm^{-1}.[18]

The second compound is tricyclopentadienylscandium. Although this compound was first synthesized in 1954,[2] its crystal structure was not done until 1972 by Atwood.[10b] The compound was found to be a polymer with each scandium coordinated with two pentahapto Cp ions and to two others through essentially only one carbon atom. Thus, the compound contains both terminal pi-Cp ligands and scandium bridging sigma-Cp ligands.

The type of organolanthanide compound where all of the ligands are sigma-bonded encompasses a much greater variety than the previous group. The earlier literature recounts several attempts to prepare sigma-bonded derivatives of the lanthanides, though none of the results were reproducible.[5,19] Finally, in 1968 F. A. Hart, et al. reported what appeared to be the first purely sigma bonded derivatives of the lanthanides.[8] The following compounds were prepared by reacting anhydrous metal halides with the appropriate alkyl or aryl lithium reagent in THF or THF/ether:

  (6)   triphenylscandium
  (7)   triphenylyttrium
  (8)   lithium tetraphenyllanthanate
  (9)   lithium tetraphenylpraseodymate
 (10)   tris(phenylethynyl)scandium

These compounds were characterized by ir, elemental analysis for the metal, Michler's ketone test, and reaction with HgCl$_2$. It is of interest to note that in the reaction with HgCl$_2$, only 65-76% of phenyl mercuric chloride could be isolated. Elemental analysis for carbon were generally low, often less than could be isolated as the mercuric chloride derivative, and it was suggested that the formation of carbides accounted for this. All of these compounds are air sensitive (pyrophoric) and essentially insoluble in all solvents except THF. This insolubility precluded better characterization of these compounds.

Hart, et al. also described the reaction of metal halides with methyl lithium. However, they were unable to separate the products from LiCl and they could only speculate on their composition. Reactions with mercuric chloride resulted in the isolation of CH$_3$HgCl.

In 1972, Hart and co-workers reported the synthesis and single crystal X-ray structure of tetrakis(tetrahydrofuran)lithium tetrakis-(2,6-dimethylphenyl)lutetiate.[10a] This compound was prepared in an analogous manner to the other sigma-bonded compounds prepared by Hart and provides conclusive evidence for their existence. The lutetiate derivative is the first crystallographically characterized

Table I

Chemical and Physical Properties of Sigma-Bonded Derivatives of the Lanthanides

Compound	Prep	Color	MP	Reactions	Ref.
$Sc(C_6H_5)_3$	2a	yellow-brown	dec. 140°	$+ HgCl_2 \rightarrow C_6H_5HgCl$ $+ CO_2 \rightarrow C_6H_5COOH$ $+ (C_6H_5)_2C=O \rightarrow (C_6H_5)_3COH$	8b
$Sc(CH_3)_3$	2a	orange-yellow	darkens 140° in vacuo	$+ HgCl_2 \rightarrow CH_3HgCl$	8b
$Sc(C\equiv CC_6H_5)_3$	2a	dark brown	stable to 250° in vacuo	$+ CO_2 \rightarrow C_6H_5C\equiv CCOOH$	8b
$[ScCp_2(C\equiv CC_6H_5)]_2$	3	yellow	dec. 254°		8c
$(ScCp_3)_n$	3	straw	240°	$+ FeCl_2 \rightarrow FeCp_2$	11,12
$Y(C_6H_5)_3$	2a	brown		$+ HgCl_2 \rightarrow C_6H_5HgCl$	8b
$Y(CH_3)_3$	2a	orange-yellow	dec. 140° in vacuo	$+ HgCl_2 \rightarrow CH_3HgCl$	8b
$Li[La(C_6H_5)_4]$	2b	dark brown		$+ HgCl_2 \rightarrow C_6H_5HgCl$	8b
$La(CH_3)_3$	2a	brown		$+ HgCl_2 \rightarrow CH_3HgCl$	8b
$Li[Pr(C_6H_5)_4]$	2b	dark brown		$+ HgCl_2 \rightarrow C_6H_5HgCl$	8b
$[Li(THF)_4][Lu(2,6-Me_2C_6H_3)_4]$	2b	colorless			10a
$[Li(THF)_4][Yb(2,6-Me_2C_6H_3)_4]$	2b	dark brown			10a
RMI (M = Yb, Eu)	1	brown solutions*		$C_6H_5MI + (C_6H_5)_2C=O \rightarrow$ $(C_6H_5)_3COH$ $(M = Yb, Sm)$	7

RMI (M = Sm)	1	blue green solutions* THF		$(C_6H_5)Mi + (CH_3)_3SiCl \rightarrow$ $(CH_3)_3Si(C_6H_5)$ (M = Yb, Eu, Sm)	7
Sm(Ind)$_3$·THF	3	orange-red	darkens 165–170° melts 185–200°		13
$(h^5-C_5H_5)_2$ Gd–C≡CPh	2a	yellow	279–282° d		27
$(h^5-C_5H_5)_2$ Er–C≡CPh	2a	pink	275–280° d		27
$(h^5-C_5H_5)_2$ Yb–C≡CPh	2a	orange	305–308° d		27

*These solutions also gave positive tests with Michler's ketone and other Grignard-type reactions.

example of a four coordinate lanthanide. This low coordination number is probably due to the highly hindered nature of the 2,6-dime-
thylphenyl ligand. The only other example of a f-transition metal
complex with a coordination number less than six is the three coordinate series [Ln{N(SiMe$_3$)$_2$}$_3$] (Ln = lanthanide) where the ligand
is also very large.[20]

Another interesting feature of this compound is the bond distance between Lu and carbon. These distances, between 2.42 and 2.50
Å are approximately 0.2 Å shorter than the pentahapto metal-carbon
distance calculated for Lu-(h^5-C$_5$H$_5$) bonds. The analogous ytter-
bium compound has also been prepared and was found to be isostructural with the lutetiate derivative.

All of the previously discussed sigma-bonded lanthanides have
been compounds wherein the lanthanide is in the 3+ oxidation state;
additionally, there is no change in the oxidation state during the
preparation of the compounds from the corresponding halides. In
contrast with these compounds are the reported alkyl and aryl derivatives of Sm^{2+}, Eu^{2+}, and Yb^{2+} (see equation 1);[7] these are derived
from the corresponding metals, analogous to the Grignard reaction.
The reactions were run in THF and the resulting solutions were characterized for M/I ratio, M^{2+}/M^{3+} ratio and by other analytical procedures (reaction with iodine, acidimetric titration) to demonstrate
that the solution contained predominantly RMI species. It was found
that Eu gave cleaner reactions and better yields than Yb, while Sm
was considerably less tractable -- initiation with iodine at 30° was
necessary for Sm while Eu and Yb reacted smoothly at -20° without
initiation: additionally, quite high amounts of Sm^{3+} were determined
(∿50%) to be in solution. These results parallel the stability of
the 2+ relative to the 3+ oxidation state: Eu>Yb>Sm. Partial eva-
poration of a solution of (phenyl)YbI led to the isolation of a yellow hydrolyzable solid with ∿3THF/Yb and an I/Yb ratio of 1.15:1.

Solutions of these RMI species were found to undergo "typical"
Grignard reactions: positive tests with Michler's ketone, addition
to double bonds (ketone, isocyanide) and reaction with Si-Cl bonds.

Cerium and lanthanum metals were also treated similarly to give
solutions containing M-R bonds. However, these solutions contained
mixtures of species and were not of comparable high yield as those
of Yb, Sm and Eu.

Attempts to prepare tricyclopentadienylphenyluranium originally
met with failure and were cited as evidence that the bonding in tetracyclopentadienyluranium was pentahapto for all four Cp groups;[21]
it was previously known from an X-ray determination that the bonding
in tricyclopentadienyluranium chloride was pentahapto.[22] The phenyl
compound has since been prepared by at least two groups from the reaction of tricyclopentadienyluranium chloride with phenyl lithium:

9a,9c,9d

$$(11) \quad Cp_3UCl \; + \; Li(C_6H_5) \; \longrightarrow \; Cp_3U(C_6H_5) \; + \; LiCl$$

numerous other alkyl, aryl and acetylide derivatives of tricyclopentadienyluranium (IV) have been prepared analogously:[9]

(12a) Cp_3UCl + LiR

(12b) Cp_3UCl + NaR   } $\longrightarrow$ Cp_3UR

or (12c) Cp_3UCl + RMgX

These compounds have been purified by crystallization from toluene-
hexane mixtures[9b] or by continuous extraction into hexane or pentane.[9c] This is the only system for which actinide-carbon sigma
bonds are known at present, although the thorium analogs have apparently been prepared.[23] The known sigma-bonded compounds and some
chemical and physical data are presented in Table II.

For tricyclopentadienylphenylethynyluranium, a single crystal
x-ray structure has been determined and conclusively proves the sigma-bonded nature of the fourth group.[10c]

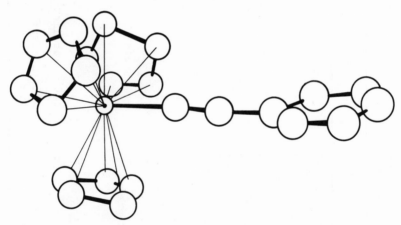

Figure I.   Structure of Tricyclopentadienylphenylethynyluranium(IV).

As in the case of the tetrakis(2,6-dimethylphenyl)lutetiate ion, the
uranium-carbon sigma-bond distance is considerably shorter than the
pentahapto carbon-uranium bond distances: approximately 0.3 Å shorter in the uranium case versus 0.2 Å for lutetium. If these shorten-
ed distances for sigma-bonds reflect covalent f orbital contributions
to the bonding, this result is not unexpected since the 5f orbitals
are more available than the 4f orbitals.

Table II

Chemical and Physical Properties of Cp_3UR Compounds

R	Color	Stability	Reactions	Ref.
Me	yellow-brown		$+ (CH_3)_2C=O \rightarrow NR$	9b,26
n-butyl	dark red-brown			9b,26
allyl	dark brown	$t_{1/2} \approx$ 20-96 hours at 72° in toluene	$+ CH_3OH \rightarrow (Cp_3)_3UOCH_3$	9b
neopentyl	dark red-brown			9b
C_6F_5	dark brown			9b
$\underline{i}-C_3H_7$	dark red-brown			9b
C_6H_5	green		$+ H_2O \rightarrow C_6H_6 + Cp_3U^+$	9a,9c
$\underline{p}-MeC_6H_4$	yellow-green			9c,9d
$C\equiv CC_6H_5$	yellow-green	183-185°		9c
$C\equiv C-H$	yellow-green	melts with partial decomp.		27
$CH_2-\underline{p}-MeC_6H_4$				26
$CH_2C_6H_5$				26
Ferrocenyl	brown	180° (sublime)		31
$Cp_3UFcUCp_3$	green	190° decomp.		31

The question of covalency for both lanthanide and actinide organometallic compounds is one of both longstanding and current interest and can only be discussed briefly here. Results for pi-bonded
ligands suggest, as does the considerable bond shortening for the M-C
sigma-bonds, that the actinides possess considerably more covalent
character than the lanthanides and, in fact, that the lanthanides,
at most, possess very small covalent contributions to the bonding.
For example, it is known that the Cp groups in tricyclopentadienyluranium chloride do not react rapidly with water (although the chloride is ionic) and do not react with ferrous chloride to give ferrocene,[2] while the lanthanide cyclopentadienides do.[12] Similar behavior has been found for "uranocene", which is stable towards water,[4]
while $[Ln(COT)_2]^-$ are found to be very hydrolytically unstable:[24]
additionally,

(13) $[Ce(COT)_2]^- + UCL_4 \xrightarrow{\text{THF}} U(COT)_2$ (87%, immediate
reaction)

The use of Mossbauer spectroscopy (the isomer shift) to determine covalency in the Np(IV)-Cp bonds for $NpCp_4$ and $NpCp_3Cl$[25] suggests that this would be useful for determining covalent contributions for a series of R groups in Cp_3NpR versus Cp_3NpCl; although
no Cp_3NpR compounds have been reported, it can be expected that their
synthesis would be straightforward.

The protons of the sigma-bonded moieties exhibit extraordinar-
ily large upfield chemical shifts: (τ) +197 for the methylene protons of the benzyl group,[26] +202 for the methyl and +334 for the
methylene of the sigma-bonded allyl.[9b] For the protons of the pi-
bonded cyclopentadienyl groups, shifts of $\sim$+10-13 ppm upfield from
benzene are found. Although the original report[9c] of the chemical
shift order of p>o>m is incorrect and follows that found for the
phenyl in the phenylacetylide derivative of o>m>p,[27] these chemical
shifts can be separated into contact and psuedocontact terms and
used to measure the extent of 5f electron delocalization and covalency.[23] All of the sigma-bonded moieties have been found to have
upfield shifts except the acetylide derivative.[27]

The allyl derivative exhibits an A_4X pattern at RT and above
but at -85° an A_2BCD pattern for a monohaptoallyl is obtained, the
molecule exhibiting fluxional behavior.[9b] This behavior differs
from that found for tetrallylthorium[28] and tetrallyluranium:[29] these
compounds exhibit a piallyl structure at low temperatures, the uranium exhibiting anti and syn methylene protons split by the methine
proton, and the thorium changing to a dynamic structure at $\sim$+80°.
Since tetracyclopentadienyluranium requires more forcing conditions
[30] than the preparation of the sigma-bonded derivatives, it is not
unreasonable that the pi-bonded allyl in tricyclopentadienylurani-
umallyl exists at higher temperatures wherein the coordination number is expanded and that decreasing an already "small" coordination

number in the tetraallyl compounds in unfavorable and therefore the
sigma-bonded derivative requires higher temperatures.

These compounds have surprisingly high thermal stability and
seem to undergo decomposition via a path different from those normally associated with sigma-bonded organometallic compounds -- β hydrogen abstraction or homolytic bond scission: The major decomposition product of the n-butyl derivative is n-butane and not 1-butene
(<2%) or n-octane (<1%).[9b] Additionally, reference to Table II indicates very little difference in stability between compounds with
or without β hydrogens.

Marks, et al. have proposed that decomposition takes place via
hydrogen abstraction from a cyclopentadienyl group.[23] The mass spec-
tra of both $(\underline{h}^5\text{-}C_5H_5)_3U(C_6H_5)$ and $(\underline{h}^5\text{-}C_5H_5)_3U(C_6D_5)$ show the loss of
a C_5H_6 fragment,[28] again demonstrating hydrogen transfer from a cy-
clopentadienyl ring.

Attempts to prepare sigma-bonded derivatives from $(Ind)_3UCl$ under conditions similar to those for Cp_3UCl,[9c,9d] have not been successful: in the case of $\underline{p}\text{-}CH_3C_6H_4$ high yields of bitolyl have been
obtained.[27]

Two novel binuclear sigma-bonded compounds have been prepared
in a similar manner to Cp_3UR compounds: p-bis(tricyclopentadienyluranium)benzene and 1,1'-bis(tricyclopentadienyluranium)ferrocene.[27]
The first compound was synthesized with the expectation that magnetic coupling, via the organic bridge, might be found for the unpaired 5f electrons. However, the RT magnetic moment shows no "abnor-
mality" from other sigma-bonded derivatives (e.g., the phenyl) and
low temperature magnetic studies need to be undertaken before magnetic coupling should be excluded; the ferrocene derivative, as
would be expected, shows no deviation either.

A new series of organolanthanide complexes, $(Cp)_2Ln\text{-}C{\equiv}CPh$ (Ln =
Gd, Er, Yb; Ph = phenyl) has been prepared. For the Gd and Er derivatives, they are the first well characterized compounds to contain a metal-carbon sigma bond. Visible spectra of the Er complex
indicate at least some localized interaction between the metal and
the phenylacetylide moiety.[31]

PI-BONDED ORGANOMETALLIC DERIVATIVES
OF THE LANTHANIDES AND ACTINIDES

Several series of complexes have been prepared in which one or
more organic ligands (indene,[13b] cyclopentadiene,[32] and cyclooctatetrene[33-36]) are π-bonded to a lanthanide or actinide element. In
addition, there are three isolated examples of uranium complexes
containing the π-allyl,[37,38] and π-benzene[39] moieties.

Trisindenyl derivatives of the lanthanides are synthesized via the following reactions:[1]

$$LnCl_3 + 3NaC_9H_7 \xrightarrow{THF} Ln(C_9H_7)_3 \cdot THF + 3NaCl$$

Ln = La, Sm, Gd, Tb, Dy, and Yb

The organic moiety is π-bonded to the central metal atom; the bond retains appreciable ionic character for all complexes except Sm. A covalent bonding mode has been proposed in the latter case (cf. discussion under σ-bonded complexes).

Lanthanide complexes[3a] with cyclopentadiene ligands are of the general formulations: $Ln(Cp_3)_3$, Ln = La, Ce, Pr, Nd, Sm, Eu, Gd, Tb, Dy, Ho, Er, Tm, Yb and Lu; $LnCl(Cp)_2$, Ln = Sm, Gd, Dy, Ho, Er, Yb and Lu; $LnCl_2(Cp) \cdot 3$ THF, Ln = Sm, Eu, Gd, Dy, Ho, Er, Yb and Lu; and $Ln(Cp)_2$, Ln = Sm, Eu and Yb. The C_5H_5 ligands are viewed as π-bonded to the metal, although the chemical reactivity and spectral and magnetic properties are indicative of an ionic bonding mode.

Actinide complexes[32] with cyclopentadiene are of the general formulations: $Ac(Cp)_3$, Ac = U, Pu, Am, Cm, Bk and Cf; $Ac(Cp)_4$, Ac = Th, Pa, U, Np; $X-U(Cp)_3$, X = F, Cl, Br and I; and $R-U(Cp)_3$, R = C_6-H_5, $CH_2C_6H_5$, C_8H_5, $\underline{n}-C_4H_9$, CH_3, etc. As with the lanthanide complexes the Cp ligand is π-bonded. The latter group of complexes has been of recent interest because of the formation of metal-carbon σ bonds.

The degree of covalency for $Cl-U(Cp)_3$ is greater than that of the lanthanide cyclopentadienyl derivatives as evidenced by the failure of the uranium complex to react with $FeCl_2$ to form ferrocene. The transuranium complexes are more covalent in character than the lanthanides, but less covalent than the uranium derivative.[32b]

The complex $(Cp)_3U(C_3H_5)$ was studied by Marks et al.[38] and found to be a π-bonded allyl species. The nmr spectra were interpreted as a σ ⇄ π ⇄ σ interconversion. At room temperature, the allyl group shows a dynamic A_4X spectrum. On lowering of the temperature to 179°K, the allyl resonance collapses and an A_2BCD pattern characteristic of a monohapto allyl linkage is frozen out. It is estimated that the trihapto bonding configuration is no higher than about 8-9 kcal/mol in energy above the monohapto configuration. By comparison, the trihapto bonding mode is lower in energy for $U(C_3H_5)_4$.[37]

The last extensive series of π-bonded ligands involves complexes with cyclooctatetrene[33-36,42,43] which are of the general formulation: $K[Ln(C_8H_8)_2]$, Ln = La, Ce, Pr, Nd, Sm, Gd and Tb; $[Ln(C_8H_8)Cl \cdot 2$ THF$]_2$, Ln = Ce, Pr, Nd and Sm; $Ac(C_8H_8)_2$, Ac = Th, Pa, U, Np and Pu; and $U(C_8H_7R)_2$, R = C_2H_5, $\underline{n}-C_4H_9$, vinyl and cyclopropyl. The X-ray crystal structure has been done for $[Ce(C_8H_8)Cl \cdot 2$ THF$]_2$.[44] The

cerium atoms of the dimer unit are bridged asymetrically by the chlorine atoms.   The C_8H_8 ring is planar and has aromatic C–C bond
lengths.   The other dimeric complexes are believed isomorphic with
the Ce species.   The bisoctatetrene species are sandwich type complexes with planar rings and D_{8h} symmetry.[45,46]

Spectral and chemical data indicate the lanthanide species are
more ionic than the actinide analogues.[36]   The ligands are labile as
demonstrated by the following reactions:

$$K[Ce(C_8H_8)_2]$$

$$or \qquad\qquad + \ UCl_4 \ \longrightarrow \ U(C_8H_8)_2$$

$$K[Sm(C_8H_8)_2]$$

Uranocene does not undergo ligand exchange even at elevated temperatures.[42]   A mixture of 1,1'-diethyluranocene and 1,1'-dibutylurano-
cene was refluxed in diglyme for a period of five hours.   Upon examination of the reaction mixture, there was no observable amount of
scrambled sandwich complex.   For uranocene, the metal-ring bond must
be substantially covalent in nature.   The ease of hydrolysis and
products obtained are also indicative of the ionic nature of the
C_8H_8 ligand for the lanthanide species.   In contrast, uranocene is
stable to water.

The Mossbauer spectrum of $Np(C_8H_8)_2$, absorption spectra of U-
$(C_8H_8)_2$, $Np(C_8H_8)_2$ and $Pu(C_8H_8)_2$ and the magnetic properties of the
three actinide species show significant departure from those observed for principally ionic actinide complexes.[34]   In the case of
neptunocene, the large isomer shift observed is indicative of additional shielding of the 6s shell.   This suggests a strong electronic contribution from the ligand to metal orbitals.   There are strong
absorptions in the visible region for these complexes which are a
factor of 10 or more stronger than the intensities of bands resulting from normal 5f–5f transitions.   The mechanism that permits observation of these Laporte-forbidden transitions is the mixing in
of higher energy orbitals.

Further evidence of covalency is found in the observed magnetic moments; the experimental values are lower than those calculated.
Consideration of covalent contributions (i.e., reduction of orbital
size) results in calculated values more closely in agreement with
observed magnetic moments.

The reaction of allyl Grignard and uranium(IV) at low temperatures yields a π-complex,

$$4C_3H_5MgBr + UCl_4 \xrightarrow[-30°]{} U(C_3H_5)_4 + 4MgBrCl$$

The IR of the tetraallyl species does not indicate a σ-allyl structure; rather there are absorption bands characteristic of a π-allyl-
ic group. The complex is thermally unstable above -20° and decomposes with the violent evolution of gas. The allyl groups are readily cleaved with water to give propylene, 1,5-hexadiene, and cyclohexane.

The last example of a π-complex is $(\pi-C_6H_6)UAl_3Cl_{12}$ which has
been studied by X-ray analysis.[39] The complex is synthesized by
the following reaction:

$$UCl_4 \ + \ AlCl_3 \ \xrightarrow[]{C_6H_6} \ (C_6H_6)UAl_3Cl_{12}$$

The uranium atom is coordinated to three $AlCl_4$ tetrahedral units
through U-Cl-Al bridge bonds and to the benzene ring. The U-C mean
distance was found to be 2.91 ± 0.01 Å. By comparison, a U-C single bond is 2.19 Å and the U-C distances in $U(C_8H_8)_2$ and $ClU(Cp_3)_3$
are 2.74 Å and 2.65 Å, respectively. The lengthening of the U-C
bond in $(C_6H_6)UAl_3Cl_{12}$ as compared to the other π-bonded species is
a result of the highly electronegative chlorine atoms also bonded
to the uranium. The IR gives no indication of the π-nature of the
benzene-uranium bond.

REFERENCES

1. G. Wilkinson and J. M. Birmingham, J. Amer. Chem. Soc., 76, 6210
   (1954).
2. L. T. Reynolds and G. Wilkinson, J. Inorg. Nucl. Chem., 2, 246
   (1956).
3. a. G. Hayes and J. L. Thomas, Organometal. Chem., Rev. A 7, 1
   (1971).
   b. H. Gysling and M. Tsutsui, Advan. Organometal. Chem., 9, 361
   (1970).
4. A. Streitwieser, Jr. and U. Mueller-Westerhoff, J. Amer. Chem.
   Soc., 90, 7364 (1968).
5. G. Gilman, R. G. Jones, E. Bindschadler, D. Blume, G. Karmas,
   G. A. Martin, Jr., J. F. Nobis, J. R. Thirtle, H. L. Yale and
   F. A. Yoeman, J. Amer. Chem. Soc., 78, 2790 (1956).
6. V. M. Pletz, J. Gen. Chem., U.S.S.R., 20, 27 (1938).
7. a. D. F. Evans, G. V. Fazakerley and R. F. Phillips, Chem. Comm.,
   244 (1970).
   b. D. F. Evans, G. V. Fazakerley and R. F. Phillips, J. Chem.
   Soc., 1931 (1971).
8. a. F. A. Hart and M. S. Saran, Chem. Comm., 1614 (1968).
   b. F. A. Hart, A. G. Massey and M. S. Saran, J. Organometal.
   Chem., 21, 147 (1970).
   c. R. S. P. Coutts and P. C. Wailes, J. Organometal. Chem., 25,
   117 (1970).

9. a. G. Brandi, M. Brunelli, G. Lugli, A. Mazzei, N. Palladino, U. Pedretti and F. Salvatori, Proc. Third Int. Symp., Inorg. Chim. Acta, E10 (1970).

   b. T. J. Marks and A. M. Seyam, J. Amer. Chem. Soc., 94, 6545 (1972).

   c. A. E. Gebala and M. Tsutsui, Chem. Lett., 775 (1972).

   d. A. E. Gebala and M. Tsutsui, J. Amer. Chem. Soc., 95, 91 (1973).

10. At least three single crystal X-ray diffraction studies have been carried out of compounds in which sigma bonds are present:
   a. S. A. Cotton, F. A. Hart, M. B. Hursthouse, and A. J. Welch, Chem. Comm., 1225 (1972).

   b. J. L. Atwood and K. D. Smith, J. Amer. Chem. Soc., 95, 1488 (1973).

   c. J. L. Atwood, F. Hains, Jr., M. Tsutsui and A. E. Gebala, Chem. Comm., 452 (1973).

11. A. F. Reid and P. C. Wailes, Inorg. Chem., 5, 1213 (1966).

12. J. M. Birmingham and G. Wilkinson, J. Amer. Chem. Soc., 78, 42 (1956).

13. a. M. Tsutsui and H. J. Gysling, J. Amer. Chem. Soc., 90, 6880 (1968).

   b. M. Tsutsui and H. J. Gysling, J. Amer. Chem. Soc., 91, 3175 (1969).

14. J. H. Osiecki, C. J. Hoffman and D. P. Hollis, J. Organometal. Chem., 3, 107 (1965).

15. R. B. King and M. B. Bisnette, Inorg. Chem., 3, 796 (1964).

16. F. A. Cotton, A. Musco and G. Yagupsky, J. Amer. Chem. Soc., 89, 6136 (1967).

17. J. L. Atwood, J. H. Burns and P. G. Laubereau, J. Amer. Chem. Soc., 95, 1830 (1973).

18. E. A. Jeffery and T. Mole, J. Organometal. Chem., 11, 393 (1968).

19. H. Gilman and R. G. Jones, J. Org. Chem., 10, 505 (1945).

20. D. C. Bradley, J. S. Ghotra and F. A. Hart, Chem. Comm., 349 (1972).

21. Y. Hristidu, Thesis, University of Munich, 1963, as cited in R. von Ammon, R. B. Kanellakopulas and R. D. Fisher, Chem. Phys. Lett., 4, 553 (1970).

22. C. H. Wong, T. M. Yen and T. Y. Lee, Acta Crystallogr., 18, 340 (1965).

23. T. J. Marks, J. R. Kolb and A. M. Seyam, 165th National Meeting of the American Chemical Society, Los Angeles, Calif., No. INOR-50 (1973).

24. F. Mares, K. Hodgson and A. Streitwieser, Jr., J. Organometal. Chem., 24, C68 (1970).

25. D. G. Karraker and J. A. Stone, Inorg. Chem., 11, 1742 (1972).

26. F. Calderazzo, Inorg. Chem. Acta, Plenary Lecture, ICCC, XIV, 21 (1972).

27. M. Tsutsui, N. Ely and A. E. Gebala, J. Amer. Chem. Soc., in press.

28. G. Wilke, B. Bogdanovic, P. Hardt, P. Heimbach, W. Keim, M.

Kroner, W. Oberkirch, K. Tanaka, E. Steinrucke, D. Walter and H. Zimmermann, Angew. Chem., 78, 157 (1966).

29. N. Paladino, G. Lugli, U. Pedretti, M. Brunelli and G. Giacometti, Chem. Phys. Lett., 5, 15 (1970).

30. M. L. Anderson and L. R. Crisler, J. Organometal. Chem., 17, 345 (1969).

31. M. Tsutsui and N. Ely, J. Amer. Chem. Soc., in press.

32. Complexes containing the C_5H_5 ligand have been reviewed, e.g.
    a. G. Hayes and J. L. Thomas, Organometal. Chem., Rev. A 7, 1 (1971).
    b. G. T. Seaborg, Pure Appl. Chem. 30, 539 (1972).

33. A. Streitwieser, Jr. and N. Yoshida, J. Amer. Chem. Soc., 91, 7528 (1969).

34. D. G. Karraker, J. A. Stone, E. R. Jones, Jr. and N. Edelstein, J. Amer. Chem. Soc., 92, 4841 (1970).

35. A. Streitwieser, Jr., U. Mueller-Westerhoff, G. Sonnichsen, F. Mares, D. G. Morrell, K. O. Hodgson and C. A. Harmon, J. Amer. Chem. Soc., 95, 8644 (1973).

36. K. O. Hodgson, F. Mares, D. F. Starks and A. Streitwieser, Jr., J. Amer. Chem. Soc., 95, 8650 (1973).

37. G. Lugli, W. Marconi, A. Mazzei, N. Paladino and U. Pedretti, Inorg. Chim. Acta, 3, 253 (1969).

38. T. J. Marks, A. M. Seyam and J. R. Kolb, J. Amer. Chem. Soc., 95, 5529 (1973).

39. M. Cesari, U. Pedretti, A. Zazzetta, G. Lugli and W. Marconi, Inorg. Chim. Acta, 5, 409 (1971).

40. J. L. Atwood, J. H. Burns and P. G. Laubereau, J. Amer. Chem. Soc., 95, 1830 (1973).

41. G. Brandi, M. Brunelli, G. Lugli and A. Mazzei, Inorg. Chim. Acta, 7, 319 (1973).

42. A. Streitwieser, Jr. and C. A. Harmon, Inorg. Chem., 12, 1102 (1973).

43. D. A. Starks and A. Streitwieser, Jr., 167th National A.C.S. meeting, Los Angeles, California, Abstract INOR 148 (April 1-5, 1974).

44. K. O. Hodgson and K. N. Raymond, Inorg. Chem., 11, 171 (1972).

45. A. Zolkin and K. N. Raymond, Inorg. Chem., 91, 5667 (1969).

46. K. O. Hodgson and K. N. Raymond, Inorg. Chem., 11, 3030 (1972).

MECHANISMS OF HOMOGENEOUS CATALYTIC HYDROGENATION AND RELATED PROCESSES

Jack Halpern

Department of Chemistry, The University of Chicago, Chicago, Illinois 60637

Homogeneous catalytic processes such as hydrogenation, hydroformylation and oligomerization are typically multistep processes as exemplified by the widely accepted mechanism of the $Rh(PPh_3)_3Cl$-catalyzed hydrogenation of olefins depicted below ($L=PPh_3$):[1]

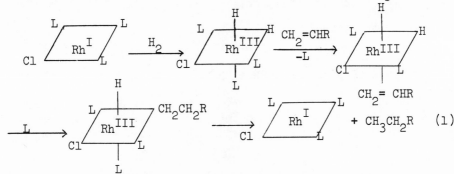

$$(1)$$

Such processes are further characterized by the fact that the catalytic complexes and intermediates are typically labile species which coexist in several forms related through dissociative equilibria, for example,[2,3]

$$(2)$$

Such lability is widespread in homogeneous catalytic systems and indeed can frequently be recognized as an essential feature of

catalytic activity, a requirement of which is the facile incorporation of reactants (e.g., olefins or CO) into the coordination sphere,
and elimination of products from the coordination sphere, through
substitutional or dissociative processes. The widely recognized importance of "coordinatively unsaturated" species in catalysis is
closely linked to this theme.

Although attempts have been made to deduce the mechanisms of
such catalytic processes through kinetic studies of the overall process, the complexity of the systems renders the results of such studies of questionable reliability and imposes severe limitations on
the completeness and degree of detail of the mechanistic schemes that
are derived in this way.[1,4] In favorable cases the mechanisms may
be more reliably elucidated by separate examination of each of the
component steps and species of the reaction sequence. In such studies it is essential to combine detailed kinetic measurements with
corresponding characterization of species through appropriate spectroscopic and other physical measurements. Useful supplementary
mechanistic information may also be provided by the study of reactivity patterns associated with systematic variation of substituents
on the substrates as well as ligands; isotopic tracers; kinetic isotope effects; and stereochemical studies.

Such detailed mechanistic studies will be described for two
classes of reactions involving $Rh(PPh_3)_3Cl$, namely:

1. The $Rh(PPh_3)_3Cl$-catalyzed hydrogenation of olefins (Eq. 1);
2. The decarbonylation of aldehydes by $Rh(PPh_3)_3Cl$ (Eq. 3).[5,6]

$$RCHO + Rh(PPh_3)_3Cl \longrightarrow RH + Rh(PPh_3)_2(CO)Cl + PPh_3 \qquad (3)$$

The detailed mechanistic information derived from such studies
provides significant new insights into the factors influencing catalytic activity and into the optimization of such activity.

CATALYTIC HYDROGENATION OF OLEFINS

Detailed studies, the results of which will be described, have
established that predominant pathway of the $Rh(PPh_3)_3Cl$-catalyzed
hydrogenation of olefins (in the presence of excess PPh_3 to suppress
the accumulation of the dimer $Rh_2L_4Cl_2$) corresponds to the mechanistic scheme depicted by Figure 1.[2] Attention is directed to the following features of this scheme.

1. The kinetics of the hydrogenation of $RhClL_3$, i.e.,

$$RhClL_3 + H_2 \longrightarrow RhClH_2L_3 \qquad (4)$$

are described by the two-term rate-law,

$$\frac{-d[RhClL_3]}{dt} = \left[k_2 + \frac{k_1 k_4}{k_{-1}[L] + k_4[H_2]} \right] [H_2][RhClL_3] \qquad (5)$$

where, $k_2 = 4.8\ M^{-1}\ sec^{-1}$, $k_1 = 0.7\ sec^{-1}$, $k_{-1}/k_4 = 1.0$ and $k_4 >$ $10^4\ M^{-1}\ sec^{-1}$, in benzene at $25°$.[7] Up to 0.1 M PPh$_3$ the predominant pathway is, thus, not the direct hydrogenation of RhClL$_3$ (Reaction 4) but, rather, the dissociative pathway corresponding to,

$$RhClL_3 \underset{k_{-1}}{\overset{k_1}{\rightleftharpoons}} RhClL_2 + L \qquad (6)$$

$$RhClL_2 + H_2 \xrightarrow{k_4} RhClH_2L_2 \qquad (7)$$

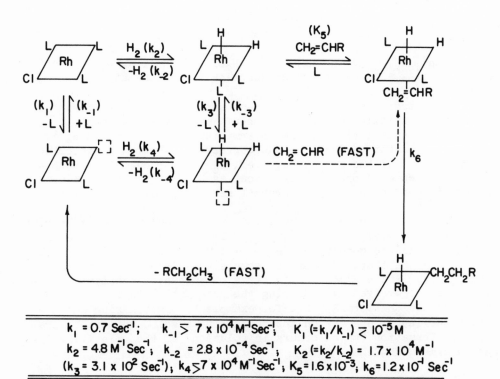

$k_1 = 0.7\ Sec^{-1}$; $k_{-1} \gtrsim 7 \times 10^4\ M^{-1}Sec^{-1}$, $K_1\ (= k_1/k_{-1}) \lesssim 10^{-5}\ M$

$k_2 = 4.8\ M^{-1}Sec^{-1}$; $k_{-2} = 2.8 \times 10^{-4}\ Sec^{-1}$; $K_2\ (= k_2/k_{-2}) = 1.7 \times 10^4\ M^{-1}$

$(k_3 = 3.1 \times 10^2\ Sec^{-1})$; $k_4 \gtrsim 7 \times 10^4\ M^{-1}Sec^{-1}$, $K_5 = 1.6 \times 10^{-3}$; $k_6 = 1.2 \times 10^{-1}\ Sec^{-1}$

Figure 1. Mechanistic scheme for the Rh(PPh$_3$)$_3$Cl-catalyzed hydrogenation of olefins. The values listed refer to the hydrogenation of cyclohexene in benzene at $25°$.

$$\text{RhClH}_2\text{L}_2 + \text{L} \xrightarrow{\text{fast}} \text{RhClH}_2\text{L}_3 \qquad (8)$$

It is significant that although the stable forms of the reactant and product are the tris-phosphine complexes (i.e., RhClL_3 and RhClH_2L_3, respectively) the preferred route for the oxidative addition of H_2 is that involving the coordinatively unsaturated bis-phosphine species, RhClL_2 and RhClH_2L_2. This implies, not unexpectedly, that rates of oxidative addition reactions are more susceptible to steric influences than the stabilities of either the reactants or products. (Note that even at low PPh_3 concentrations, rapid interception of RhClL_2 by H_2 according to Eq. 7 prevents dimerization to form $\text{Rh}_2\text{Cl}_2\text{L}_4$).

2. Over a wide range of conditions, hydrogenation of RhClL_3 to RhClH_2L_3 by the above pathways is fast compared to the subsequent steps in the olefin hydrogenation scheme which are, accordingly, rate-determining. The rate-law for catalytic hydrogenation is thus described by,

$$\frac{-d[\text{H}_2]}{dt} = \frac{k_6 K_5 [\text{Rh}]_{\text{Total}} [\text{RCH=CH}_2]}{K_5 [\text{RCH=CH}_2] + [\text{L}]} \qquad (9)$$

where, $[\text{Rh}]_{\text{Total}} \sim [\text{RhClH}_2\text{L}_3] + [\text{RhClH}_2\text{L}_2(\text{RCH=CH}_2]$, $K_5 = 1.6 \times 10^{-3}$ and $k_6 = 1.2 \times 10^{-1}$ sec^{-1} for cyclohexene in benzene at 25°.[8]

3. In studies on a series of p-substituted styrenes it was found that the influence of electron-withdrawing substituents is to increase the stabilities of the intermediate olefin complexes, i.e., K_5 ($\rho \sim + 1.1$) but to decrease the corresponding rates of insertion rearrangements of the complexes, i.e., k_6 ($\rho \sim -0.8$). The two opposing influences approximately cancel each other so that the net effect of substituents on the overall rate is small.

4. The limiting rate of catalytic hydrogenation attainable with this system is given by, $k_6[\text{Rh}]_{\text{Total}}$. This rate is approached under readily realizable conditions, e.g., $\sim$ 1 M olefin and $< 10^{-3}$ M PPh_3. It is concluded that, contrary to some suggestions previously advanced, very little advantage in terms of enchanced catalytic activity is to be derived by modifying the catalyst so that it is predominantly in the form RhClL_2, for example by immobilizing and isolating the latter species on the surface of a solid support to prevent the formation (which occurs in solution) of the relatively inactive dimer, $\text{Rh}_2\text{Cl}_2\text{L}_4$.

5. Analysis of the various factors that influence the catalytic activity of complexes of the type RhClL_3 suggests that such activity is, in fact, close to optimal for L = PPh_3.

DECARBONYLATION OF ALDEHYDES

The results of kinetic and spectral measurements to be described support the following mechanistic scheme for the decarbonylation of aldehydes by $RhClL_3$ (Eq. 3).[2]

$$RhClL_3 + RCHO \xrightleftharpoons{K_6} Rh(RCHO)ClL_2 + L \qquad (10)$$

$$Rh(RCHO)ClL_2 \xrightarrow[\text{slow}]{k_7} \left[\begin{array}{c} H \\ \diagdown \\ O=C{-}RhClL_2 \\ \diagup \\ R \end{array} \right] \longrightarrow$$

$$\left[\begin{array}{c} H \\ \diagdown \\ R{-}RhClL_2 \\ \diagup \quad | \\ C \\ O \end{array} \right] \rightarrow RhClL_2(CO) + RH \qquad (11)$$

The rate-law is, accordingly, described by Eq. 12, where $K_6 = 7 \times 10^{-4}$ and $k_7 = 2 \times 10^{-3}$ sec^{-1}, for propionaldehyde in o-dichlorobenzene at 25°.

$$\frac{-d[RhClL_3]}{dt} = \frac{k_7 K_6 [RCHO][Rh]_{Total}}{K_6[RCHO] + [L]} \qquad (12)$$

The limiting rate of decarbonylation, i.e., $k_7[Rh]_{Total}$, is approached under readily attainable conditions, for example, [RCHO] $\sim$ 1M and [L] $< 10^{-3}$ M. The same limiting rate applies to decarbonylation by $Rh_2Cl_2L_4$ and is also expected to apply to "isolated" $RhClL_2$ species such as might be achieved by immobilization on a solid support.

Recognition that the species, $Rh(RCHO)ClL_2$, which is involved in the rate-determining oxidative addition step is "coordinatively unsaturated" with respect to such oxidative addition, led to attempts to achieve enhanced activity through incorporation of an additional appropriate ligand into the coordination shell. This was accomplished with a variety of ligands, the most effective for this purpose proving to be dimethylformamide (DMF). Thus the decarbonylation of aldehydes by $Rh_2Cl_2L_4$ (Eq. 13) in the presence of DMF was found to proceed according to the mechanism depicted by Eq. 14 and 15, where $K_8 \sim 1$ M and $k_9 = 0.2$ M^{-1} sec^{-1} for propionaldehyde in o-dichlorobenzene at 25°.

$$2RCHO + Rh_2Cl_2L_4 \longrightarrow 2RH + 2RhClL_2(CO) \qquad (13)$$

$$Rh_2Cl_2L_4 + 2DMF \xrightleftharpoons{k_8} 2RhClL_2(DMF) \qquad (14)$$

$$\text{RhClL}_2(\text{DMF}) + \text{RCHO} \xrightarrow[\text{slow}]{k_9} \left[\begin{array}{c} \text{H} \\ \text{O=C} \\ \text{R} \end{array} \diagdown \text{RhClL}_2(\text{DMF}) \right] \longrightarrow$$

$$\text{RhClL}_2(\text{CO}) + \text{RH} + \text{DMF} \qquad (15)$$

The limiting rate of decarbonylation under these conditions is given by $k_9[\text{Rh}]_{\text{Total}}[\text{RCHO}]$ and rates substantially in excess of those attainable in the absence of DMF (i.e., $k_7[\text{Rh}]_{\text{Total}}$) are readily achieved.

The influence of other ligands in this context and the systematic application of ligand variation to the optimization of reactivity and catalytic activity will be discussed.

ACKNOWLEDGMENT

The contributions of Dr. C. S. Wong and Dr. T. Okamoto who performed most of the experiments discussed in this paper are gratefully acknowledged. The research was supported by the National Science Foundation.

REFERENCES

1. J. A. Osborn, F. H. Jardine, J. F. Young and G. Wilkinson, J. Chem. Soc. (A), 1711 (1966).
2. C. S. Wong, Ph.D. Dissertation, The University of Chicago, 1973.
3. P. Meakin, J. P. Jesson and C. A. Tolman, J. Amer. Chem. Soc., 94, 3240 (1972).
4. S. Siegel and D. Ohrt, Inorg. Nucl. Chem. Lett., 8, 15 (1972).
5. M. C. Baird, C. J. Nyman and G. Wilkinson, J. Chem. Soc. (A), 348 (1968).
6. K. Ohno and J. Tsuji, J. Amer. Chem. Soc., 90, 99 (1968).
7. J. Halpern and C. S. Wong, J.C.S. Chem. Comm., 629 (1973).
8. J. Halpern and T. Okamoto, unpublished results.

DISCUSSION

DR. PETTIT: You never wrote out in detail the structure that you think is the aldehyde rhodium complex. How does the aldehyde coordinate?

DR. HALPERN: I suppose through the oxygen, but we do not know that. The same question arises with respect to acetonitrile. The complexes are very weak. We cannot isolate the aldehyde complex even though it rearranges slowly enough to be observed in solution.

DR. COLLMAN: Do you observe it spectroscopically?

DR. HALPERN: Yes. What we see when we add aldehyde to a solution of tristriphenylphosphine chloro rhodium is a fast spectral change associated with the aldehyde complex equilibrium and then a slow change characterized by an isobestic feature in which we go from the bistriphenylphosphine chloro rhodium aldehyde complex to the carbonyl.

DR. IBERS: Is it possible to observe the vibrational spectrum? Can you see the C-O stretch?

DR. HALPERN: I suppose yes in principle. Unfortunately, the solubility of tristriphenylphosphine chloro rhodium in benzene, the solvent we like to work in, is not more than about 10^{-2} molar. Thus, it would not make a very good candidate for infrared spectral studies. To examine such systems by NMR one needs a higher concentration and frequently the tri-para-tolyl compound which is somewhat more soluble is used. The particular system we chose is one that does not lend itself to infrared characterization.

DR. OSBORN: In cationic rhodium and iridium catalysts which we made some years ago, we can isolate adducts of ketones and perhaps Jim Ibers would like to do the structure of one of these.

DR. IBERS: We would prefer a carbon dioxide compound.

DR. HALPERN: That reminds me - your reference to cationic complexes. I made the comment in passing, that in many respects triphenylphosphine is an optimal ligand for this catalyst system. I was thinking of a system which contains a halide. There are other ways in which you can modify the catalyst. One is by moving the halide to form a cationic complex. This opens up a new possibility that you have explored but to which none of what I said relates directly.

DR. WHITESIDES: What is the conclusion as to overall rate determining step under the conditions that normally are used in synthetic chemistry for hydrogenation?

DR. HALPERN: At least for cyclohexene, it is the rearrangement of the hydride olefin complex because the rate that we calculate for that step is in good agreement over a wide range of conditions, with the observed catalytic rate. For styrene the observed rate is higher than we calculate for the path I have shown, but it does approach it under many conditions. The additional contribution may well be due to the other (i.e. "olefin") pathway - but for cyclohexene such a contribution appears to be negligible.

DR. WHITESIDES: Do you have any idea as to how you can increase the rate of that rearrangement?

DR. HALPERN: What we would like is to modify the complex in such a way that the rearrangement is faster as long as that is the rate determining regime. I think that is what John Osborn has accomplished with the cationic complexes. That is one way perhaps of destabilizing the olefin complex. We think from the trends we have seen that what you want is a less stable olefin complex in order to undergo the migratory insertion rearrangement faster; but the olefin complex should not be so unstable as not to form. We have

looked, for example, at substituted arylphosphines, but we don't see
very large effects. The direction to go may be to a cationic system
where the olefin is less stably bound relative to the sigma complex.
    DR. COLLMAN: Does solvent play a role in that?
    DR. HALPERN: Not in relatively inert solvents. For example,
the difference between benzene and dichlorobenzene is very small.
If we went to more reactive solvents such as DMF I am sure we would
see larger effects but we would have a hard time understanding them.
    DR. COLLMAN: The reason that I asked is that we have some pre-
liminary kinetic data on the supported analogs. We are measuring
the turnover numbers and comparing with homogeneous systems. We can
approach within experimental error the same turnover numbers that you
get in the fastest case you can do in solution. But the additional
of a polar solvent gives about a five fold increase in rate.
    DR. HALPERN: I should emphasize that this work was designed to
learn as much as possible about the fundamental reactions, not to
optimize the system and therefore we have stuck with the given solvent. In some cases where we have seen an obvious rationale to im-
prove reactivity, we have pursued this in a systematic way. The effect of DMF on decarbonylation was one such case. In the case of hydrogenation we have not seen an obvious rationale. We could clearly
play with solvents, phosphine ligands and so on but we don't see
clearly on fundamental grounds what step we should take to optimize
the system.
    DR. WHITESIDES: We, of course, are interested in the opposite
step which is the loss of a beta hydride to give a metal hydride o-
lefin complex. It seems clear that step occurs best when you have
a cis elimination, and it is not at all clear that the best geometry
for the olefin coordinated to a metal hydride linkage is going to
bear any resemblance to that sort of thing. It seems very probable
that you have the olefin lined up this way and then the metal hydride
sticking out that way so that you would have to have a 90 degree rotation before you could have an insertion. If that is going to be
the case it is probably going to be pretty hard to muck around with
that stuff, at least from that point of view. Do you have any evidence of steric bulk, which might argue for this kind of geometry
just on the basis that the hydrides are very small and can fit the
rest of the olefin linkage over there?
    DR. HALPERN: One of my thoughts about that is that one way we
could perhaps increase the rate of migratory insertion is by increasing the steric bulk of the ligand, if we did not do it to a point of
wiping out the reactivity of the other steps. That is, maintaining
that insertion step as rate determining and still getting a high
loading of olefin. We think that in that step we are going from the
six coordinate olefin hydride complex to a five coordinate sigma
complex because we do not see a phosphine dependence of the rate.
Thus, increased steric hindrance ought to drive that step. One obvious approach that we have not yet pursued would be to make a somewhat bulkier ligand. The point you raised is of course very important in the context of the work on assymetric hydrogenation that

Knowles is doing.

DR. KUMADA: What opinion do you have about Tolman's 16 and 18 electron rule? On that basis, the species you write should be ruled out.

DR. HALPERN: As far as I know Tolman et al. agrees with us on the mechanism that involves loss of phosphine and addition of hydrogen to the bis(phosphine)rhodium monomer, i.e. a 14 - electron species. In their paper they comment that this is unexpected on the basis of the 16 electron rule and suggest that perhaps one of the phenyl rings of triphenylphosphine is bending over and keeping the rhodium happy. However, in general I think the rule is a very misleading one and that its impact is negative. My reason for saying this, is that we have all known for a very long time that the 16 and 18 electron configurations are stable and that most of the stable species belong to those configurations. We did not need any rules to remind us of that, and if that was all that the "16 and 18 electron rule" was it would not have attracted much attention. I think that Tolman's rule goes further in suggesting that one should keep in mind that these are the only configurations of importance in reaction mechanisms, even as reaction intermediates. And that, I think, is very misleading. The example I have shown you is a very good one, that the 14 electron bisphosphine rhodium chloride, which is a minor species, is terribly important, and there are many more examples. I think G. Whitesides has one in which olefin elimination involves a 14 electron intermediate. Because of their high reactivity, which in turn is linked to their low stability, such 14 - electron species will turn out very frequently to be important as intermediates in reaction mechanisms, and rather than approaching the problem of reaction mechanisms with a view that these are unlikely we should always be looking for them. For this reason I feel that the mole is highly misleading and its impact is negative. In addition to the unambiguous evidence which is accumulating for species such as intermediates, I know of at least three different X-ray structures on stable 2 coordinate d^{10} complexes of platinum(0) and palladium(0).

HOMOGENEOUS CATALYSIS OF NICKEL COMPLEXES. OLIGOMERIZATION, POLYMERIZATION AND SOME RELATED REACTIONS OF DIOLEFINS

Junji Furukawa and Jitsuo Kiji[*]

Department of Synthetic Chemistry, Kyoto University, Kyoto 606, Japan

ACID-PROMOTED REACTIONS OF DIOLEFINS ON Ni(0) COMPLEXES

It has already been reported that the cyclodimerization of butadiene proceeds smoothly under the influence of some nickel complexes to give 2-methylenevinylcyclopentane (I) ((a), Eq. 1).[1] In this reaction a ligand-containing nickel(0) complex, combined with protic acid, has been proved to be an active species.[2] Not only the cyclodimerization but also following reactions of diolefins occur in the presence of the catalysts; (b) trans-1,4-polymerization,[3] (c) amination of 1,3-diolefin and norbornadiene,[4] (d) allyl transfer,[5] (e) a novel dimerization of norbornadiene (Eq.2),[6] and (f) cycloaddition of butadiene to strained ring olefins to form an exo-methylene- and methyl-substituted four-membered cyclic compound (Eq. 3).[7]

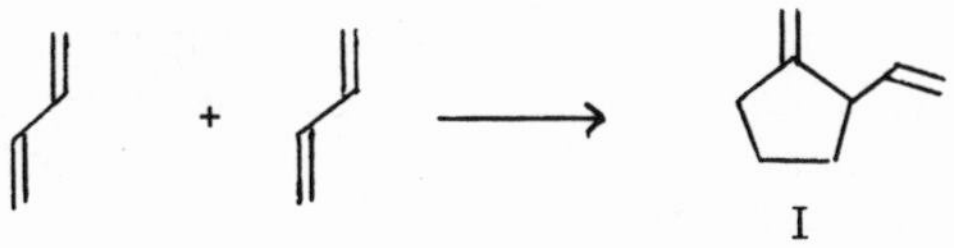

$$(1)$$

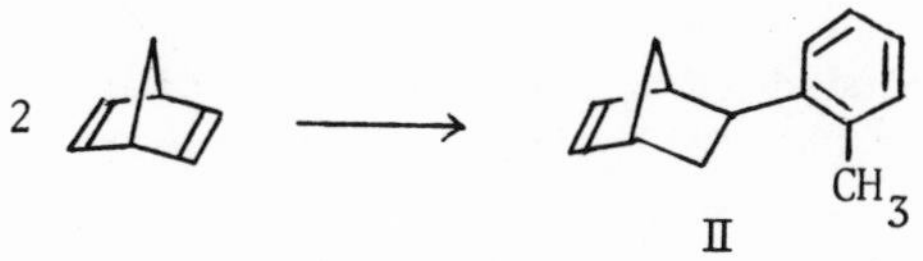

$$(2)$$

$$(3)$$

119

This catalytic system is characteristic in a point that a controlled amount of protic acid is involved in the system and thus different from the conventional ligand-containing nickel catalyst.[8] Consequently, these reactions can be characterized as an acid-promoted reactions on zerovalent nickel complexes. Figure 1,2 and 3 show how these reactions are influenced by a catalytic amount of protic acid. Figure 1 shows the dimerization and polymerization of butadiene by tetrakis(phosphite)nickels $Ni[P(OR)_3]_4$ in alcohol.

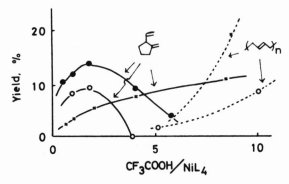

Figure 1. Reaction of butadiene by NiL_4—CF_3COOH in iso-propanol. ●, L = $PPh(OEt)_2$; o,L = $P(OEt)_3$; x,L = $P(OCH_2CH_2Cl)_3$. Conditions: NiL_4, $0.25 \times 10^{-3}M$; butadiene, $10^{-2}M$; 80°C for 8 – 10 hr.

Addition of CF_3COOH to the system promotes the formation of I, but further addition retards it and promotes the 1,4-trans-polymerization. Amination of 1,3-diolefin by bis(cyclooctadiene)nickel $Ni(COD)_2$/n-Bu_3P or by $Ni[P(OEt)_3]_4$ is also promoted by CF_3COOH. Added acid enhances the reaction of 1,3-diolefin with amine (Figure 2). Dimerization of norbornadiene by $Ni(COD)_2$/n-Bu_3P is interesting to note. In the absence of protic acid, the conventional 2+2 cycloaddition product is formed in high yield. When one equivalent of CF_3COOH is added to this system, a dramatic change of the reaction can be observed. The cyclodimer no longer is formed but o-tolyl-2-norbornene (II) is formed in moderate yield (ca. 50%). Further addition of CF_3COOH promotes the reaction of amine with norbornadiene.[9] In this way the evidence of the effect of the added acid has been obtained and it has been concluded that the active species of the catalyst is well-characterized by the following equilibrium:[10]

$$NiL_x \ + \ HX \ \rightleftharpoons \ HNiL_x \cdot X \qquad\qquad (4)$$

The reaction courses of the diolefins, e.g., the cyclodimerization or 1,4-trans-polymerization of butadiene, are related to an equilibrium state of the catalytic system (Eq. 4).

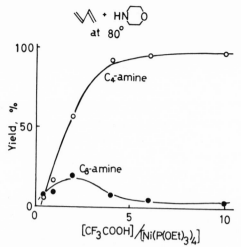

Figure 2. Effect of CF_3COOH on the reaction of butadiene with morpholine. ●, Octadienyl amine; o, butenyl amine. Conditions: NiL_4, 0.25 x 10^{-3}M; butadiene 10^{-2}M; at 80°C for 8 hr.

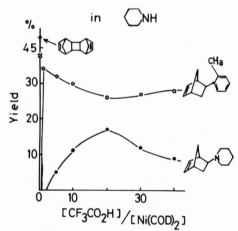

Figure 3. Effect of CF_3COOH on the reaction of norbornadiene in piperidene. Conditions: $Ni(COD)_2$, 0.05 x 10^{-3}M; n-Bu_3P, 0.2 x 10^{-3}M; norbornadiene, 2.x 10^{-3}M; at 96°C for 15 hr.

A series of these investigations originated from the mechanistic study on the cyclodimerization (a). Concerning the cyclodimerization, the cylization mechanism as shown in Equation 5 has been proposed. The cyclization of this type is well-characterized in terms of "intramolecular insertion" of one double bond and the formation of five-membered cyclic compounds is a general reaction.[11,12]

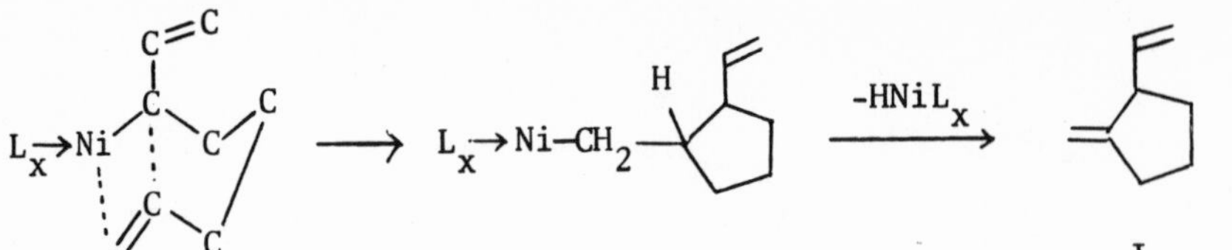

(5)

While the olefin-dismutation reaction has been discussed,[13] examples of cycloaddition reactions giving a four-membered cyclic compound are scarce and only a few cyclic compounds have been isolated as the stable products.[14],[15]

It has also been found that the cycloaddition reaction (f) can be characterized as an intramolecular insertion of one double bond. The catalysts, which are effective for the cyclodimerization (a), are effective for the reaction (f). Consequently, it appeared to us that this reaction might be an acid-promoted one. Indeed, π-crotyl complex (IV)[16] reacts with dicyclopentadiene (DCPD) smoothly and IIIa is formed in high yield. This fact leaves little doubt that

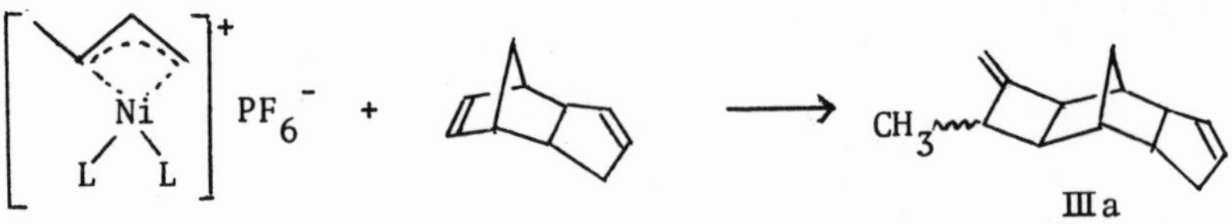

$L = P(OEt)_3$

the cycloaddition reaction (f) proceeds through a π-crotyl complex which is formed by the reaction of butadiene with the nickel hydride (Eq. 4), because of the reaction of the nickel hydride with 1,3-diolefin is known to give π-allyl complex.[16] The cycloaddition product (IIIa) is formed more rapidly than I as shown in Figure 4. The formation of I is very slow even in the absence of DCPD. Therefore, the formation of I is so sufficiently slow that the cycloaddition reaction is possible with a small amount of the Ni-H species, namely, the formation of IIIa is possible under the conditions, where the equilibrium of the reaction (Eq. 4) lies far to left. A driving

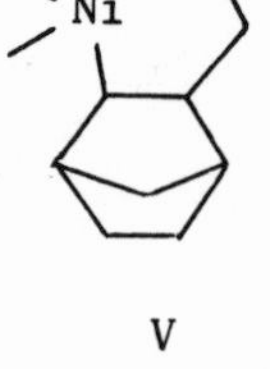

V

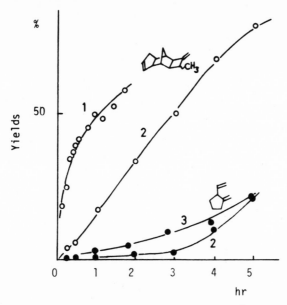

Figure 4.   Rates of reaction of $[\pi\text{-}C_4H_7NiL_4]^+PF_6^-$ with DCPD (1), of butadiene with DCPD by $(n\text{-}Bu_3P)_2NiBr_2$—$NaBH_4$ (1:1) (2), and of butadiene by $(n\text{-}Bu_3P)_2NiBr_2$—$NaBH_4$ (1:1) (3).   Conditions: In ethanol at 80°C.

force of the ring closure to the four-membered cyclic compound may be attributed to the strain of the carbon chain of a reaction intermediate (V) of the π-allyl complex with the bicyclic compound.[17-20]

CONCLUSION

The conclusion we have reached in these studies is that these reactions can be controlled by adding a controlled amount of protic acid or that the regulation of the reactivity in the reaction of diolefins is possible by adjusting the acidity of the system. The reactions described above are summarized in Scheme 1.

ACKNOWLEDGEMENT

The contribution of Messrs. S. Yoshikawa, K. Yamamoto, E. Sasakawa, and S. Nishimura who performed most of the experiments in this paper are gratefully acknowledged.

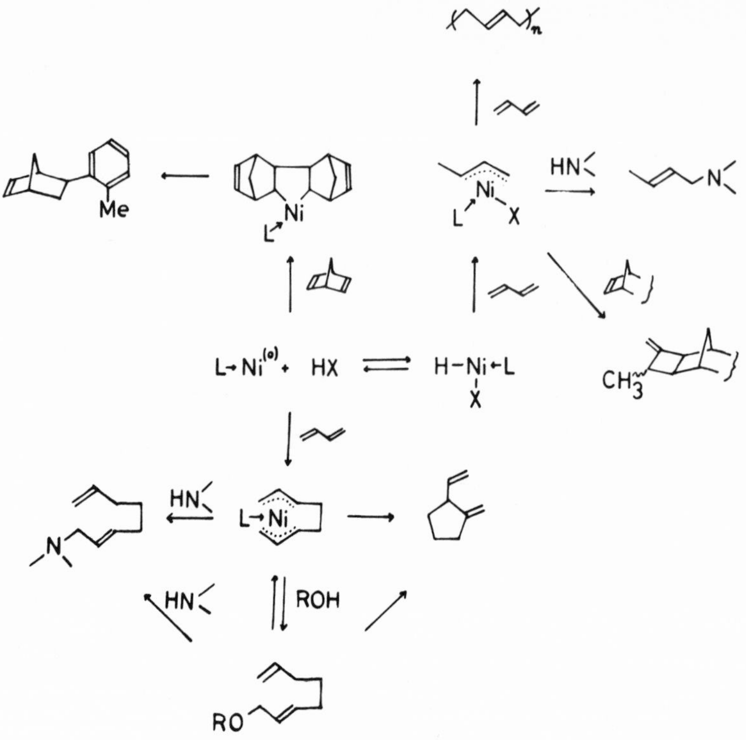

Scheme 1

REFERENCES

1. J. Kiji, K. Masui and J. Furukawa, Tetrahedron Letters, 2561
   (1970); Bull. Chem. Soc. Japan, 44, 1956 (1971).
2. J. Kiji, K. Yamamoto, S. Mitani, S. Yoshikawa and J. Furukawa,
   Bull. Chem. Soc. Japan, 46, 1791 (1973).
3. J. Furukawa, J. Kiji, T. Konishi, K. Yamamoto, S. Mitani and
   S. Yoshikawa, Makromol. Chem., 174, 65 (1973).
4. J. Furukawa, J. Kiji, S. Mitani, S. Yoshikawa, K. Yamamoto and
   E. Sasakawa, Chem. Letters, 1211 (1972); J. Kiji, K. Yamamoto,
   E. Sasakawa and J. Furukawa, Chem. Comm., 770 (1973); J. Organometal. Chem., in press.
5. J. Furukawa, J. Kiji, K. Yamamoto and T. Tojo, Tetrahedron, 29,
   3149 (1973).
6. S. Yoshikawa, K. Aoki, J. Kiji and J. Furukawa, ibid., 30, 405
   (1974).
7. S. Yoshikawa, S. Nishimura, J. Kiji and J. Furukawa, Tetrahedron
   Letters, 3071 (1973).
8. P. Heimbach and R. Traunmuller, "Chemie der Metall-Olefin-Kom-
   plexe", Verlag Chemie, Weinheim/Bergstr., 1970, p. 113.

9. J. Kiji, S. Nishimura, S. Yoshikawa, E. Sasakawa and J. Furukawa, to be submitted to Bull. Chem. Soc. Japan.
10. C. A. Tolman, J. Amer. Chem. Soc., 92, 4217 (1970).
11. K. Ziegler, "Organometallic Chemistry", H. Zeiss, Ed., Reinhold
    Publishing Corp., New York, N. Y., 1960, p. 194.
12. G. P. Chiusoli and L. Casser, Angew. Chem., 79, 177 (1967).
13. F. D. Mango and J. H. Schachtschneider, J. Amer. Chem. Soc., 91,
    1030 (1969).
14. P. R. Coulson, J. Org. Chem., 37, 1253 (1972).
15. L. G. Cannel, J. Amer. Chem. Soc., 94, 6867 (1972).
16. C. A. Tolman, ibid., 92, 6777 (1970).
17. M. C. Gallazzi, T. L. Hanlon, G. Vitulli and L. Porri, J. Organ-
    ometal. Chem., 33, C45 (1971).
18. M. Zocchi, G. Tiechi and A. Albinati, ibid., 33, C47 (1971).
19. R. P. Hughes and J. Powell, ibid., 60, 387 (1973).
20. J. A. Sadowinick and S. J. Lippard, Inorg. Chem., 12, 2659 (1973).

DISCUSSION

DR. COLLMAN: I notice a striking similarity which I never thought about before, to the form of your butadiene dimer. Two butadienes complex with nickel and you can write this as a metallocyclopentane which has two vinyl groups on it. That intermediate is the same form as the intermediates that many people write for the dismutation reaction and I wonder if there is some kind of relationship between the dismutation and this chemistry.

DR. KIJI: I think there is a great difference between our system and the conventional dimerization intermediate built by Wilke, or the dismutation reaction. In our case a controlled amount of protic acid is present in the system but in the other case no protic acid is present. I think there is a further difference after protolysis, where our intermediate is stabilized towards cyclization.

DR. WHITESIDES: Does the butadiene dimerization reaction work with isoprene, and if so, what does it give?

DR. KIJI: No, no appreciable reaction occured with isoprene.

HOMOGENEOUS CATALYTIC ACTIVATION OF HYDROCARBONS

G. W. Parshall

Central Research Department, E. I. du Pont de Nemours
and Company, Experimental Station, Wilmington, Delaware
19898

A historical trend has been evident in the reactions of C-H
bonds with soluble transition metal complexes over the past decade.
The general progression in terms of reactions of increasing difficulty has been I. Ortho-metallation of aromatic rings, II. Substitution of benzene C-H, III. Intramolecular attack on aliphatic C-H,
IV. Alkane activation. The work from our laboratory has involved
the first three of these topics. Our primary tool has been the study of exchange between D_2 and the C-H bonds of organic compounds.
Although this is not a generally useful synthetic reaction, it provides a convenient tool for the detection of rapid reversible interactions of potential catalytic importance.

ORTHO METALLATION

In the late 1960's, a number of reactions were observed in which
the central metal atom of a complex attacks an ortho C-H bond of an
aryl ring in one of the ligands bonded to the metal.[1] One of the
first systems that we studied was a cobalt hydride complex (1) which
contained triphenyl phosphite ligands.[2] As shown in Figure 1, the
cobalt hydride 1 reacted with deuterium gas at 100° to generate a
cobalt deuteride complex in which the ortho carbons of the phosphite
ligands bear deuterium atoms rather than hydrogen. At the time that
this chemistry was first carried out, we suspected that the reaction
involved an intermediate such as 2 in which one of the aryl rings
had become bonded to the cobalt through a sigma bond. We were not
able to isolate 2 at the time, although we did isolate rhodium and
ruthenium analogs.

Recently Dr. Gosser,[3] in our laboratories, has isolated 2, both

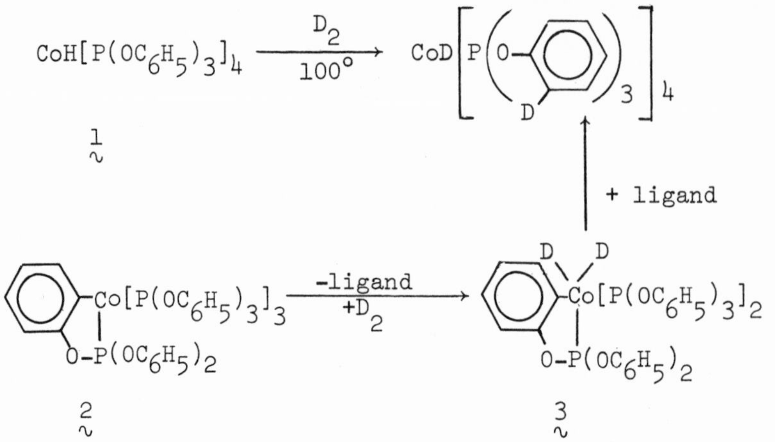

Figure 1. Exchange between D_2 and the ortho C-H bonds of coordinated triphenyl phosphite.

by the careful pyrolysis of 1 and by an independent synthesis from π-cyclooctenyl-1,5-cyclooctadiene-cobalt. We were very pleased to find that 2, the proposed intermediate in the exchange reaction, re- acts with deuterium gas to give the ortho-deuterated triphenyl phos- phite complex. However, much to our surprise, the reaction with 2 occurs at room temperature in contrast to a temperature of 80-100° required for the deuterium exchange with 1. This result, as well as the enhanced catalytic properties of the ortho-metallated compound, can probably be explained by a greater rate of dissociation of the triphenyl phosphite ligands from 2 as compared to 1. The ligand dissociation process frees a coordination site for the oxidative ad- dition of deuterium to give the intermediate aryl cobalt deuteride (3). Subsequent reductive elimination of Co-C and Co-D ligands gen- erates the observed exchange product (after acquisition of another phosphite ligand from solution).

In studies in our laboratories as well as elsewhere, many other examples of this ortho-metallation reaction have been observed. It has now been detected for triphenylphosphine or triphenyl phosphite complexes of all the group VIII transition metals.

BENZENE C-H EXCHANGE

The observation of the intramolecular exchange process led us to consider the question of whether benzene would undergo a compar- able exchange with deuterium in the presence of transition metal complexes. Although this process had been observed with heteroge-

neous catalysts such as metallic platinum as early as 1937,[4] it was
not known for soluble catalysts.

In an exploration of the reaction of benzene with D_2 in the
presence of metal polyhydrides, we soon discovered two families of
catalysts for exchange. One group was found in our studies of the
chemistry of the early transition metals. These compounds, MH_3-
$(C_5H_5)_2$ (M = Nb, Ta), catalyze the reaction of benzene and substituted benzenes with deuterium to give deuterated benzenes. Similar
reactions were found to occur with a second group of compounds, the
phosphine-substituted polyhydrides such as $IrH_5(PMe_3)_2$, $ReH_5(PPh_3)_3$
and $TaH_5(DMPE)_2$ (DMPE = tetramethylethylene diphosphine).

In a careful study of this reaction,[6] Dr. Klabunde found that
the exchange is promoted by electron-withdrawing substituents on the
benzene ring and is strongly susceptible to steric effects. In competitive experiments between substituted benzenes, he found the following reactivity sequence: p-difluorobenzene > fluorobenzene > benzene > toluene = anisole > p-xylene. This reactivity sequence is
valid for all the complexes studied except $TaH_3(C_5H_5)_2$. This hydride shows no discrimination amongst the substrates studied except
that p-xylene reacted much more slowly than did the other aromatic
compounds. This discrimination is probably due to a steric effect
because in p-xylene all the aromatic C-H bonds are ortho to a methyl
group. The aliphatic hydrogens of the methyl groups did not exchange
with deuterium.

All of our experimental observations seem consistent with a
mechanism in which an aryl C-H bond reacts with coordinatively unsaturated metal complex by oxidative addition as shown in Figure 2.

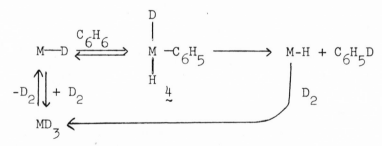

Figure 2. Oxidative addition and reductive elimination sequences in
exchange.

The key intermediate is the reactive species, M-D or M-H, produced
by dissociation of deuterium or hydrogen from a polyhydride complex
(other ligands omitted). Oxidative addition of a benzene C-H bond
generates the intermediate 4 which can undergo reductive elimination

to give a deuterated benzene. Reaction of the M-H product with D_2 can regenerate the MD_3 complex to start another exchange cycle.

Sound precedent for the formation of arylmetal hydride intermediates such as 4 is found not only in the intramolecular exchange
studies described above but also in recent work by Green.[7] Hydrogen elimination from $WH_2(C_5H_5)_2$ in the presence of benzene gave a
stable phenyltungsten hydride, $WH(C_6H_5)(C_5H_5)_2$, analogous to 4.

Recent work in our laboratories by Drs. Tebbe and Klabunde and
Mrs. Seiwell has indicated that many other types of transition metal complexes catalyze exchange of benzene C-H bonds by similar mechanisms. Although the polyhydride species are the most effective
catalysts, a preformed hydride function is not essential.

INTRAMOLECULAR ALIPHATIC C-H INTERACTIONS

Several reports of intramolecular interactions of metal atoms
with aliphatic C-H bonds, analogous to ortho-metallation in aromatic systems, have appeared in recent years. Most of these observations have dealt with platinum systems such as those studied by
Bernard Shaw and his coworkers.[8] In a typical experiment, they observe a stoichiometric reaction in which HCl is eliminated between
Pt-Cl and C-H bonds. Typically a five-membered ring is formed as in

Dr. Klabunde came across evidence of a catalytic attack on aliphatic C-H bonds in the course of reexamination of a ruthenium system[2] that we had studied several years ago. The hydride complex,
$RuHCl(P\emptyset_3)_3$, reacts with D_2 to produce $RuDCl(P\emptyset_3^D)_3$ in which $\emptyset^D$ represents an ortho-deuterated phenyl group. However, the complex al-
so catalyzes ortho-deuteration of any excess triphenylphosphine added to the solution. Apparently the rate of exchange between free
and coordinated phosphine is sufficiently rapid to equilibrate the
two species in the course of a deuteration experiment.

When excess diphenyl-n-propylphosphine (5) was added to a benzene solution of RuHCl(PØ$_3$)$_3$ being treated with D$_2$, the anticipated
exchange of the ortho hydrogens of 5 occurred.[9] However, quite unexpectedly, some deuterium also appeared in the α, β and γ positions
of the propyl group. Exchange was fastest for the methyl group. In
fact, on a statistical basis, CH$_3$ exchange was as rapid as ortho
exchange and proceeded at a measurable rate even at 20°.

In a series of kinetic runs under exchange conditions spanning
the temperature range 20-160°, Dr. Klabunde obtained approximate
activation energies of 6 kcal for exchange of the ortho and γ hydrogens and 13 kcal for the α hydrogens. β-Exchange was very slow even
at elevated temperatures.

We were inclined to assume a mechanism (Figure 3) analogous to
that for ortho metallation of an aryl group. Oxidative addition of
a γ-C-H bond as shown to produce a cyclic structure followed by reductive elimination would accomplish exchange. However, as in the
studies of oxidative addition reported by Dr. Osborn at this meeting,

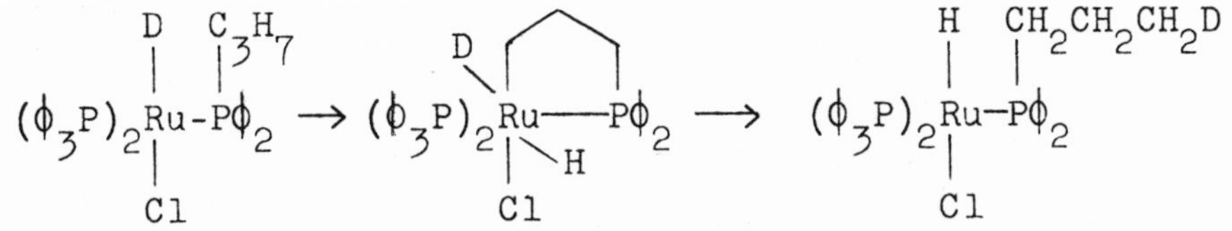

Figure 3. Scheme for aliphatic HD exchange.

there are alternative oxidative addition pathways which might be
labelled "concerted" or "radical". Dr. Klabunde's experiment to
distinguish between the two paths resembles one used by Dr. Osborn.
He prepared and resolved the optically active phosphine 6 and then
subjected it to our standard D$_2$ exchange experiment in the presence
of RuHCl(PØ$_3$)$_3$. One would anticipate that a concerted exchange process at the γ-carbon would lead to deuterium substitution without
loss of optical activity while a radical process would lead to racemization. Surprisingly, in the course of four weeks at 108°, no
significant exchange occurred at the γ carbon (although deuterium

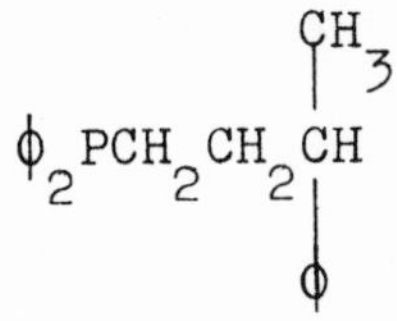

6

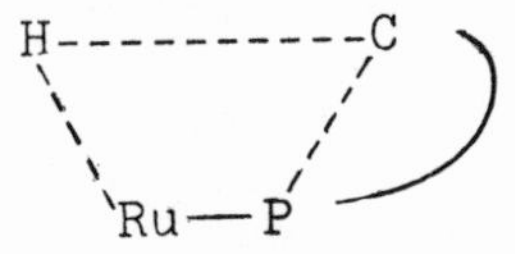

7

did appear at the α, β, δ and ortho positions). Only 17% loss of
optical activity occurred. The absence of reaction is probably assignable to steric hindrance at the γ-carbon in $\underset{\sim}{6}$.

    Perhaps the most important point of point of these experiments
is that a transition metal is able to react with a C-H bond at a
saturated carbon, perhaps via a transition state such as $\underset{\sim}{7}$. This
result suggests that any C-H bond can oxidatively add to a sufficiently reactive transition metal atom. Thus, the activation of alkanes by homogeneous catalysts, the fourth step in the historical
progression that I referred to in my introduction, may be attainable.
Recent reports[10] by Shilov and others provide preliminary evidence
that such is the case.

REFERENCES

1. G. W. Parshall, Accounts Chem. Res., 3, 139 (1970).
2. G. W. Parshall, W. H. Knoth and R. A. Schunn, J. Amer. Chem. Soc.,
   91, 4990 (1969).
3. L. W. Gosser, to be published.
4. A. Farkas and L. Farkas, Trans. Faraday Soc., 33, 827 (1937).
5. E. K. Barefield, G. W. Parshall and F. N. Tebbe, J. Amer. Chem.
   Soc., 92, 5234 (1970).
6. U. Klabunde and G. W. Parshall, ibid., 94, 9081 (1972).
7. M. L. H. Green and P. J. Knowles, J. Chem. Soc., A, 1508 (1971).
8. A. J. Cheney, B. E. Mann, B. L. Shaw and R. M. Slade, ibid., A,
   3833 (1971).
9. U. Klabunde, to be published.
10. M. B. Tyabin, A. E. Shilov, and A. A. Shteinman, Doklad. Akad.
    Nauk SSSR, 198, 381 (1971); R. J. Hodges, D. E. Webster and P. B.
    Wells, J. Chem. Soc., A, 3230 (1971).

DISCUSSION

    DR. MARKS: In your deuteration of the cyclopentadienyl rings
with the bimolecular mechanism, if that works you should be able to
deuterate ferrocene and some other things.
    DR. PARSHALL: Yes, one would think so.
    DR. MARKS: Some people have seen activation of alkanes, I be-
lieve.
    DR. PARSHALL: Yes, Wells and Shilov have done a fair amount of
work with platinum catalyzed activation of alkanes. It is a very
complex system. There is some question as to whether or not this is
really homogeneous. It may well be, but it looks much more complicated than a simple mononuclear complex. One characteristic of the
Pt systems that we don't see in any of our systems is what is termed

multiple exchange - several exchanges each time the metal atom comes in contact with the substrate. Ours go one hydrogen at a time, whereas both the heterogeneous and the Wells and Shilov systems give multiple exchange.

DR. HALPERN: I am struck by the fact that the two metals that have thus far been implicated in the activation of saturated hydrocarbons, i.e., platinum (II) and ruthenium (II), are both metal ions that we have concluded from our work on H_2 activation, activate H_2 not by oxidative addition but by electrophilic attack leading to heterolytic splitting. In such cases we think the metal interacts only with one hydrogen atom and that the proton that comes off in the process is being accepted by some base, i.e. $M^{n+} + H_2 + B \rightarrow MH^{(n-1)+} + BH^+$. In that type of reaction pattern which is very distinctive, we see the role of the base quite clearly. If we make the medium more basic the activity goes up, implying that the base is involved in the rate determining step. If that were true also for the carbon hydrogen activation, I would visualize that process as one in which the metal interacts with the carbon but not necessarily with the hydrogen. In fact, I would visualize this as one in which some proton acceptor, such as chloride ion is acting as the hydrogen acceptor and therefore I would be inclined to draw the transition state somewhat differently than you did. Now I realize that this is not terribly relevant to the present situation, but it is to the arguments that are made relating structures of stable species to the products of reaction which are sometimes formed. For example, in the paper to which you referred, the parallel is drawn between a structure in which the hydrogen (not the carbon) is apparently pushed toward the metal, and, as I recall, it is suggested that this may be important for the activation of a carbon hydrogen bond. The point is made in connection with some work of Shaw in which a metal carbon bond is formed in a similar system. Well, it is not clear to me how pushing the hydrogen up to the metal helps metal-to-carbon bond formation. I see these as two separate kinds of processess that have become confused. There is a tendency for such confusion at an early stage of a subject, such as that which we witnessed in connection with the activation of dihydrogen. When a new phenomenon such as activation of H_2 (or N_2!) by metal ions is first discovered one tends to think of all hydrogen activating reactions as being similar. Gradually as one learned more about it it became apparent that H_2 activation can occur in several distinctly different ways and that is was misleading to try to interpret all the systems in terms of a uniform mechanism. I think the same thing is true here and emphasizing the importance of metal-hydrogen proximity may not be relevant at least for certain cases of C-H activation.

DR. BERGMAN: In that system where you measured the 6 kilocalorie per mole activation energy, could you tell us what the activation entropy is?

DR. PARSHALL: I do not have the entropy value here.

DR. BERGMAN: What was the temperature range? Roughly about 140 degrees?

DR. PARSHALL: Yes, about 20 degrees up to about 160 degrees.

DR. BERGMAN: So at about 100° if the slope is really that shallow the entropy must be terrifically negative.

DR. HALPERN: That is fairly characteristic of oxidative addition reactions. Even the oxidative addition of H_2 to iridium (I), one of the situations which we think of as non polar, has activation entropies in the range of -20 e.u. The activation entropies of the oxidation of alkyl halides are very negative, -40 -50 e.u. with correspondingly low activation energies.

DR. BERGMAN: Is the reason you get minus 40 or 50 e.u. because you are combining a large number of steps?

DR. HALPERN: No, there we think it is because we are looking at a nucleophilic displacement of a halide. All the parameters of that reaction parallel closely those for quaternary ammonium salts.

DR. BERGMAN: So the negative entropy is due to the generation of a lot of charge in a non polar solvent?

DR. HALPERN: In that particular case, yes.

DR. BERGMAN: Is it possible that so much charge is really being generated?

DR. HALPERN: In CH I would be surprised.

METALLOCYCLES AS INTERMEDIATES IN ORGANOTRANSITION-METAL REACTIONS

Robert H. Grubbs[*], Dale D. Carr and Patrick L. Burk

Department of Chemistry, Michigan State University, East Lansing, Michigan 48824

Metallocycles have been proposed as intermediates in a number of transition metal catalyzed reactions of organic compounds.[1,2,3] In most cases it is difficult to distinguish between a non-concerted, stepwise reaction proceeding through a metallocyclic intermediate and a concerted "symmetry controlled" reaction.[2,3,4,5] Consequently, stable tetramethylene metallocycles have been prepared and their structure and reactivities have been examined. These compounds should provide models for reactions which may involve metallocycles. Evidence for metallocycles in transition metal catalyzed olefin reactions, especially in the olefin metathesis reaction, has also been sought.

The reaction of 1,4-dilithiobutane with transition metal halides produces tetramethylene metallocycles.[6,7,8] This reaction was initially suggested for the reaction of tungsten and has more recently been reported for titanium[7] and platinum.[7,8]

$$L_nMCl_2 + \underset{Li}{\overset{Li}{\Big[\quad\Big]}} \rightarrow L_nM\bigcirc$$

The titanium complex decomposes rapidly at room temperature while the platinum complex requires temperatures of ca. 120° for appreciable decomposition rates. The platinum metallocycles decomposed with half-lives that were > 10^4 that of acyclic analogs.[8] The platinum complex was sufficiently stable for an X-ray structure determination.[7]

We have recently prepared a series of nickel phosphine metallo-

135

cycles which are of intermediate stability ($t_{1/2} \approx$ 12 hr at 25°) and are extremely oxygen sensitive. Our preliminary results demonstrate the range of reactions open to tetramethylene metallocycles.

These complexes decompose at room temperature to yield butene, ethylene and cyclobutane as products. As indicated in Table I, the ratios of these products were dependent on the structure of the other ligands.

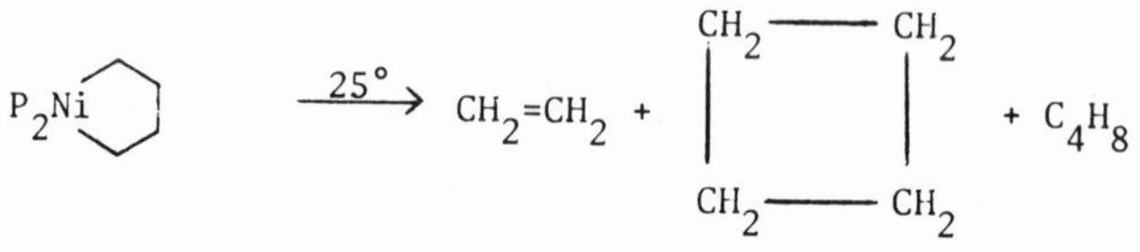

$$P_2Ni \quad \xrightarrow{25°} \quad CH_2{=}CH_2 \; + \; \begin{matrix} CH_2 \text{——} CH_2 \\ | \qquad\quad | \\ CH_2 \text{——} CH_2 \end{matrix} \; + \; C_4H_8$$

Table I

P	C_2H_4	Cyclobutane	C_4H_8 Linear
ϕ_3P(II)	2.1%	.13%	98%
$\phi_2(CH_3)P$(III)	34	5	61
diphos(IV)	95%	5%	--

The ratio of the products and the rate of the reactions were increased by the addition of coordinating solvents or olefins to the reaction solution. The most consistent data to date for the triphenylphosphine complex (II) is given in Table II.

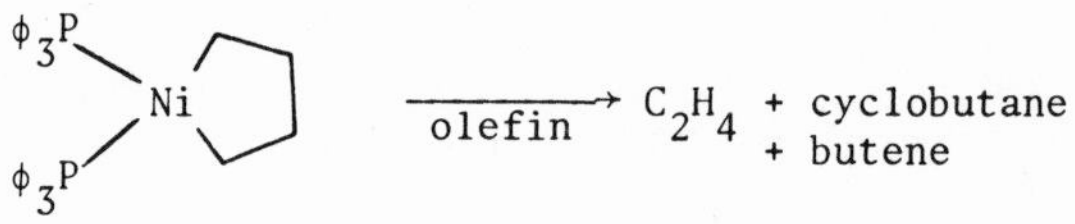

$$\begin{matrix} \phi_3P \\ \quad\searrow \\ \qquad Ni \\ \quad\nearrow \\ \phi_3P \end{matrix} \quad \xrightarrow{\text{olefin}} \quad C_2H_4 + \text{cyclobutane} + \text{butene}$$

Table II

Olefin	Olefin/II	Ethylene	Cyclobutane	Butene
butylvinyl ether	16	10	43	47
acrylonitrile	16	72	28	--
maleic anhydride	16	73	23	4
tetracyanoethylene	16	77	23	--

Although our data is not complete enough for a detailed analysis of the factors controlling the mode of decomposition, it is apparent the activation energies for the different modes of decomposition are similar (3-4 kcal). Consequently, small changes in the electronic factors of the ligands easily alter the ordering of the energy levels of the related transition states. As seen above, any of the three products can become the major product depending on the choice of ligands.

In some special cases tetramethylene metallocycles can be prepared from olefins[6] or strained cyclobutanes.[2] These reactions are formally the reverse of the reactions indicated above.

McDermott and Whitesides observed that the reaction between bistitanocenedinitrogen and ethylene produced a reasonable yield of a complex which gave the same reactions as the unstable but well characterized metallocycle (V) produced from titanocene dichloride and 1,4-dilithiobutane. Fraser, et al,[9] observed that the strained

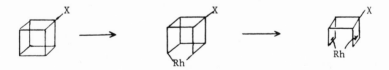

olefin, norbornene reacted with iridium to produce a stable metallocycle.

Cubane reacts with rhodium(I) complexes to produce metallocyclic intermediates which rearrange to a diolefin complex. The careful

$$\text{(cube with X, top-right corner)} \longrightarrow \text{(cube with X, Rh)} \longrightarrow \text{(cube with X, Rh)}$$

work of Halpern and coworkers[2] demonstrated that metallocycles were intermediates in the metal catalyzed ring openings of a number of cubane derivatives.

The very important olefin metathesis reaction[10] was examined to determine the probability of the intermediacy of metallocyclic species in this reaction. These studies have progressed along two lines;

reactions were carried out which should produce metallocycles under
conditions as close to the standard metathesis conditions as possible, and a standard metathesis solution was probed by examining side
products from the reaction and attempting to trap intermediates by
the rapid quenching of such solutions. The earlier observation that
the reaction of 1,4-dilithiobutane-2,3-d$_2$ produced a mixture of mo-
nodeutero, dideutero and non-deuterated ethylene indicated that an
intermediate was produced in this reaction which gave reactions required for the intermediate in the olefin metathesis reaction.[1]
Stereochemical and solvent studies suggested that the rearrangement
in this reaction must have occurred before the intermediate decomposed to olefin. These studies support the non-concerted metallocyclic mechanism for the metathesis reaction.

Before considering methods of detecting metallocyclic intermediates in a metathesis reaction, some characteristics of the WCl$_6$/
RLi reaction were investigated.

It was observed that WCl$_6$/2RLi solutions in aromatic solvents
at the concentrations reported in the original work, were not homogeneous when ethylene and a number of other olefins were used as the
reactants. To determine if the soluble complexes were the active
catalysts, a series of metathesis reactions of an ethylene-d$_2$/ethy-
lene mixture in which only the volume of solvent was varied were
carried out. The percent metathesis per unit time increased as the
concentration decreased (volume of solvent increased) from .1 $\underline{M}$ to
.025 $\underline{M}$. This result is only consistent with metathesis taking place
in solution. Addition of two equivalents of lithium octoxide to the
tungsten hexachloride, before the addition of the alkyl lithium activation, gave a more homogeneous, more active solution.

The mechanism suggested for the metathesis reaction requires a
steady state concentration of the metallocycle (A).

Since transition metal alkyl bonds are cleaved by acids to produce hydrocarbons, treatment of a metathesis solution with acid
should produce a low yield of a hydrogenated dimer of the starting
olefin. The amounts of this material would give an indication of
the concentration of (A) at equilibrium. A 0.1 $\underline{M}$ solution of tungsten hexachloride in benzene under an atmosphere of ca. 6 equivalents
of ethylene was treated with 2 eq. of octyl Li.[11,12] After .5 hr,
the gas above the mixture was sampled, and it was found to contain
about 1% of products made up of propene, cyclopropane, butane and
butene. The source of these gases and their relationship to the metathesis reaction and its mechanism as suggested earlier[2] remains

to be determined. The reaction mixture was then treated with $4N$. DCl in D_2O. The yield of butane from the reaction increased by 30%. The resulting butane was collected by g.l.c. and analyzed by mass spectroscopy. These data are presented in Table III. Comparison of the fragmentation pattern of this sample to authentic samples of the isomeric butanes-d_2 demonstrated that at least 80% of the butane-d_2 showed the same fragmentation pattern as the 1,4-isomer. The observation that the 1,4-isomer was the major product is most consistent with the acid cleavage of an intermediate metallocycle.[13] For example, the reaction[8] of VI with DCl produced good yields of butane-d_2. As can be seen from Table III the mass spectrum of the bu-

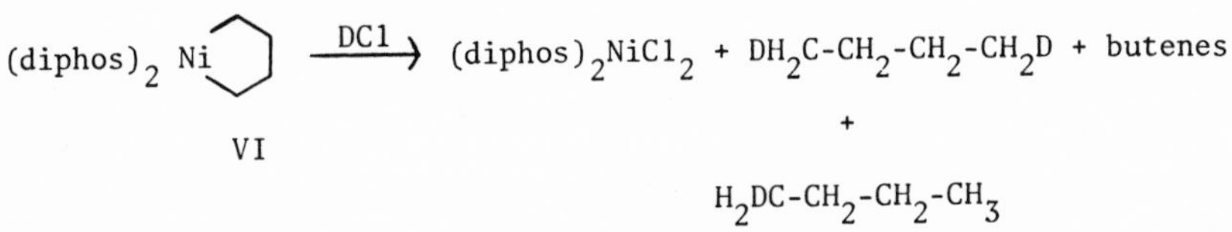

$$(diphos)_2 Ni \bigcirc \xrightarrow{DCl} (diphos)_2 NiCl_2 + DH_2C-CH_2-CH_2-CH_2D + butenes$$

VI

$$+$$

$$H_2DC-CH_2-CH_2-CH_3$$

Table III

Butane m/e	60	59	58	45	44	43
D_0	0.01	0.54	12.3	0.05	3.33	100
1,4-D_2[a]	7.8	8.1	7.0	7.2	100	86
1,2-D_2[b]	19.2	3.9	1.3	92.8	100	30.6
2,3-D_2	10.3.	2.3	0.7	100	19.8	20
W/C_2H_4+DCl	12.6	14.9	6.09	18.32	100	46.1
VI+DCl	13.0	9.2	3.4	18.8	100	46.1

a) from 1,4-dilithiobutane + D_2O

b) from 1-butene + $D_2/(\phi_3 P)_3 ClRh$

c) from 1,4-dilithiobutane-2,3-d_2 + H_2O

tane derived from VI was very similar to that of the butane obtained from deuteron quenching of the metathesis reaction.

A calculation based on the yields of deuterobutane obtained in the above reactions indicated that at least .05% of the tungsten in the metathesis reaction was in the metallocyclic form at equilibrium.

This is equivalent to a 10^{-3} $\underline{M}$ concentration of metallocycle (A) at equilibrium and represents a reasonable concentration for a steady state intermediate.

A similar reaction was run using a catalyst prepared from tungsten hexachloride, 2 eq of lithium n-octoxide and 2 eq of butyllithium. This reaction mixture shows about twice the catalytic activity of the simple reaction mixture which does not contain an alcohol activator for the metathesis of deuterated ethylenes. This reaction on quenching with deuterated acid produced an amount of butane-d_2 which corresponded to at least .1% of the tungsten in the reaction mixture being in the metallocyclic form at equilibrium. These trapping experiments which were necessary in the study of the metathesis reaction require much more careful investigations since: a) only small, although reasonable, amounts of the trapped products were obtained, b) there is the possibility that the butane-d_2 was produced from an acid promoted-metal catalyzed ethylene dimerization reaction, and c) the conditions required to observe the reaction products are at much lower olefin to metal ratio and are more concentrated solutions than the normal metathesis reaction.

However, there does appear to be a direct relationship between the metathesis activity of a solution and the yield of butane-d_2 obtained on quenching with deuterated acid. In addition to the observed relationship pointed out above, it was found that a .5 $\underline{M}$ solution of tungsten under the usual metathesis conditions, gave slow metathesis and no observable butane-d_2 on deuteron quenching.

The amount of observed butane-d_2 also falls well within the limits set by the observed E_a of 6-7 kcal/mole and the $\Delta S^{\ddagger}$ of -42 eu.[14]

Although the possible reaction pathways of stable model metallocyclic systems have been reasonably well established, conclusive evidence for the intermediacy of a metallocyclic intermediate in the olefin metathesis reaction will require much more careful, difficult work. However, the experimental evidence to date supports the intermediacy of such complexes in this reaction.[14]

REFERENCES

1. R. Grubbs and T. K. Brunck, J. Am. Chem. Soc., 94, 2538 (1972).
2. L. Cassar, P. E. Eaton and J. Halpern, J. Am. Chem. Soc., 92, 3515 (1970).
3. T. Katz and S. Cerefice, J. Am. Chem. Soc., 91, 6520 (1969).
4. F. D. Mango and J. H. Schachtschneider, J. Am. Chem. Soc., 89, 2484 (1967).
5. G. S. Lewandos and R. Pettit, Tetrahedron Lett., 780 (1971).
6. J. McDermott and G. M. Whitesides, J. Am. Chem. Soc., 96, 947 (1974).
7. C. Biefeld, H. A. Eick, and R. Grubbs, Inorg. Chem., 12, 2166 (1973).
8. J. McDermott, J. F. White and G. M. Whitesides, J. Am. Chem. Soc. 95, 4451 (1973).

9. A. R. Fraser, P. H. Bird, S. A. Beaman, J. R. Shapley, R. White
   and J. A. Osborn, J. Am. Chem. Soc., 95, 597 (1973).
10. N. Calderon, Accounts Chem. Res., 5, 127 (1972).
11. J. L. Wang and H. R. Menapace, J. Org. Chem., 33, 3794 (1968).
12. The optimum metathesis rate is obtained when the W/RLi ratio
    (for octyl lithium) is 2.
13. Under similar conditions, a 1:1 mixture of 1,2-dideuteroethy-
    lene, ethylene has changed to a 1:2:1 mixture of 1,2-dideutero-
    ethylene, deuteroethylene and ethylene.
14. We acknowledge the support of the Petroleum Research Fund ad-
    ministered by the American Chemical Society for this research.

DISCUSSION

DR. MANGO: I have three questions. The first one is regarding
your addition of deuterium halide to quench the reaction. You have
in there very low valent tungsten which is probably heterogeneous
and covered with ethylene. You also have your active catalyst which
may or may not be in the form of a metallocycle - it could be a bis-
ethylene complex. Once the deuterium halide adds to low valent W it
becomes, or it could become, a very active dimerization catalyst.
If it goes through a normal 1,2 addition to ethylene you will end up
not with butene, but, if you assume 2 moles of HX add, tungsten hexachloride and butane. It should be dideutero [1-4] butane and you
could get in addition some 1,3 isomer. I am wondering whether or
not a mixture of 1-4 and 1-3 would not be consistent with the mass
spec data which you have?

DR. GRUBBS: That is possible. The thing that you have to assume
there is that this tungsten alkyl, that if you encourage the dimerization, you added one DCl to the tungsten and got to the point where
you have the alkyl (butyl) function on there that did not have to
wait around for the second DCl.

DR. MANGO: These are both very fast processes because both low
valent catalytic species are incredibly sensitive to things like water or anything polar. The basic statement that I would like to make
is regarding the general value of a certain model, a certain metal
system to totally different metals; for example we have metathesis
of simple olefins which involves tungsten, molybdenum and rhenium
and essentially nothing else.

DR. GRUBBS: Iridium functions that way in special cases.

DR. MANGO: The examples that are frequently sited to justify a
metallocycle are 1) primarily non catalytic in themselves, 2) do not
metathesize anything. The rhodium work of Halpern's is stoichiometric. Osborn's interesting iridium metallocycle is again not a catalytic system. It does not metathesize olefins, nor catalytically
give cyclobutane. You get around 30% yield of cyclobutane dimer after treating it with triphenyl phosphine. With the platinocycle you

can get a cyclobutane, I understand, but in the published work the product is primarily the butene. But in no case do you get the products one would associate with either of the important 2+2 cycloaddition reactions unless you go to special conditions. This nickelocycle of yours does give you interesting products, for example, cyclobutane. But can one infer catalytic significance to this? You have to come back to the fact that nickel does not metathesize simple olefins. And, although nickel in a low valent state has been looked at by many workers for a long time it has never taken ethylene to cyclobutane. If we are going to argue that metallocycles play an important role in the chemistry of nickel then one should expect someone to observe cyclobutane products or metathesis. Every catalytic chemist, at least in industry, who prepares a low valent metal, and particularly nickel, examines the catalytic chemistry of ethylene and cyclobutane is one of the things you simply do not see. You see a lot of butene 1, a lot of dimerization, and oligomerization. You can argue a metallocycle for that, but if you do, then you would have to rationalize why you are getting cyclobutane when you actually make the metallocycle and thermally decompose it, yet in the catalytic system where you are getting turnover numbers up in the millions you see little or no cyclobutane.

DR. COLLMAN: Let me interject a piece of hearsay; I was told by a person at Phillips petroleum that they had found almost all of the transition elements to give some metathesis.

DR. MANGO: I have seen their reported work and I doubt it. They take their substrate, like propylene and then go to a high enough temperature to get some ethylene and butene. They go up to 700 degrees at times, and at these temperatures you may expect everything. They claim this is disproportionation, and thus claim almost every element on the periodic chart. But they are not getting it. A second interesting point is that Phillips did not discover this. It was discovered about five years earlier by Peters, I believe from Standard of Indiana.

SEQUENTIAL MULTISTEP REACTIONS CATALYZED BY POLYMER-BOUND HOMOGENEOUS Ni, Rh, AND Ru CATALYSTS

Charles U. Pittman, Jr.[*] and Larry R. Smith

Department of Chemistry, University of Alabama, University, Alabama 35486

Homogeneous catalytic reactions employing transition metal complexes have become increasingly important in recent years, as exemplified by the production of acetaldehyde from ethylene and air by the Wacker process.[1] In many cases, however, homogeneous catalysts are very expensive and must be separated from the products by costly separation steps. These two factors have in part, limited the scope of practical application of homogeneously catalyzed reactions. One way to achieve easy recovery of the catalyst from product, while simplifying the purification of products, is to anchor the catalyst to immobile supports such as silica, ceramic, or polymeric materials. Such studies of "heterogenized" homogeneous catalysts are beginning to receive increased attention.[2] Industrial groups at Mobile,[3] ESSO,[4] Monsanto,[5] and ICI[6] as well as the academic groups of Bailer,[7] Grubbs,[8,9] Pittman,[10,11,12] and Collman[13] have anchored several organotransition metal complexes to substrates, via phosphine and nitrogen complexation, and studied their catalytic reactivity.

No studies have been reported where more than one transition metal catalyst have been anchored to the same substrate and then used sequentially in the same reactor to effect multistep reactions. However, this concept has been applied by Mosbach[14] who bound both hexokinase and glucose-6-phosphate isomerase to the same polystyrene support. Using this system glucose was converted, sequentially to glucose-1-phosphate and then isomerized to glucose-6-phosphate. Thus, the product from the first enzymatic reaction became the substrate for the second. Could this concept be applied to "heterogenized" homogeneous transition metal catalysts? In this paper we demonstrate that the answer is yes.

Binding two, or more, catalysts to the same substrate raises

several questions. Will each catalyst react as it does individually
or will the catalysts interfere with one another? Can such systems
be recycled? Where catalysts do react or interfere with each other,
can this be avoided by anchoring in such a way that contact is prevented? Finally, will one catalysts intervene in the reaction path
of the second under the conditions used for that second catalyst?

　　Swellable polystyrene beads (Bio Rad SX-1, 1% divinylbenzene,
14,000 mol. wt. exclusion limit) with a high degree of internal mobility[9,10,12,13] were chosen for use in this study. Catalysts were
anchored to these polymers (as outlined in scheme 1) by phosphine
exchange. In these resin beads, catalyst sites bound to the same
bead can come into contact with each other. However, binding one
catalyst to a given batch of resin and the second catalyst to another batch of resin, followed by mixing the beads from both batches
together, allows both catalysts to be used in the same reactor with
contact between them prevented. For example, beads containing bound
$(PPh_3)_2Ni(CO)_2$, 4, could be readily mixed with beads containing bound
Wilkinsons' catalyst, 6. No contact between the Ni and Rh could occur in that case. Alternatively, by anchoring both catalysts to the
same polymer beads (as in 5) contact between catalysts sites is not
necessarily prevented.

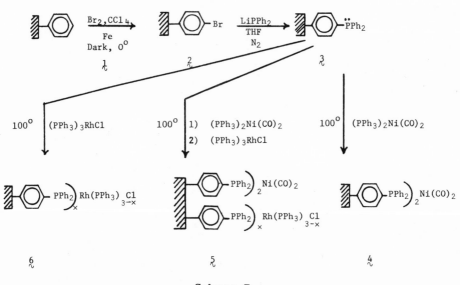

Scheme I

　　Scheme II summarizes the catalysts which have been anchored.
The representation of 4, 5, and 6 in schemes I versus II points out
that the individual metal sites can be multiplied bound to phosphine
sites of the swollen mobile polymer or singly bound depending on the

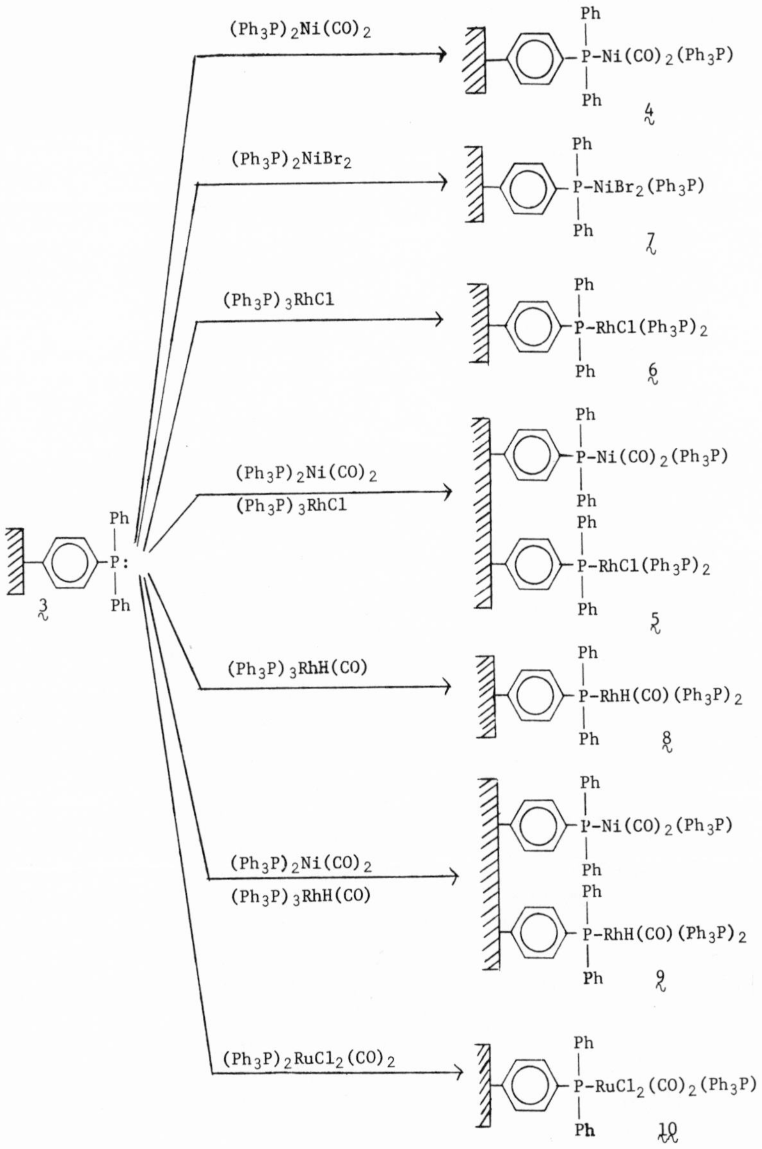

$(Ph_3P)_2Ni(CO)_2$

$(Ph_3P)_2NiBr_2$

$(Ph_3P)_3RhCl$

$(Ph_3P)_2Ni(CO)_2$
$(Ph_3P)_3RhCl$

$(Ph_3P)_3RhH(CO)$

$(Ph_3P)_2Ni(CO)_2$
$(Ph_3P)_3RhH(CO)$

$(Ph_3P)_2RuCl_2(CO)_2$

Ph
|
P-Ni(CO)$_2$(Ph$_3$P)
|
Ph   4

Ph
|
P-NiBr$_2$(Ph$_3$P)
|
Ph   7

Ph
|
P-RhCl(Ph$_3$P)$_2$
|
Ph   6

Ph
|
P-Ni(CO)$_2$(Ph$_3$P)
|
Ph
Ph
|
P-RhCl(Ph$_3$P)$_2$
|
Ph   5

Ph
|
P-RhH(CO)(Ph$_3$P)$_2$
|
Ph   8

Ph
|
P-Ni(CO)$_2$(Ph$_3$P)
|
Ph
Ph
|
P-RhH(CO)(Ph$_3$P)$_2$
|
Ph   9

Ph
|
P-RuCl$_2$(CO)$_2$(Ph$_3$P)
|
Ph   10

Ph
|
P:
|
Ph   3

Scheme II

stoichiometry employed during anchoring. In this study, excess polymer-bound phosphine was employed and the metals were multiply che-
lated. For example, the P/Ni ratio of 4 was 2.15, close to the limiting value of two for structure 4 in Scheme I verses that of three
for structure 4 in Scheme II. This accounts for our observation that
after many catalytic recycle steps, no loss of metal to solution was

observed.

CATALYTIC REACTIONS EMPLOYING ANCHORED CATALYSTS 4-10

Cyclooligomerization of 1,3-butadiene by $(Ph_3P)_2Ni(CO)_2$, previously studied by Wilke,[14,15] was repeated to determine yields under our experimental conditions to compare with polymer-bound catalytic reactions. The products observed (eq. 1) using anchored catalyst 4 (4-vinylcyclohexene, (Z,Z)-1,5-cyclooctadiene, and (E,E,E)-1,5,9-cyclododecatriene) and the product distribution compared quite well with those using $(PPh_3)_2Ni(CO)_2$ homogeneously and with those reported.[14,15] Reactions stopped short of completion, showed that

cis-1,2-divinylcyclobutane was a product using the polymer-bound catalyst. Reactions 3 and 4 (Table I) also showed no difference between catalysts bound to SX1 or SX2 beads (2,700 mol. wt. exclusion limit). This was understandable since 1,3-butadiene is sterically small, therefore the polymer's pore size presents no problems within this range. The anchored catalyst could be repeatedly recycled until the maximum molar turnover of 1100-1200 was reached. The reactions rate using the anchored catalyst 4 at 115° was close to that of the unbound catalyst at 90°. Diffusion into the resin accounts for the difference. When 4 was ground into a fine powder, thereby reducing diffusion effects, the rate using the anchored catalyst at 115° compared to the unbound catalyst at 108°.

Anchored $(PPh_3)_2NiBr_2$ (i.e. polymer 7), upon reduction with two moles of $NaBH_4$ in THF- ethanol, gave an active anchored Ni(0) species which selectively catalyzed the linear oligomerization of butadiene to (E,E)-1,3,6-octatriene in 90% yield and over 98% isomeric purity. The molar turnover was 1500. This contrasts with the results of Kiji[16] who obtained a complex mixture of 2-methylenevinylcyclopentane, and 1,3,6-, 2,4,6-, and 1,3,7-octatriene with $(PPh_3)_2NiBr_2$ in EtOH THF was necessary to swell the polymer. The $NaBH_4$ -7 catalysts

did not oligomerize isoprene or cooligomerize isoprene and butadiene.

After reduction of 7 with $NaBH_4$, an excess of the bound hydroformation catalyst, $(PPh_3)_3RhH(CO)$, (i.e. 8) was added. With these

Table I

Results of 1,3-butadiene Cyclooligomerization by Homogeneous
and Polymer-bound $(Ph_3P)_2Ni(CO)_2$ [a]

#	Catalyst	mmol Catalyst	mmol ⟋	Overall Yield	DVCB	4-VCH	1,5-COD	1,5,9-CDT
1	$(Ph_3P)_2Ni(CO)_2$ [b]			100	0	26.6	64.3	8.9
2	$(Ph_3P)_2Ni(CO)_2$	0.156	188.5	100	0	25.3	63.2	11.5
3	4a [d]	0.179	185.0	26.0[c]	33.6	34.2	20.0	11.2
4	4b [d]	0.155	185.0	21.9[c]	22.6	33.1	32.9	11.4
5	4e [d]	0.171	115.0	90.0	0	33.3	55.6	11.1
6	recycle	0.171	115.0	42.3	0	31.8	57.9	11.3
7	4e	0.171	55.5	92.0	0	28.7	61.2	10.0
8	recycle	0.171	55.5	90.8	0	31.7	56.4	11.9
9	recycle	0.171	55.5	81.7	0	29.2	59.7	11.1
10	recycle	0.171	55.5	11.2	0	27.6	61.5	10.9

[a] All reactions are run 24 hrs. in benzene or THF (20 ml), 90° for homogeneous reactions, 110°–115° for polymer-bound catalysts, conducted in sealed bombs under 1,3-butadiene pressure.
[b] Wilke's results.[14,15] [c] Reactions stopped short of equilibrium to analyze for divinylcyclobutane.
[d] Analyses of the polymers gave for 4a, 2.39% P, 2.10% Ni; for 4b, 1.75% P, 1.82% Ni; for 4e, 2.32% P and 2.10% Ni. SX-2 beads were used in 4b.

two catalysts present, oligomerization followed by hydroformylation
within the same reactor was successfully effected! Linear oligomerization was carried out as before, and then the reactor was pres-
surized to 500 psi with H_2/CO (1/1) to effect selective hydroformylation of the terminal bond of 11 at 65°. Overall, an 82% yield of
the normal and branched aldehydes 12 and 13 were obtained in a 3.4
ratio.

$$\text{7} \quad \xrightarrow[100°]{\text{NaBH}_4} \quad \xrightarrow[\text{1/1, 500psi}]{\text{H}_2:\text{CO}} \quad \underset{\text{major}}{12} \quad + \quad \underset{\text{CHO}}{13} \qquad (3)$$

Sequential cyclooligomerization of butadiene, followed by hydrogenation of its oligomers, was next effected using 1) $(PPh_3)_2Ni$-
$(CO)_2$ and $(PPh_3)_2RhCl$ in solution, 2) a mixture of the beads of polymer-anchored catalysts 4 and 6, and 3) catalyst 5 containing both
species bound to the same resin beads. The reactions were carried
out in benzene, cyclooligomerizations at 110-115° and hydrogenations
at 50° and 350 psi H_2. Butadiene was added, the oligomerizations
were conducted, and then the reactor cooled to 50° followed by pres-
surization to 350° psi (Table II). Recycling the polymer beads, 4
plus 6 or 5, proved feasible with no loss of activity up to the molar turnover limit of 4. Separate experiments indicated $(PPh_3)_2Ni$-
$(CO)_2$ (or its anchored forms) does not act as a hydrogenation catalyst and $(PPh_3)_3RhH(CO)$ does not act as an oligomerization catalyst.
Furthermore, the catalysts do not interfere or react with each other
as proved using polymer 5 or dissolved $(PPh_3)_2Ni(CO)_2$ and $(PPh_3)_3$-
RhCl. The product distribution of alkanes directly reflected the
distribution of polyenes resulting from the oligomerization step.

$$\xrightarrow[\substack{(PPh_3)_3RhCl \\ \text{or} \\ 4 + 6 \\ \text{or} \\ 5}]{\substack{(PPh_3)_2Ni(CO)_2 \\ \\ 110°, \\ 24\ hr\ \ then \\ 50°,350\ psi \\ H_2\ 4\ hr}} \qquad (4)$$

14   15   16

Next cyclooligomerization was sequentially followed by hydroformylation (eq. 5). Three catalyst systems were employed: 1) dissolved $(PPh_3)_2Ni(CO)_2$ and $(PPh_3)_3RhH(CO)$, 2) a mixture of resin beads
of these anchored catalysts (4 and 8) and 3) these two catalysts anchored to the same beads (i.e. 9). The cyclooligomerizations as before, were carried out at 110-115° for 24 hours using the anchored
catalysts while the hydroformylation proceeded in 85-100% yield at

Table II

Results of Reaction of 1,3-butadiene with Homogeneous and Polymer-bound $(Ph_3P)_2Ni(CO)_2$ and $(Ph_3P)_3RhCl$

#	Catalyst	mmol	Catalyst	mmol	mmol //	%Conversion	% 4-VCH	% 1,5-COD	% 1,5,9-CDT	% Alkane[d]
7	$(Ph_3P)_2Ni(CO)_2$	0.157	-	-	-	-	23.6[a]	-	-	0
8[e]	-	-	$(Ph_3P)_3RhCl$	0.054	128	0	0	0	0	0
9	$(Ph_3P)_2Ni(CO)_2$[b]	0.157	$(Ph_3P)_3RhCl$[c]	0.108	187	98.3	20.8	66.4	12.8	99.3
10	4a	0.179	6[f]	0.083	181	84.7	22.3	54.9	20.5	86.4
11	4a	0.179	6[f]	0.083	55	92.2	22.7	61.3	16.0	86.3
12	recycle	0.179	recycle	0.083	55	91.8	23.1	60.8	16.1	84.5
13	recycle	0.179	recycle	0.083	55	81.3	22.8	59.7	17.5	83.2
14	recycle	0.179	recycle	0.083	55	21.6	23.9	58.6	17.5	81.9
15	5[f]	0.413	5[f]	0.300	55	92.3	24.9	60.7	14.4	84.1
16	recycle	0.413	recycle	0.300	55	92.1	25.1	60.5	14.4	82.8

[a] mmol 4-VCH added to reaction at 350 psi H_2. [b] Homogeneous oligomerization 90°, 24 hrs., polymer-bound oligomerizations 110°-115°, 24 hrs. [c] Hydrogenations at 50°, 350 psi H_2. [d] Total percent of the cyclo-oligomerization products which were hydrogenated to alkanes. [e] No H_2 pressure. [f] polymer 5 analyzed for 2.49% Ni and 3.09% Rh while 6 contained 1.83% P and 1.70% Rh.

70° and 18 hours. The hydroformylation was selective. Only the exocyclic double bond of 4-vinylcyclohexene was hydroformylated and all endocyclic double bonds were inert under a 1:1 H_2 to CO atmosphere at 500 psi. Table III summarizes these reactions.

$$ (5) $$

The ratio of normal to branched aldehyde (18/17) was about 4.3 using both the soluble catalysts (Table IV) and the combination of polymer beads of 4 and 8. However, using 9, where both catalysts were anchored to the same polymer, resulted in a ratio of 5.8. The explanation for this change rests with a change in the excess P/Rh ratio which changed from 1.7 in 8 to 6.1 in 9. Pittman and Hanes previously demonstrated that the normal to branched product ratio varied from 2.6 to 4.3 as the excess bound-phosphine to rhodium ratio increased from 2.3 to 4.3 using anchored catalysts of structure 8. Recycling both bound catalyst systems resulted in no loss of activity until the maximum turnover value for $(PPh_3)_2Ni(CO)_2$ was reached.

Recently, Fahey[17,18] reported that 1,5-cyclooctadiene and 1,5,9-cyclododecatriene were hydrogenated to their respective monoenes with remarkable selectivity by $(PPh_3)_2RuCl_2(CO)_2$. Monoene yields greater than 97 and 92% were obtained respectively in the presence of a large excess of PPh_3 (P/Ru ratio of 11.1 to 20 were optimum). Monoenes were obtained in only a slight excess over alkanes when no added PPh_3 was present. Thus excess PPh_3 was required for high selectivity.

In order to see if the internal mobility of bound phosphine groups on 1% divinylbenzene-crosslinked polystyrene resins was sufficient to promote this selectivity, this Ru derivative was anchored such that the P to Ru ratio was 14.8 (polymer 10). 1,5-Cyclooctadiene reductions were then conducted in benzene at 165°, 150 psi H_2, for 15 hours without added triphenylphosphine. A cyclooctene to cyclooctane ratio of 55 to 45 was obtained. Thus, anchored phosphine sites do not encounter the bound Ru at a rate fast enough to influence the mechanism in the manner that excess added PPh_3 did. Under these conditions the mobility of polymer-bound 2,2,6,6-tetramethyl-4-piperidinol-1-oxyl was shown by esr to have a correlation time (τ) of at least 1 x 10^{-10} sec. which was as fast as that of the same

Table III

Results of Reaction of 1,3-butadiene with Homogeneous
and Polymer-bound $(Ph_3P)_2Ni(CO)_2$ and $(Ph_3P)_3RhH(CO)$

#	Catalyst	mmol	Catalyst	mmol	mmol	Total Yield	% 4-VCH	% 1,5-COD	% 1,5,9-CDT	% Aldehydes[d]	1°/2°
1	$(Ph_3P)_2Ni(CO)_2$	0.157	-	-	-	-	14.8[a]	-	-	-	-
2	-	-	$(Ph_3P)_3RhH(CO)$[g]	0.109	55	0	0	0	0	0	-
3	$(Ph_3P)_2Ni(CO)_2$[b]	0.157	$(Ph_3P)_3RhH(CO)$[c]	0.109	185	96.1	22.2	57.5	20.3	85.9	4.21
4	4a[e]	0.179	8[f]	0.140	105	89.5	28.5	45.0	26.5	89.3	4.18
5	recycle	0.179	recycle	0.140	55	38.1	27.2	49.6	23.2	88.7	4.27
6	4a	0.179	8[h]	0.140	55	96.7	23.1	58.2	19.7	91.1	4.22
7	recycle	0.179	recycle	0.140	55	92.1	24.1	57.9	18.0	89.3	4.31
8	recycle	0.179	recycle	0.140	55	42.6	27.1	50.4	22.5	86.2	4.58
9	9[h]	0.186	9[h]	0.137	75	88.4	24.7	57.4	17.9	84.1	5.81
10	recycle	0.186	recycle	0.137	75	74.3	25.1	59.1	15.8	83.2	5.72

[a] mmol 4-VCH added to reaction at 500 psi H_2, CO 1:1. Homogeneous oligomerizations for 24 hrs. at 90°.
[c] Homogeneous hydroformylations 70°, 6 hrs., 500 psi H_2, CO 1:1. [d] Aldehydes resulting from 4-vinylcyclohexene.
[e] Polymer-bound oligomerizations at 110°-115°, 24 hrs. [f] Polymer-bound hydroformylations at 70°, 18 hrs.,
500 psi H_2, CO 1:1. [g] No H_2, CO pressure. [h] Polymer 8 analyzed for 2.01% P, 1.44% Rh and polymer 9 for
4.06% P, 1.09% Ni, and 1.41% Rh.

Table IV

Results of Hydroformylation of 1,3-butadiene Cyclooligomers
by Homogeneous and Polymer-bound $(Ph_3P)_3RhH(CO)$

#	Catalyst	mmol Catalyst	Olefin	mmol Olefin	% Aldehyde	$1°/2°$
1	$(Ph_3P)_3RhH(CO)$	0.109	4-VCH	14.8	100	4.29
2	$(Ph_3P)_3RhH(CO)$	0.109	1,5-COD	15.1	0	-
3	$(Ph_3P)_3RhH(CO)$	0.109	1,5,9-CDT	15.9	0	-
4	8	0.140	4-VCH	14.8	100	4.35
5	recycle	o.140	4-VCH	14.8	96.4	4.28
6	recycle	0.140	4-VCH	14.8	96.4	4.30

In benzene (20 ml), 65°-70°, H_2, CO 1:1 total pressure 500 psi.
Homogeneous reactions run 4 hrs., polymeric catalysts 12-18 hrs.

spin lable bound to uncrosslinked polystyrene.[19] However, the trans-
lational motion of bound phosphine moieties must be greatly restricted relative to free dissolved PPh_3 or high selectivity would have
been observed.

Next an eighteen to twenty mole excess of added PPh_3 was added
to Ru - containing polymer 10 in benzene solution. At 165° and 150
psi of H_2, 1,5,9-cyclododecatriene was selectively reduced to cyclo-
dodecene in 95% yield, 1,5-cyclooctadiene to cyclooctene in 90%
yield, and 4-vinylcyclohexene to 4-ethylcyclohexane in 88% yield.
Once again the bound catalyst system could be recycled repeatedly
with no loss of activity or selectivity. No ruthenium was leached
from the polymer during these reactions. These results are summarized in Table V.

Studies of the sequential, one pot cyclooligomerization of butadiene followed by selective reduction of the oligomers to their corresponding monoenes were conducted using 1) soluble $(PPh_3)_2Ni(CO)_2$
and $(PPh_3)_2RuCl_2(CO)_2$ and 2) using a mixture of beads of polymer 4
and 10. The Ru catalyst did not promote any side reaction during
cyclooligomerization and the Ni catalyst was inert to the hydrogenation conditions, although it did decompose at 165-170°. Both soluble and anchored catalyst combinations gave high yields of sequential cyclooligomerization (98 and 92%) and selective reduction (85-

(6)

Table V

Results of Hydrogenation of 1,3-butadiene Cyclooligomers by Homogeneous and Polymer-bound $(Ph_3P)_2RuCl_2(CO)_2$

#	Catalyst	mmol	Polyene	mmol	Alkane	Alkene	Diene	Triene
a	$(Ph_3P)_2RuCl_2(CO)_2$	0.130	1,5,9-CDT	1.10	1.9	99.7	0.3	0.3
a	$(Ph_3P)_2RuCl_2(CO)_2$	0.130	1,5-COD	1.10	6.1	93.4	0.8	-
1[b]	$(Ph_3P)_2RuCl_2(CO)_2$	0.090	1,5,9-CDT	12.3	1.8	97.1	0.6	0.5
2	$(Ph_3P)_2RuCl_2(CO)_2$	0.090	1,5-COD	16.3	3.9	93.8	3.3	-
3	$(Ph_3P)_2RuCl_2(CO)_2$	0.090	4-VCH	16.3	5.7	87.7	6.8	-
4[c]	10a[e]	0.053	1,5-COD	16.3	5.7	90.2	4.1	-
5	recycle	0.053	1,5-COD	16.3	5.4	90.5	4.1	-
6	recycle	0.053	1,5-COD	16.3	5.5	90.6	3.9	-
7	10a[e]	0.053	4-VCH	16.3	5.4	88.2	6.6	-
8	10a[e]	0.053	1,5,9-CDT	12.3	2.1	95.1	2.0	1.8
9[a]	10b[e]	0.097	1,5-COD	16.3	45.9	54.1	0.0	-
10	recycle	0.097	1,5-COD	16.3	43.8	56.2	0.0	-

[a] Fahey's results. [b] Homogeneous reactions at 145°, 5 hrs. [c] Polymer-bound reactions at 165°-170°, 15 hrs. Homogeneous reactions conducted at 150 psi H_2, 0.262 g, 1.0 mmol Ph_3P added in 20 ml benzene. [d] No excess free Ph_3P added, except the excess bound to the polymer. [e] Polymer 10a contained 1.88%P and 1.27%Ru while 10b contained 5.03%P and 0.98%Ru.

Table VI

Results of Reaction of 1,3-butadiene with Homogeneous
and Polymer-bound (Ph₃P)₂Ni(CO)₂ and (Ph₃P)₂RuCl₂(CO)₂

#	Catalyst[a]	mmol	Catalyst[a]	mmol	mmol	Total Yield	%VCH[f]	%COD[f]	%CDT[f]	%19[d]	%20[d]	%21[d]
11[b]	-	-	L₂RuCl₂(CO)₂	0.090	100	0	0	0	0	0	0	0
12[c]	L₂Ni(CO)₂	0.157	L₂RuCl₂(CO)₂	0.090	162	97.2	21.8	68.8	9.4	85.3	89.1	98.2
13	L₂Ni(CO)₂	0.157	L₂RuCl₂(CO)₂	0.090	55	98.1	23.6	64.1	13.3	86.1	88.7	94.6
14	recycle	0.157	recycle	0.090	55	91.2	22.9	65.3	11.8	85.8	84.9	91.7
15	recycle	0.157	recycle	0.090	55	56.2	23.9	62.8	13.3	86.4	87.2	92.2
16[e]	4a	0.179	10a	0.053	55	92.3	24.1	56.8	19.1	84.2	88.9	91.8
17	recycle	0.179	recycle	0.053	55	0	0	0	0	0	0	0

[a] L represents Ph₃P. [b] No hydrogen pressure. [c] Homogeneous oligomerization at 90°, 24 hrs., hydrogenations at 140°, 4 hrs., 150 psi H_2, in 20 ml benzene with 0.262 g. Ph₃P added. [d] Percent of respective polyene hydrogenated to monoene; 19 is 4-ethylcyclohexene, 20 is Z-cyclooctene, and 21 is E-cyclododecene. [f] VCH is 4-vinylcyclohexene, COD is (Z,Z)-1,5,cyclooctadiene, CDT is (E,E,E)-cyclododecatriene.

90% for both systems). The yields and product distributions (Table VI) were similar, but the rate of hydrogenation was slower using anchored catalyst 10 than when $(PPh_3)_2RuCl_2(CO)_2$ was used. This required using 165° to achieve hydrogenation rates with 10 comparable to those at 145° using the homogeneous catalyst. Since $(PPh_3)_2Ni(CO)_2$ decomposes above 155°, it was not possible to recycle beads 4 and 10, but the homogeneous catalyst was readily recycled by removing solvent and products by vacuum distillation and carrying the products through the reaction again. The homogeneous catalyst combination was still able to effect selective hydrogenation upon recycling. This was because 145° was the upper limit that $(PPh_3)_2Ni(CO)_2$ encountered.

In conclusion, it has been demonstrated that homogeneous organometallic catalysts can be used sequentially in multistep reactions within the same reactor. Furthermore, their polymer-anchored analogs were used similarly and gave yields and product distributions similar to the soluble catalyst. This success suggests many multistep syntheses can be sequentially carried out in the same reactor using bound homogeneous catalysts. The following studies are in progress.[20]

ACKNOWLEDGEMENTS

The Office of Naval Research and the National Science Foundation (Grant No. GH-37566) are thanked for their generous support of this work.

REFERENCES

1. G. Szonyi, "Recent Homogeneously Catalyzed Commerical Processes", Homogeneous Catalysis, Vol. 70, ACS Advances in Chemistry Series, American Chemical Society, 1968.
2. For a review see J. C. Bailar, Cataly. Rev., 0000 (1974).

3. W. O. Haag and D. D. Whitehurst, "New Heterogeneous Catalysts for Carbonylation and Hydroformylation", Meeting of the Catalysis Society, Houston, Texas, 1971; German Patent 1,800,379 (1969).

4. A. A. Oswald, L. L. Murrell, and L. J. Boucher, Olefin Hydroformylation Catalysts with Rhodium and Cobalt Phosphine Complexes Covalently Anchored to Silica", Presented at the 167th National Meeting of the American Chemical Society, April, 1974, see abstracts pp 162 and 155.

5. J. Rony, J. Catalysis 14 (1969) 142.

6. K. G. Allum, R. D. Hancock, S. McKenzie, and R. C. Pitkethly, "Supported Transition Metal Complexes as Heterogeneous Catalysts" presented at the Vth International Congress on Catalysis, West Palm, Florida, see abstracts 30-1.

7. H. S. Bruner and J. C. Bailar, Jr., J. Amer. Oil Chemists Soc., 49, 533 (1972).

8. R. H. Grubbs and L. C. Kroll, J. Amer. Chem. Soc., 93, 3062 (1971).

9. R. H. Grubbs, L. C. Kroll, and E. M. Sweet, J. Macromol. Sci.-Chem., A-7, 1047 (1973).

10. G. O. Evans, C. U. Pittman, Jr., R. McMillan, R. T. Beach, and R. Jones, J. Organometal. Chem., 67, 295 (1974).

11. C. U. Pittman, Jr. and G. O. Evans, Chem. Tech., Sept. 560 (1973).

12. C. U. Pittman, Jr. and R. M. Hanes, Ann. N. Y. Acad. Sci., 0000 (1974).

13. J. P. Collman, L. S. Hegedus, M. P. Cooke, J. R. Norton, G. Dolcetti, and D. N. Marquant, J. Amer. Chem. Soc., 94, 1789 (1972).

14. B. Bogdanovic, P. Heimbach, H. Kroner, G. Wilke, G. Brandt, and E. B. Hoffman, Ann. Chem., 727, 143 and 169 (1969).

15. P. Heimbach, P. W. Jolly, and G. Wilke, Advances in Organometal. Chem., 8, 29 (1970).

16. J. Kiji, K. Yamamoto, S. Mitani, S. Yoshikawa and J. Furukawa, Bull. Chem. Soc. Jap., 46, 1971 (1973).

17. D. R. Fahey, J. Org. Chem., 38, 80 (1973).

18. D. R. Fahey, ibid, 38, 3343 (1973).

19. Personnel Communication from S. L. Regen.

20. C. U. Pittman, Jr. and G. Wilemon, work in progress.

CONCERTED REACTIONS OF ORGANIC LIGANDS ON TRANSITION METALS

Rowland Pettit

Department of Chemistry, The University of Texas at
Austin, Austin, Texas 78712

INTRODUCTION

The development of the Woodward-Hoffman Rules[1] has provided a
basis for the understanding of concerted reactions occurring in purely organic systems. Within the framework of this theory concerted
reactions may be classified into two groups i.e. "allowed" and "forbidden" processes; and associated with the latter group there is an
added energy barrier in the reaction coordinate of the process which
generally renders these processes to be of such a high energy that
alternative non concerted processes of lower energy are favored if
reaction occurs at all. However, when such "forbidden" reactions are
conducted with the organic species simultaneously coordinated to a
transition metal then it appears that the concerted reaction may then
become "allowed". Theoretical arguments, based on the correlation or
orbital symmetries of the reactants and products, have been developed
to substantiate the idea of the changeover of the process from "forbidden" to "allowed";[2] these will not be reproduced here, but rather
I shall concentrate on the experimental results of three separate reactions which illustrate the phenomenon.

The Disrotatory Ring Opening of Cyclobutene-Metal Complexes

In earlier experiments it was found that silver ions catalyzed
the isomerization of strained cyclobutene derivatives to the corresponding butadiene derivatives. For example, dibenzotricycloocta-
diene 1 is completely isomerized to dibenzocyclooctatetraene 2 in
the presence of catalytic amounts of $AgBF_4$ in 10 seconds at $0°$; in
contrast, the condition required to effect the purely thermal isomerization of 1 to 2 is heating for five hours at $180°$.[3]

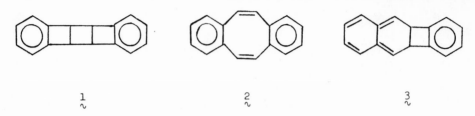

1   2   3

When the catalyzed isomerization is conducted in the presence
of maleic anhydride there is isolated the Diels Alder adduct of the
o-xylylene derivative 3 with maleic anhydride. We conclude therefore that the role of the silver iron in the overall formation of 2
is to effect a facile disrotatory cyclobutene-butadiene ring opening which converts 1 to 3; the latter compound by a normal Cope Rearrangement affords 2. In a series of related compounds e.g. ben-
zotricyclooctadiene and syn- and anti-tricyclooctadiene it was found
that the rate of the silver catalyzed conversion to the corresponding
cyclooctatetraene derivative was related to the amount of strain energy released in the process.

In order to provide evidence that the basic reaction involved
the ring opening of a cyclobutene metal olefin complex (equation 1)
the iron tetracarbonyl complexes of syn- and anti-tricyclooctadiene

$$ M \longrightarrow \quad\quad M \quad\quad \cdots\cdots\cdots (1) $$

(4 and 5 respectively) were prepared.[4] Each complex is readily converted to isomeric iron tircarbonyl complexes of bicyclooctatriene
(6 and 7) upon heating in hexane. There is no significant change in
the rate of conversion upon changing the solvent from hexane to methanol hence an ionic process is not indicated; furthermore, the reactions are inhibited by addition of carbon monoxide.

Fe(CO)$_4$ ⇌ Fe(CO)$_3$ → Fe(CO)$_3$

4   8   6

Fe(CO)$_4$ ⇌ Fe(CO)$_3$ → Fe(CO)$_3$

8   9   7

We consider that the reactions involve thermal loss of a CO ligand to generate the olefin-iron tricarbonyl complexes 8 and 9 and
that these undergo disrotatory cyclobutene ring opening reaction
leading to the butadiene-Fe(CO)$_3$ complexes 6 and 7 in which the effective atomic number of iron is maintained at that of Krypton.
These results support the assertion that strained cyclobutene-ole-
fin metal complexes readily isomerize to butadiene-metal complexes
via a disrotatory ring opening process as depicted in equation (1).

2Π+6Π Cycloaddition Reactions Between
Ligands Coordinated to an Iron Atom

Both 2Π+6Π concerted cycloaddition (equation 3) are calculated
to exothermic processes; however, the first of these is an "allowed"
process and many examples are known (i.e. the Diels Alder Reaction)
but the latter is a "forbidden" process and no examples of such reactions appear to have been reported. In accordance with the belief

that "forbidden" reactions may become "allowed" with involvement of
a transition metal atom, we have attempted to accomplish 2+6 cycloaddition between the organic reactants simultaneously coordinated to
a metal (equation 4).

Solutions of cycloheptatriene-iron tricarbonyl 10 in tetrahydrofuran, upon irradiation in the presence of methylacetylene dicarboxy-
late at 0°, afford the diene iron tricarbonyl complex 11.[5] The formation of complex 11, the structure of which was determined by X-ray
analysis, reveals that 2+6 cycloadditions has occurred and that the
acetylene moeity had added to the triene on the same face to which
the Fe(CO)$_3$ group is attached; presumably then at some point in the
reaction the acetylene was bonded to the iron atom.

The reaction between triene-Fe(CO)$_3$ complexes and acetylenes

10

11

appears to be a fairly general one. The diethylene ketal of tropone
iron tricarbonyl, upon similar irradiation in the presence of acetylene carboxylic ester, gave rise to a complex of similar structure to 11. When a solution of the tropone-Fe(CO)$_3$ complex was irradiated at -78° in tetrahydrofuran, the lamp then shut off and the
acetylene derivative added, there was obtained after warming to room
temperature the free 2+6 adduct 12 containing no iron atom. Similar
irradiation of solution of cyclooctatetraene iron tricarbonyl followed by addition of acetylene dicarboxylic ester or diphenyl ace-

12

13

14

tylene afforded the 2+6 adducts 13 (R=COOCH$_3$ and ø). Likewise
cyclooctatreine-iron tricarbonyl and diphenylacetylene react to yield
the hydrocarbon 14.

Since the adducts can be produced by addition of the acetylene
to the irradiated solutions after the lamp has been turned off, it
is clear that the new carbon bonds which are formed in the adducts
result from a thermal not photochemical process. A reasonable mechanism could then seem to be the following. The light reacts with
the complex 10 for example to produce the monoolefin iron carbonyl
species 15(a) (L = tetrahydrofuran, M-1, N=3) and 15(b) (L = THF,
m=2, n=2). The light then plays no further role in the reaction.
A solvent ligand is thermally replaced by an acetylene to generate
the complex 16 which undergoes intramolecular concerted addition to
afford the adduct 17. Dieneiron tricarbonyl complexes of the type
17 (n=3, m=1) are stable and are isolated as such whereas the dicarbonyl monoetherate complexes (n=2, m=2) are thermally unstable and
result in the isolation of the uncomplexed ligand.

An important point for which proof is so far lacking (in common
with many concerted processes) is that the two new carbon carbon
bonds are simultaneously formed in the reaction coordinate. Further

$\underset{\sim}{15}$ $\underset{\sim}{16}$ $\underset{\sim}{17}$

work will be needed to establish this feature, in any event the overall reaction promises to have certain synthetic utility.

Metal Catalyzed Isomerizations of Saturated Hydrocarbons

An area of considerable current interest involves the mechanism
of transition metal catalyzed isomerization of strained saturated
hydrocarbons. Results to date would suggest that in some cases at
least the reactions proceed in a concerted manner since no evidence
can be obtained (such as through trapping experiments) to indicate
the intermediacy of ionic or radical species. For example syn tri-
cyclooctane ($\underset{\sim}{18}$) undergoes complete isomerization to a mixture of
1,5 cyclooctadiene ($\underset{\sim}{19}$) bicyclooctene ($\underset{\sim}{20}$) and tetrahydrosemibullval-
ene $\underset{\sim}{21}$ within five minutes upon treatment in methanol containing
$AgBF_4$. Anti-tricyclooctane $\underset{\sim}{22}$, on the other hand, shows no reaction
after treatment with $AgBF_4$ for five days in refluxing aqueous acetone.[6]

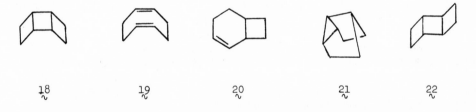

$\underset{\sim}{18}$ $\underset{\sim}{19}$ $\underset{\sim}{20}$ $\underset{\sim}{21}$ $\underset{\sim}{22}$

Oxidative addition of a low valent transition metal to a highly strained ring system is a well established phenomenon and in many
cases this would seem to be the initial step in the reaction where-
by such metals induce strained hydrocarbons to undergo rearrangement
or reaction with other substrates. But such initial oxidative addition is not indicated in the behavior of 18 and 22 with silver
ions. We consider it more likely that the Ag^+ ion forms an initial,
though of course very weak, complex with a strained bridging bond in
18 and that this system then undergoes concerted rearrangement to
analogous complexes of 19, 20 and 21 thence to the formation of the
free hydrocarbons. It is of interest to note that whereas these

products are formed rapidly in the presence of Ag^+ in each case, in the absence of the metal, their formation would constitute a "forbidden" process by any concerted reaction.

We have also found that silver ions catalyze the isomerization of quadricyclane to norbornadiene and, contrary to the case with syn-tricyclooctane, when the reaction is conducted in the presence of methanol a small amount of the methyl ether 23 is produced.[7] The production of the ether is readily accounted for in the following scheme: the silver ion reacts with norbornadiene to afford 24, via

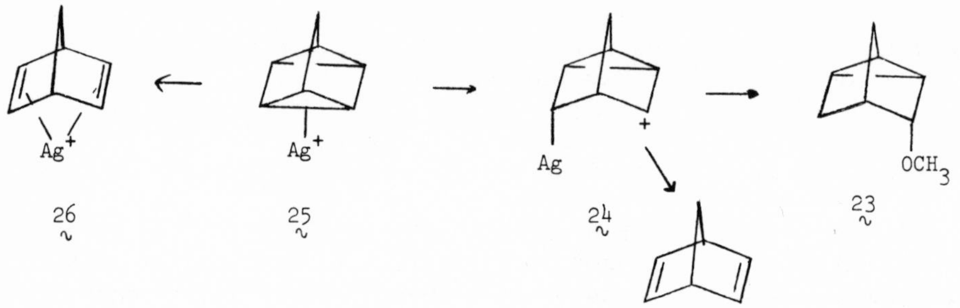

$$\underset{26}{Ag^+} \quad \longleftarrow \quad \underset{25}{Ag^+} \quad \longrightarrow \quad \underset{24}{Ag^+} \quad \longrightarrow \quad \underset{23}{OCH_3}$$

an initial complex 25; 24 reacts then with methanol to yield, after protolysis of the C–Ag bond, the methyl ether. The norbornadiene could also conceivably be produced from 24 by elimination of Ag^+ as indicated. However, in a series of experiments conducted in an inert solvent containing varying amounts of methanol it is found that the ether: norbornadiene ratio is independent of the methanol concentration. We conclude therefore that the ion 24 is not a significant source of norbornadiene and propose that a competing reaction of the complex 25, leading directly to the norbornadiene silver complex 26 in a concerted manner, is operative.

It appears that no one mechanistic scheme can account for all of the metal catalyzed rearrangements of various strained hydrocarbons; whether or not the metal will undergo initial oxidative addition, possibly followed by formation of distinct ionic species, or whether concerted reactions are involved will depend both on the nature of the metal and the nature of the hydrocarbon.

ACKNOWLEDGEMENT

The author thanks the National Science Foundation and the Robert A. Welch Foundation for financial support.

REFERENCES

1. R. B. Woodward and R. Hoffman, 1970, "The Conservation of Orbital Symmetry", Academic Press, New York, N. Y.
2. R. Pettit, H. Sugahara, J. Wristers, and W. Merk, Discuss. Faraday Soc., 47, 71 (1969); F. D. Mango, Advan. Catal., 20, 291 (1969) and references therein, M. J. S. Dewar, Angew. Chem. Int. Ed. Engl., 10, 761 (1971).
3. W. Merk and R. Pettit, J. Amer. Chem. Soc., 89, 4788 (1967).
4. W. Slegeir, R. Case, J. S. McKennis and R. Pettit, ibid., 96, 286 (1974).
5. Submitted for publication.
6. J. Wristers, L. Brener and R. Pettit, J. Amer. Chem. Soc., 92, 7499 (1970).
7. Submitted for publication.

DISCUSSION

DR. HALPERN: I want to start at a position in basic agreement with Rowland, namely that all of the various mechanisms (I think you have indicated four), are certainly possible ones. There has never been any question in anybody's mind that the theoretical considerations that suggest that transition metals can modify the selection rules for concerted reactions are sound. The issue has always been that there are alternative processes that have to be considered and the question, then, is deciding in each particular case what is going on. This makes each system subject to individual elucidation, and I think there remain specific situations in which we may still disagree on mechanism. The silver catalyzed rearrangements of saturated systems are probably in that class. These involve the technical details of each particular reaction, and that is not the point I want to address.

There is, however, one general point I want to make. In addition to real disagreements about particular mechanisms (or reactions) that have pervaded this field, there have also been situations that have evoked the appearance of disagreements that are not real, and that really have a semantic basis. I think it is important to recognize those, and for that purpose, I would like to use the board for a moment. I think wherever arguments about concerted and non-concerted processes turn up, whether they involve metals or not, one gets into the problems of defining the terms concertedness and nonconcertedness. What I mean is illustrated by considering the suprafacial 2 plus 2 cycloaddition which is a forbidden reaction. What we have meant when we said that this is a stepwise process (in the presence, say, of rhodium) is that there is a step in which the intermediate II type is formed, and therefore, going from I to III is stepwise process. This corresponds in fact to the stepwise process

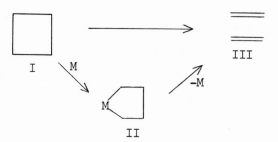

that you would have breaking one bond and then the second one - if
you just leave the catalyst out. The way you relate the catalytic
mechanism to a non-catalytic one is writing exactly the same mechanism, and just leaving the catalyst out. If you did that you would
break one bond and then break another bond.

Lets consider, in the same context, the disrotatory cyclobutene
opening. Formally, the relationship of the disrotatory cyclobutene-
butadiene rearrangement to the suprafacial 2 plus 2 cycloaddition reaction is simply that the in the cyclobutene opening, we are carrying through an extra bond we broke above, and we make the same kind
of rearrangements, but all the way, we are carrying through an extra
bond. If we wanted, then, to extend this mechanistic scheme, whether it is right or wrong, to this process in a purely formal way,
the way I would do it is to write exactly the same intermediate, but
carry it through an extra formal bond.

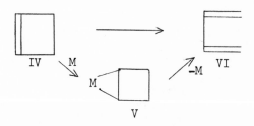

If I did that I would simply be consistently using the same valence-
bond representation as I did earlier. In fact, whereas everyone
uses valence bond representation in I → II people generally do not
use valence bond representation, in IV → V. Instead, the way everyone would represent IV → V is

Now, this is your iron butadene structure. There is no question, at least in my mind, that the process which you have described, which is going from V to VII is, in fact, a concerted process. Similarly, it has always seemed to me that the step II → III is a concerted one. Oxidative addition is an allowed concerted process and I see no reason to question that it proceeds by a concerted mechanism. The apparent disagreement implied between your demonstration of a concerted organometallic rearrangement (i.e. V → VII) and our claims that I → III (via II) is stepwise, really, is not a real disagreement because the process that you have described (i.e. V → VII) corresponds formally to just the second step of the sequence I → II → III not to the overall reaction I → III. I do not think anyone would question that the step II → III is a concerted process. The difference between the two situations is that the formation of such an intermediate (i.e. I → II) where you are dealing with a sigma bond is likely to be slow compared with the subsequent rearrangement (II - III), and, therefore you do not see the intermediate (II). On the other hand formation of a metal-olefin π-complex is frequently a fast process, especially if one is dealing with a labile system, so that the organometallic intermediate (i.e. V) and its subsequent rearrangement are readily observed directly.

DR. PETTIT: Can I make one comment on that? It involves these chalk lines here. I am going to end up agreeing with you wholeheartedly, I think. As I drew that silver complex, I drew a line like one traditionally drawn for an olefin pi-complexes (VII) and then, you take the other extreme, and you draw the oxidative addition: VIII

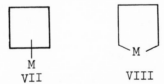

VII   VIII

There is no difference, really, between the two representations, by the metal bond in VII. I am meaning interaction between the sigma bond, and the sigma bond is IX,

IX   X

and presumably, it tries to react with some S - orbital on the metal of the appropriate symmetry. There is going to be an antibonding orbital associated with this strained sigma bond, which is going to be X,

XI

and this is going to interact with the metal d orbital XI. This
representation here we have two sigma bonds; we will call IX sigma
one, and X sigma two. IX is going to interact with an S, X is going to interact with a P orbital.

It is exactly the same. There is going to a whole gradation. There
are going to be a million intermediates.

DR. HALPERN: What you have really pointed out here is that a
concerted oxidative addition is an allowed process and it is. The
reason I am inclined not to believe too much in the reality of intermediate complexes of metals with C-C bonds is that in cases where
one examines comparable directly observable oxidative addition processes, say oxidative addition of H_2, they are fairly slow processes.
At the same time, one fails to see any evidence of recognizable stable intermediates, even only sufficiently stable to, say, increase
the solubility of hydrogen. Hydrogen is not a bad model for a saturated hydrocarbon-electronically, anyway.

DR. PETTIT: Silver forms a complex with cyclopropane. If we
draw it like XI, or like XII,

I do not think there is any difference. They are just chalk lines
on the blackboard.

DR. BERGMAN: I think that there is one other possibility here
I would like to ask you about. I tend to agree that one should not
try to argue about where one draws these lines. The important distinctions are where the atomic nucleu are in the given intermediate
you are talking about. I would say the only kinds of things one
would want to argue about are, lets say for a cyclobutane metal complex, is whether you are talking about the metal situated symmetrically below the ring or on the edge, and if it is on the edge, the
lines do not matter. What is important is where the metal nucleus
is, compared to the other atoms.

DR. PETTIT: One day molecular orbital theory will be able to
handle metals that will be like Jack has drawn here, and find out
where this energy minimum is, with respect to the position of the
nucleus. We cannot do it now.

DR. BERGMAN: I just think that that is the kind of controversy
that is a real one, and the other one, as to where you draw the
lines when all the nuclei are in the same place, as you say, is not
the sort of thing that could be profitable. The other question I

would like to ask Jack, and that is, when you drew the analogy between the cyclobutane complex you formed with a metal, and the cyclobutane, you put the metal like so

then when you dit it with cyclobutene you put the double bond; and
drew a metal ion:

I just want to say that the other possibility, of course, is to put
the metal like this.

DR. HALPERN: I would agree with Rowland (Pettit), that is less
likely.

DR. BERGMAN: But that seems to me a real alternative.

DR. PETTIT: It is a real alternative, but we eliminated that
possibility.

THE ROLE OF THE TRANSITION METAL IN CATALYZING PERICYCLIC REACTIONS

Frank D. Mango

Shell Development Company, P.O. Box 481, Houston, Texas 77001

We shall be concerned with pericyclic reactions that are catalyzed along their concerted pathways. In this respect, the ligand rearrangement will exactly mirror its metal-free pericyclic counterpart. Metals can, of course, catalyze pericyclic reactions along stepwise, nonconcerted pathways, but we shall not discuss these here. The focus of this paper will be to cite literature examples which appear to support earlier theoretical proposals[1] suggesting that certain transition metal complexes can totally remove orbital symmetry restrictions, allowing their ligands freedom to transform along otherwise forbidden pericyclic modes, and to do so with preservation of coordinate bonding.[2]

The more complex molecular orbital treatments of the theory have recently been reduced to simpler valence bond, resonance representations.[3] Resonance hybrids 1 and 2, for example, represent the corresponding allowed and forbidden pericyclic processes, respectively.

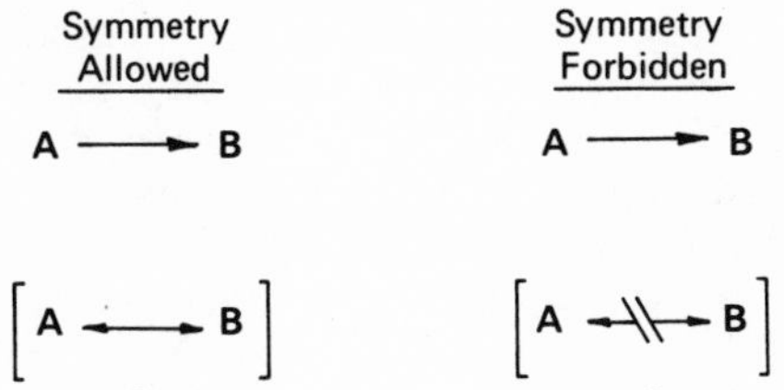

Symmetry Allowed	Symmetry Forbidden
A ⟶ B	A ⟶ B
$\left[A \longleftrightarrow B \right]$	$\left[A \nleftrightarrow B \right]$
1	2

The ground state correlation in the allowed transformation A→B translates into hybrid 1. Significantly, when the transformation A→B is

169

forbidden, B is not a significant contributor to the ground state of
A; that is, A correlates with an excited state of B and not with B
itself.

Now consider the symmetry-forbidden transformation $A \rightarrow B$ where the
reactants are fully coordinated to an appropriate transition metal
complex. Since there is now a ground state - ground state correlation between complexes, either complex is represented by 3 with its

3

appropriate mixture of contributing structures. Take, for example,
the transformation of quadricyclene to norbornadiene, the forbidden
process in eq 1. Full coordination of quadricyclene transforms it
to resonance hybrid 4. Note that the metal's focus of biscoordina-

(1)

4

tion is rotated 90°, thus preserving full coordinate bonding as the
resonance hybrid rich in the ligand quadricyclene transforms to the
one rich in norbornadiene. This can be envisioned as a smooth, concerted process as indicated in eq 2, where a slightly coordinated
quadricyclene, interacting at the outer coordination sphere (plane
A), falls to the primary coordination sphere (plane B). This process would be attended by a continuous enrichment of the contributing structure norbornadiene into resonance hybrid 4. Since the energy of norbornadiene is approximately 68 Kcal below quadricyclene,
this would constitute a relaxation of strain energy, and the relax-
ing ligand should cross a point from plane A to B where full coordination, and thus complete relaxation of strain is guaranteed.

In the catalysis of forbidden pericyclic reactions, the preferred metal systems should possess a d band occupied with sufficient metal valence electrons to give a coordinatively saturated com-

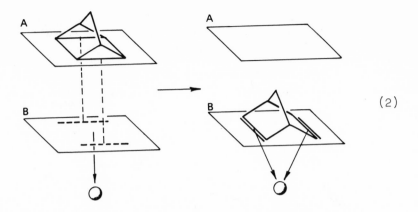

(2)

plex in which the bonding axis of the fully coordinated reacting ligand system is C_∞, C_3 or higher with respect to the nonreacting ligands. Examples of 5- and 7-coordinate systems follow:

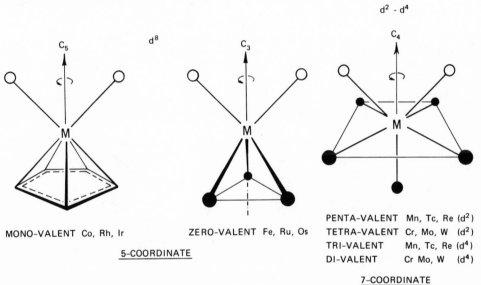

The open circles in the figures denote the coordination positions of
the reacting bonds.

The important catalytic transformations pertinent to the experimental work to be cited are summarized in eqs 3, 4, and 5. Thus, on
the appropriate metals, a bisolefin system is free to interconvert
with a fully coordinated cyclobutane. Bisacetylenes, however, are
fully restricted from a similar interconversion (i.e., eq 4) on any
single transition metal center.[3] Two, nonrestrictive-field transi-

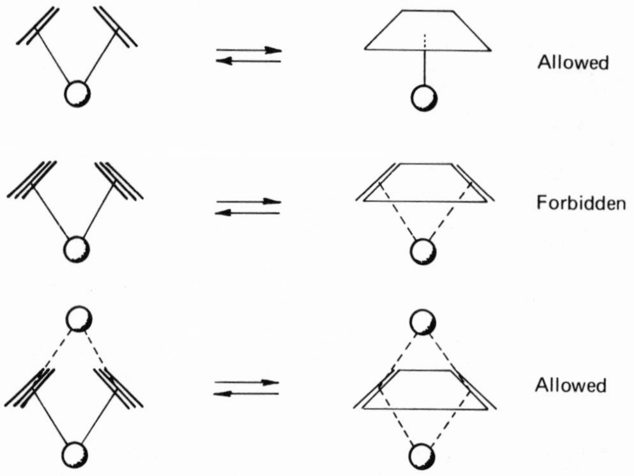

Allowed   (3)

Forbidden   (4)

Allowed   (5)

tion metal centers sharing the opposite faces of a plane containing a bisacetylene ligand system removes the symmetry restrictions to a bisacetylene - cyclobutadiene interconversion (eq 5).[1,4] As we shall see, bisacetylene coordination of this kind is not unknown and indeed may represent one important catalytic route in the catalysis of acetylenes.

Nesmeyanov, Anisimov, et. al., have shown that the diphenylacetylene complex in eq 6 undergoes the reversible dimerization indica-

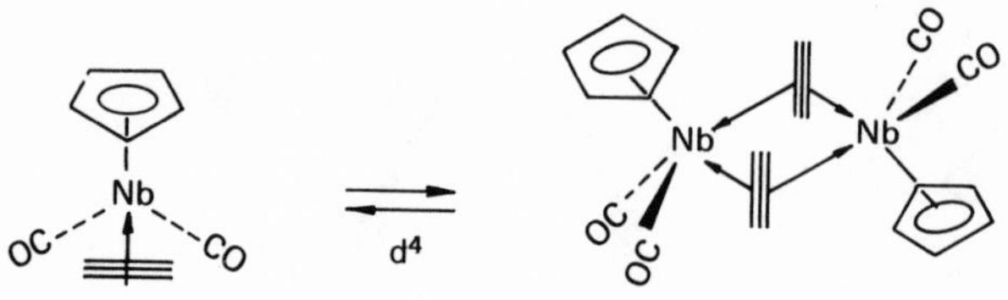

(6)

ted.[5] In a good coordinating solvent, the dimer loses CO at 80° and fuses to complex 5. Crystal analysis, confirming the indicated struc-

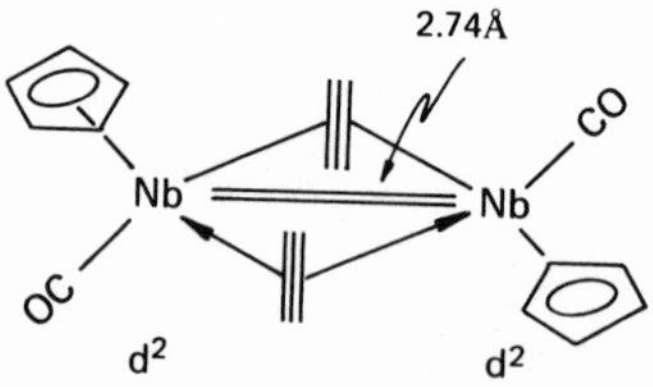

2.74Å

d^2 d^2

5

ture, showed a Nb-Nb double bond of 2.74 Å passing through the plane containing the bridging bisacetylene ligand system. Structure 5 strongly implicates the dimer in eq 6 as its precursor.

Later work on this system led to an interesting Nb complex containing three diphenylacetylene units.[6] It was formed from the bisacetylene counterpart to the monomer in eq 6 through the reactions
outlined in eq 7.

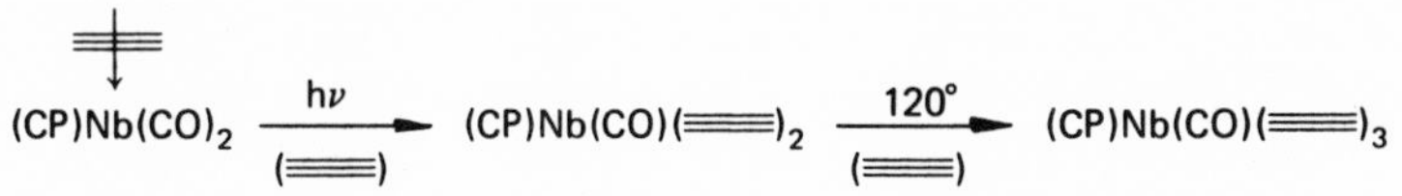

$$(CP)Nb(CO)_2 \xrightarrow[(\equiv\!\equiv)]{h\nu} (CP)Nb(CO)(\equiv\!\equiv)_2 \xrightarrow[(\equiv\!\equiv)]{120°} (CP)Nb(CO)(\equiv\!\equiv)_3 \quad (7)$$

Metallocycle 6 was suggested for the structure of the new complex. However, a later crystal analysis indicated the cyclobutadiene complex 7.[7] The facility with which the isoelectronic Nb com-

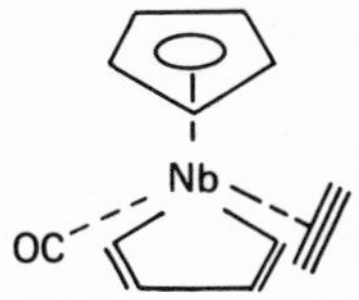

6

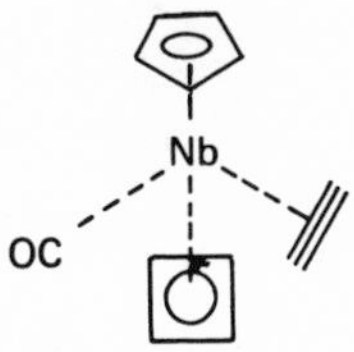

7

plex undergoes the dimerization indicated in eq 6 would strongly suggest a similar path for the bisacetylene complex. A $[2_s + 2_s]$ fusion
of bridging acetylenes as described in eq 5 is a most attractive
route to the observed product. It should be further noted that the
formation of 7 was accompanied by the appearance of considerable hexaphenylbenzene. Moreover, hexaphenylbenzene was noted as a decomposition product of 7 and not a catalytic product.

Interestingly, the vanadium analogue to the monomer in eq 6
(i.e., $(CP)V(CO)_2(Ph_2C_2)$) undergoes smooth fusion with diphenylacetylene at 100° yielding the dicarbonyl cyclobutadiene complex in eq
8.[8] The established behavior of the isostructural niobium complex

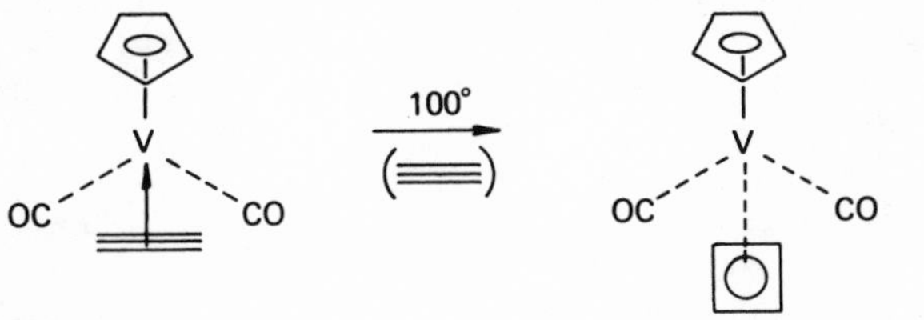

$$(8)$$

in eq 6 implicates this path as the route to the vanadium cyclobutadiene complex, namely a concerted fusion of bridging acetylene ligands as indicated in eq 5.

King has reported some intriguing cycloadditions of cyclic dia-

cetylenes.[9] $(CP)Co(CO)_2$ undergoes the general intramolecular cycl-
addition shown in eq 9. In sharp contrast to cobalt, $Fe(CO)_5$ yields

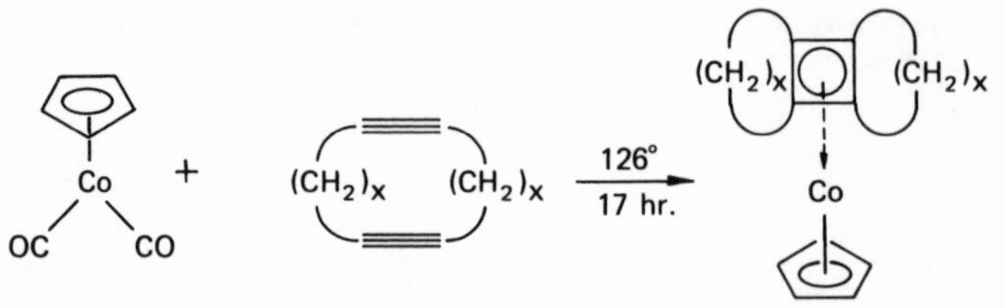

(9)

a metallocyclic product for which King reasonably assigned the structure indicated in eq 10. However, a later crystal analysis by Chin

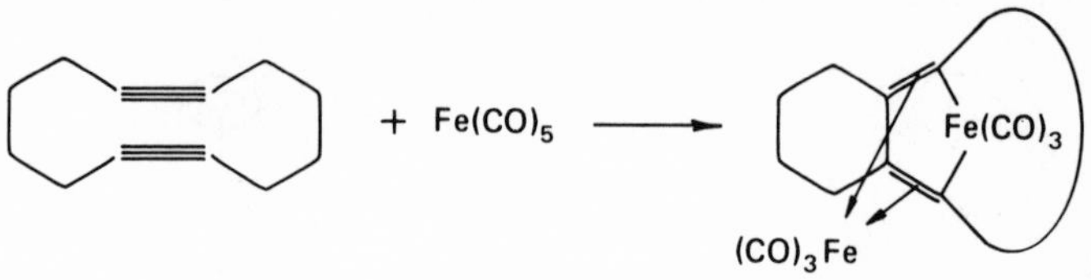

(10)

and Ban[10] established that the metallocycle in fact possessed the
"metathesized" structure 8. These authors further argued that the

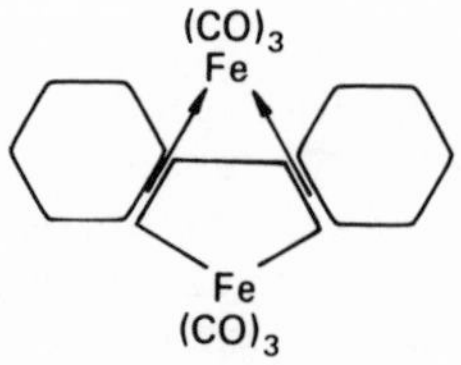

8

metallocycles were interconverting, possibly through the cyclobutadiene iron counterpart to the cobalt complex, a compound actually
isolated from the reaction in trace amounts. Moreover, this mechanism was suggested to intervene in the metathesis of acetylenes catalyzed by tungsten[11] and in olefin metathesis itself. However, mechanistic similarities between olefin metathesis and the metathesis
reported for acetylenes[11] may not be justified.[4]

A reasonable route to 8 might include the "metathesis" of a bisacetylene diiron system through the mechanism outlined in eq 5. It
is significant that the iron moieties would reasonably possess the
nonrestrictive triscarbonyl nonreacting ligand systems. Final insertion of the $Fe(CO)_3$ function would then proceed, yielding the
least sterically hindered and thermodynamically preferred metallocycle.

There is experimental support for the contention that the metallocycle is not an intermediate in the course of events leading to acetylene metathesis catalyzed by $Fe(CO)_5$. Acetylene reacts with Fe-$(CO)_5$ at high pressure yielding the acetylene counterpart to 8 (eq 11) and the cyclobutadieneiron complex in eq 12.[12] Significantly,

$$Fe(CO)_5 \;+\; HC\equiv CH \xrightarrow[110°]{9000\ atm}$$

(11)

(12)

the metallocycle was reported not to be an intermediate in the formation of the cyclobutadiene complex under the reaction conditions.

Abley and McQuillen have reported a most unusual rhodium-catalyzed hydrogenation of diphenylacetylene to _trans_-stilbene.[13] Deuteration yields the dideuterio product in eq 13. Only one deuterium is placed in the _ortho_ position of the benzene ring, arguing against

$$\text{Ph–C}\equiv\text{C–Ph} \xrightarrow{Py\ Rh\cdot DMF\cdot NaBD_4}$$

(13)

a simple exchange mechanism for the unusual placement of the second deuterium atom. The authors reasonably postulated a metal-assisted [1,3] sigmatropic rearrangement similar to that outlined in eqs 14 and 15.

Grimme has reported one of the earliest and perhaps most interesting examples of a possible metal-assisted [1,3] sigmatropic rearrangement involving a carbon-carbon bond.[14] The tricyclononatriene complex in eq 16 undergoes the [1,3] sigmatropic change indicated under surprisingly mild conditions. Significantly, the bonds of the cyclooctatriene ring remain intact. The rearranging bond, darkened in eq 16, is exo to the C-8 ring and thus not likely to be involved in an insertion step yielding a metallocycle. Retention or inversion of configuration at the bridging methylene was not determined; thus a pure metal-catalyzed [1,3] suprafacial process was not definitively established. However, the structural nature of the metal system (i.e., full coordination and a nonrestrictive ligand field)

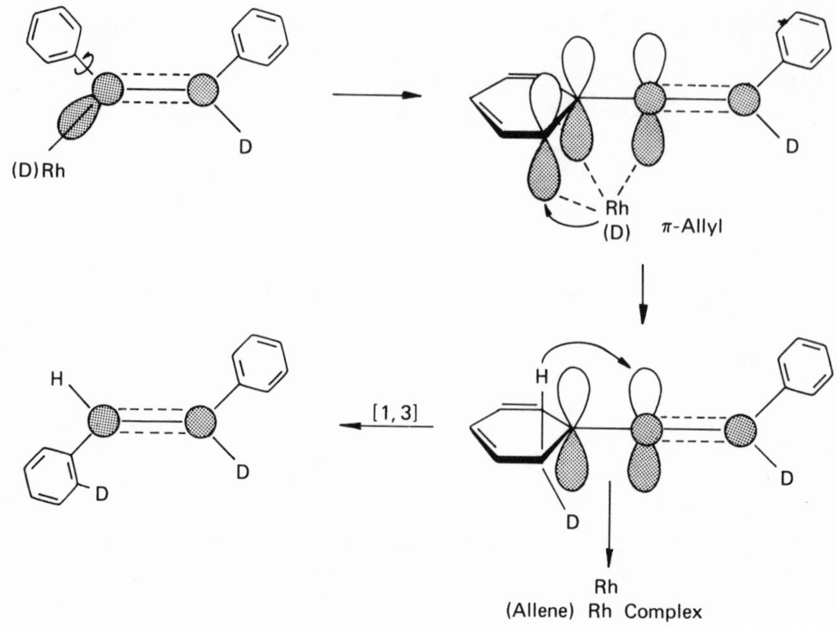

(14)

(D)Rh

Rh
(D)   π-Allyl

[1,3]

(15)

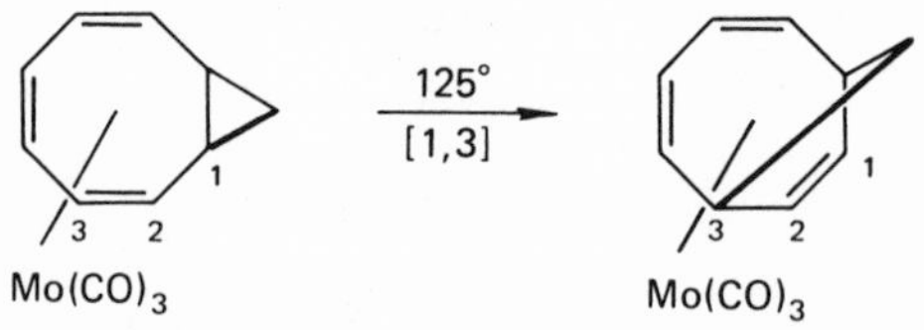

Rh
(Allene) Rh Complex

$$\xrightarrow[[1,3]]{125°}$$

(16)

Mo(CO)$_3$ Mo(CO)$_3$

makes the metal-catalyzed [1,3] suprafacial process most reasonable.

Perhaps the most remarkable example of a possible metal-cata-lyzed forbidden pericyclic process is the Cr(CO)$_3$ catalyzed valence isomerization of bullvalene (eq 17).[15] While both Mo and W carbonyls

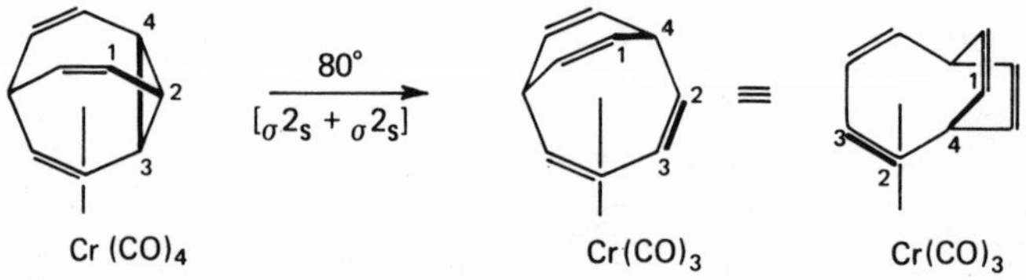

$$\xrightarrow[[_\sigma 2_s + _\sigma 2_s]]{80°}$$

≡

(17)

Cr(CO)$_4$ Cr(CO)$_3$ Cr(CO)$_3$

simply suppress the degenerate Cope rearrangement, the Cr(CO)$_4$ complex rearranges quantitatively to the indicated product. Although Aumann does not speculate on the actual role of the metal in this novel pro-cess, he does note that the ligand product is that produced through the photolytic rearrangement of bullvalene itself. Thus a symmetry forbidden pericyclic process or reaction path is clearly implicated.

Model studies of bullvalene suggest that a forbidden $[_\sigma 2_s +$ $_\sigma 2_s]$ pericyclic path of rearrangement may be involved. This process, involving the darkened bonds in eq 17, would proceed with retention at carbons 1 and 2 and inversion at carbons 3 and 4. Once again, the nonrestrictive nature of the metal carbonyl involved makes the proposed path a reasonable possibility. Finally, it should be noted that for this and similar cases, metal coordination need only focus at one of the two rearranging bonds, in this case the bridging cyclopropane bond nearest the metal. The required exchange of electron pairs between metal and ligand would in this example proceed through the bonding and antibonding components of the coordinated sigma bond.

REFERENCES

1. F. D. Mango and J. H. Schachtschneider, J. Amer. Chem. Soc., 89, 2484 (1967); ibid., 91, 1030 (1969).
2. F. D. Mango, Tetrahedron Lett., 505 (1971).
3. F. D. Mango, ibid., 1509 (1973), and references therein.
4. F. D. Mango in "Topics in Current Chemistry", Vol. 45, Springer-Verlag, Berlin, 1974, pp 65-71.
5. A. N. Nesmeyanov, K. N. Anisimov, N. E. Kolobova, and A. A. Pasynskii, Izv. Akd. Nauk SSR, Ser. Kim., 774 (1966).
6. A. N. Nesmeyanov, K. N. Anisimov, N. E. Kolobova, and A. A. Pasynskii, ibid., 1, 100 (1969).
7. A. N. Nesmeyanov, A. I. Gusev, A. A. Pasynskii, K. N. Anisimov, N. E. Kolobova, and Yu T. Struchkov, Chem. Comm., 739 (1969).
8. A. N. Nesmeyanov, K. N. Anisimov, N. E. Kolobova, and A. A. Pasynskii, Dokl. Akad. Nauk SSR, 182, 112 (1968).
9. R. B. King and A. Efraty, J. Amer. Chem. Soc., 92, 6071 (1970); 94, 3021 (1972).
10. H. B. Chin and R. Ban, ibid., 95, 5068 (1973).
11. F. Pennella, R. L. Banks and G. C. Bailey, Chem. Comm., 1548 (1968).
12. R. Buhler, R. Geist, R. Mundich and H. Plieninger, Tetrahedron Lett., 1919 (1973).
13. P. Abley and F. J. McQuillen, Chem. Comm., 1503 (1969). See also A. M. Khan, F. J. McQuillen and I. Jardine, J. Chem. Soc. (C), 136 (1967); S. Mitsui, Y. Kudo and M. Kobayashi, Tetrahedron, 25, 1921 (1969).
14. W. Grimme, Chem. Ber., 113 (1967).
15. R. Aumann, Angew. Chem., Int. Ed., 9, 801 (1970).

DISCUSSION

DR PETTIT: Can I just make a comment on that acetyline reaction
of King's. They took the acetylene I, and got out II

I   II

DR. MANGO: And a small amount of cyclobutadiene - about 1%.
DR. PETTIT: They say that this was evidence that the metallocycle that you would get initially would be expected to be III

III

and then isomerize to IV by an equivalent thing that involves a dismutation process. I read some years ago that Shannon and Ray Davis
had done the X-ray structure on this, but it was never published, unfortunately. If one takes benzocyclobutadienide tricarbonyl and just
heats it up with iron pentacarbonyl you do indeed get both V and VI.

V   VI

The point is that cyclobutadienide carbonyl complexes do on treatment
with iron carbonyl give ferroles and so the most reasonable explanation for the formation of this from the acetylene is that cyclobutadiene.
DR. MANGO: That is what I think. This avoids all of the nasty
business you have to go through, even worse than yours Bob. The metallocycle having double bonds in it undergoing the rearrangement.
I think this is a fairly straightforward reaction of cyclobutadiene
complex.
DR. GRUBBS: Isn't it known that some cobalt metallocycles can
go between metallocycle and cyclobutadiene?

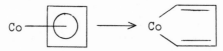

DR. MANGO: I do not know. This pertains only to iron, and the
statement was made in that paper.
DR. GRUBBS: Do you mean this thing of Rosenblum's?
DR. MANGO: No - Sugihara had cyclobutadiene cobalt tricarbonyl
and he reacted it with triphenyl phosphine, and got the metallocycle.
DR. HALPERN: I can't take exception to anything you have said in

this lecture. However, I would take issue with a point that appears
prominently in your abstract, and therefore is, in some sense, a contribution to the meeting. That is the question of the extent to
which one can regard the stoichiometric analogs of catalytic reactions as relevant to the catalytic process. If you are faced with
the problem of determining the mechanism of a catalytic reaction in
the case of an efficient catalyst it is frequently not possible to
detect the intermediates directly. The way one would like to determine unequivocally the mechanism of the reaction is by intercepting
the intermediates. Now there are two things you can do. You can
trap them in some way and of course that makes the reaction stoichiometric and not catalytic. Or, you can modify the catalyst in such a
way that the intermediate is more stable and less reactive. If you
push that to the limit, you stop it at what would be the alleged intermediate of the catalytic reaction, but you no longer have the catalyst. Now if we take the systems you are referring to, the acyl-
rhodium compound, that is what we did there:

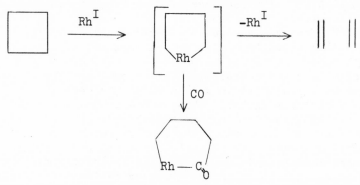

If you could insert a CO you would trap what would otherwise be an
intermediate. I agree with you that it does not prove that the catalytic reaction goes in the same way. However, at least some of
these acyl adducts upon warming do give the rearranged product. The
whole point of the experiment was to run the reaction under conditions where the adducts did not rearrange, or you would not trap
them. If you want to push the issue, then it can be pointed out
that upon workup (for example warming) the acyl adducts do in fact
give the rearranged product, for example norbornadiene from the
quadricydene adduct. One might argue, that you have got a catalytic
intermediate, and that even though you have now slowed down the decomposition of the intermediate so that it can be detected the reaction is formally still catalytic. There is another point about this
that is a very important aspect of methodology of elucidation of reaction mechanisms and explains one of the reasons why we turned to
the cubane system after other people had done experiments on quadracyclane. If you are forced to resort to analogy, you can still
make the analogy a very convincing one by saying "In one case we are
looking at a catalytic reaction, in the other case we are looking

at a stoichiometric reaction we know to be oxidative addition." Lets
start putting substituents on the substrate (and cubane is a very
good substrate for this reason, because you can put substituents on
it) and examine various patterns of the reactivity, for example, how
the rate of the stoichiometric reaction responds to substituents,
versus the rate of catalytic reaction. In the case of cubane they
were the same. Then, there are several geometrical (positional) i-
somers both for the rearrangement and for the oxidative addition
products and the distributions were the same for the two processes.
So, despite the fact that one reaction is stoichiometric and the other is catalytic, there is about a strong a case as one can make that
the stoichiometric reaction does in fact correspond to a step in the
catalytic process. The fact that one reaction is stoichiometric and
the other catalytic is not a strong argument against carrying over
the evidence from one to the other.

DR. MANGO: No, in itself, of course it is not. The additional
kinetic data you have is strong.

DR. HALPERN: But the language of the abstract makes no reference
to the real arguments about that.

DR. MANGO: What you saw was the beginning and the end. In the
middle of it, I give quite a bit of treatment and detail. I do make
the comment, though, that you do not necessarily have to coordinate
to both bonds. You can coordinate only to one, and the arguments
over this about the distinctions between simple coordination and total oxidative addition do hold. They are extremes along a continuum
and simple coordination can launch that thing right along the path.
It could be simple mono coordination in a concerted process, that is,
the one bond just begins to interact as the other one behind it begins to transform too, which would be a concerted process. In coordination to a single bond and complete oxidative addition I would
expect the same degree of specificity.

DR. HALPERN: I would not expect the response of rate to substituents to be identical in both cases. In one case the second bond
is not affected at all.

DR. MANGO: Oh, I would.

DR. HALPERN: The electronic rearrangements that go on in that
process are sufficiently different, that I think it would stretch
the credulity of most physical organic chemists to believe that substituents would not recognize that distinction.

DR. KAESZ: Didn't you have them fairly symmetrically substituted
Jack (Halpern)? In other words, what you did to one side of the molecule, you did to the other, as I remember.

DR. HALPERN: No, in the case of the rhodium, actually, we looked
at unsymmetrically substituted ones for the most part.

REACTIONS OF TRANSITION METAL DIHYDRIDES. MECHANISM OF REACTIONS
OF DIHYDRIDO-BIS(π-CYCLOPENTADIENYL)MOLYBDENUM WITH VARIOUS UNSA-
TURATED COMPOUNDS

Akira Nakamura[*] and Sei Otsuka

Department of Chemistry, Faculty of Engineering Science,
Osaka University, Toyonaka, Osaka, Japan

We have undertaken a systematic research[1] on the reactivity of
$(\pi-C_5H_5)_2MoH_2$ (abbreviated as Cp_2MoH_2) and reported the stereochemistry and mechanism of the reactions with olefins, acetylenes, and
azo compounds. Here we briefly summarize some of the results and
discuss the features which give an important insight into the unique reactivity and interesting nature of the metal-hydrogen bond.
The chemistry should have an important bearing on the nitrogen fixation by some metalloenzymes.

EDA COMPLEX FORMATION AND INSERTION REACTION[1b]

Charge transfer complexes forms between Cp_2MH_2 (M=Mo, W) and
some electron-acceptors, e.g. maleic anhydride, fumaronitrile, or
trinitrobenzene. The filled non-bonding metal orbital of A_1 symmetry
acts as a σ-donating orbital at the ground state and is involved in
the charge transfer interaction. A stronger acceptor such as TCNE
or TCNQ, causes electron transfer to give an ionic product e.g. $[Cp_2-MH_n]^+[TCNE]^-$.

Stereospecific cis-insertion of olefins,[1c] e.g. dimethyl fuma-
rate, occurs via intermediacy of somewhat excited Cp_2MoH_2 molecule
with parallel Cp rings. Some acetylenes, e.g. $CH_3O_2C-C\equiv C-CO_2CH_3$,
$CF_3C\equiv CH$, and azo compounds give insertion products through similar
mechanism.

REDUCTIVE ELIMINATION TO GIVE ALKANES, ALKENES OR HYDRAZINES

Hydrido-σ-alkyl complexes, $Cp_2MoH(R)$, thermally release alkanes

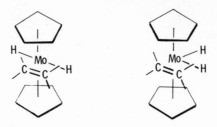

Figure 1.   Proposed Structures for Intermediates in the Olefin cis-
Insertion.

ground state
(σ-basic)

excited state
(π-basic)

electron
acceptors

>C=C<

-N=N-

-C≡C-

Cp₂Mo

Cp_2Mo

Cp_2Mo

H

H

C—C

H

H

N—N

H

H

EDA Complexes

Cp_2Mo

H

C=C

H

with the ease depending strongly on the substituents on R. α-Cyano-
or α-carbomethoxy group stabilizes the hydrido-alkyl complexes.
Stereochemistry at the α-carbon was found to be "retention" in the
case of $Cp_2MoD[CH(CO_2CH_3)CHDCO_2CH_3]$.[1c] The thermal stability of related hydrido-σ-alkenyl or -σ-hydrazino complexes is also improved
by the electron-attracting group (e.g. CO_2CH_3 or CF_3) at the α-posi-
tion.

     The reactive Cp_2Mo species reacts with various small molecules
e.g. H_2, CO,[2] N_2,[2] or larger ones containing $C=C$,[1] $-C≡C-$,[1] $-N=N-$[3]

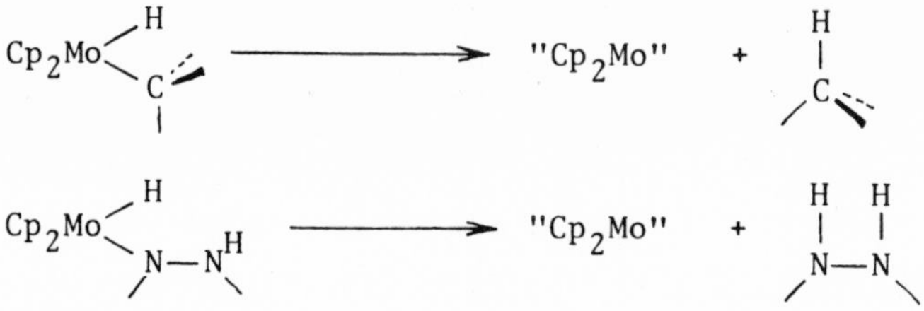

$$Cp_2Mo\begin{smallmatrix}H\\C\end{smallmatrix} \longrightarrow \text{"}Cp_2Mo\text{"} \; + \; \begin{smallmatrix}H\\C\end{smallmatrix}$$

$$Cp_2Mo\begin{smallmatrix}H\\N-N^H\end{smallmatrix} \longrightarrow \text{"}Cp_2Mo\text{"} \; + \; \begin{smallmatrix}H\;\;H\\N-N\end{smallmatrix}$$

and $-N=N-\overset{!}{C}=O$ linkages.[3] Typical examples are shown below:

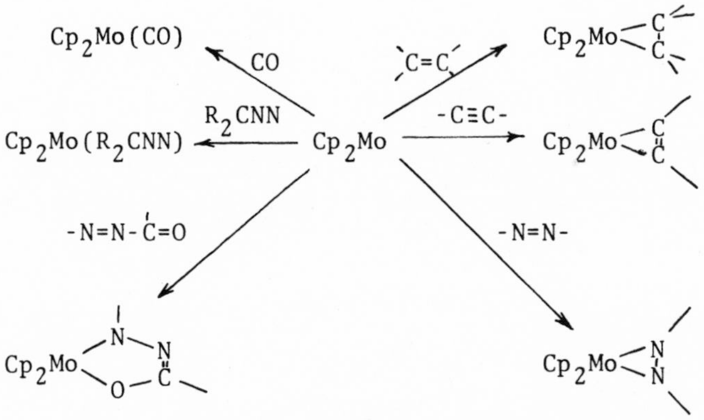

CATALYSIS

Cp_2MoH_2 catalyzes selective hydrogenation of 1,3-dienes to monoenes and α-β-unsaturated carbonyl compounds to saturated ones.

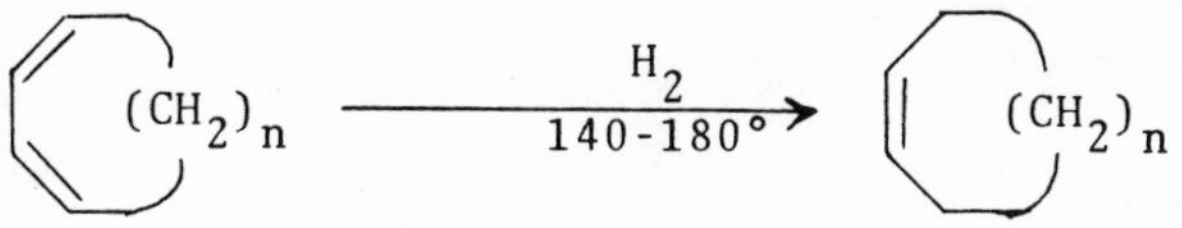

A mechanism involving 1,2-addition of Mo-H bond of the catalyst (1,2-hydrometalation) has been proposed[1d] on the basis of the reaction shown on the following page. Cp_2MoH_2 also catalyzes cis-trans isomerization of azobenzenes. The rate depends first order on the concentration of the catalyst and on that of azobenzene. A π-complexed intermediate, Cp_2MoH_2 (ArN=NAr), may be postulated.

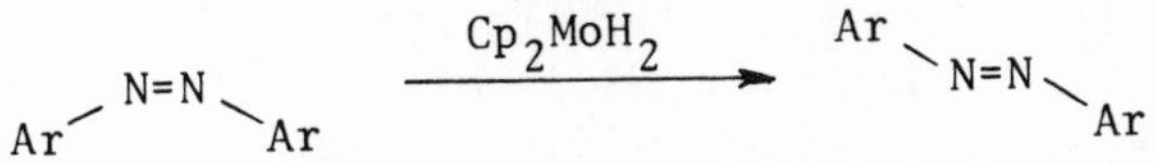

$$CH_3O_2C-C\underset{H\ \ \ \ \ H}{\overset{H\ \ \ \ \ H}{=}}C^{\diagdown}C^{\diagdown}C^{\diagup}C\diagdown CO_2CH_3$$

1,2-addn. ╱   ╱╱ 1,4-addn.

$+Cp_2MoH_2$

$$CH_3O_2C-CH\diagdown_{\underset{Cp_2MoH}{}}CH_2\diagdown C\underset{H}{\overset{H}{=}}C\diagup CO_2CH_3$$

$$CH_3O_2C\diagdown_{\underset{Cp_2MoH}{}}CH\diagup \overset{H\ H}{C=C}\diagdown CH_2\diagdown CO_2CH_3$$

↓   ↓

$$CH_3O_2C-CH_2-CH_2-C\underset{H}{\overset{H}{=}}C\diagup CO_2CH_3$$

$$CH_3O_2C\diagdown CH_2\diagup \overset{H\ H}{C=C}\diagdown CH_2-CO_2CH_3$$

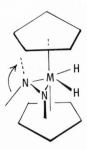

Figure 2. Proposed Structures for Intermediates in the cis-trans
Isomerization of Azobenzene.

TRANS-INSERTION OF HEXAFLUORO-2-BUTYNE

In contrast to the cis-insertion of olefins or some acetylenes,
hexafluorobutyne readily reacts with Cp_2MoH_2 even at $-78°$ to give
exclusively a trans-insertion product. Similar trans-insertion was
observed for Cp_2WH_2 and Cp_2ReH at $-30°$. Two conformational isomers
are formed from Cp_2MoH_2 or Cp_2WH_2 in a ratio ranging from 4:1 to 3:1
depending on the metal and reaction conditions. The ^{1}H and ^{19}F-nmr
spectra of these isomers indicates the structure shown on the following page. Equilibrium to a 1:1 isomer mixture occurs for M=Mo at
$30°$ and for M=W at $80°$. Bis-(methylcyclopentadienyl) analogues,
$(CH_3C_5H_4)_2MH_2$(M=Mo, W) react with hexafluorobutyne at $-70\sim-30°$ to

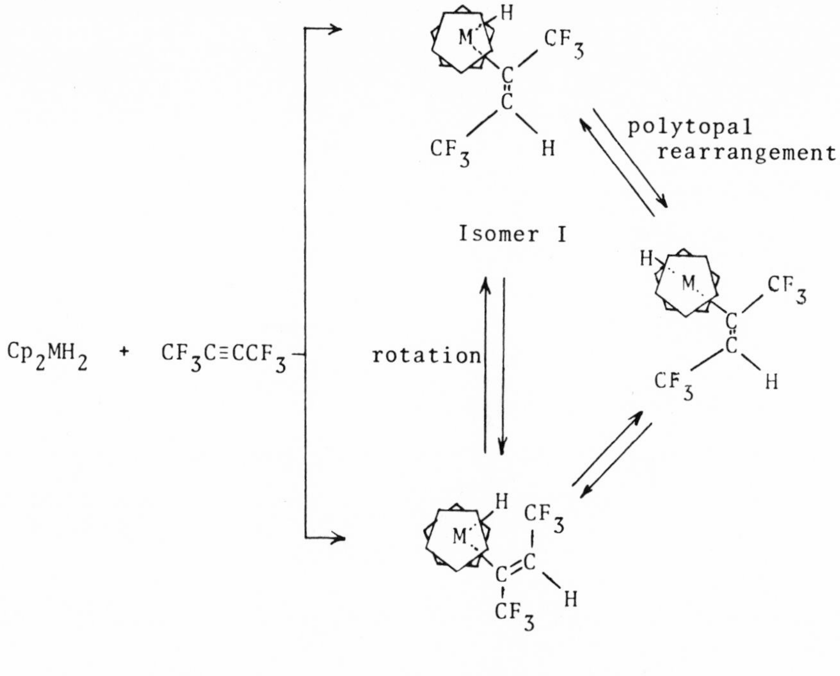

Isomer I

polytopal rearrangement

Cp_2MH_2 + $CF_3C\equiv CCF_3$

rotation

Isomer II

give only one isomer (the dimethyl analogue of Isomer I). Equilibration of the isomer occurs at ∿30° for M=Mo. When the general trend in polytopal rearrangement (i.e. 1st row>2nd row>3rd row) and similar atomic radii of Mo and W are considered, an equilibration mechanism involving polytopal rearrangement is more likely rather than the hindered rotation around the metal-carbon(alkenyl) bond.

MECHANISM FOR THE TRANS-INSERTION OF HEXAFLUORO-2-BUTYNE

Trans-insertion of disubstituted acetylenes has been observed in reaction of $HM(CO)_5$ (M=Mn, Re) with hexafluoro-2-butyne and of $HRh(CO)(PPh_3)_3$ with dimethyl acetylenedicarboxylate. Since the mechanisms for these insertion reactions have not been investigated so far and since the trans-insertion of hexafluoro-2-butyne with metallocene hydrides, such as Cp_2MH_n (M=Re, Mo, W; n=1 or 2), appears to be quite common, we examined the various possible mechanisms, i.e. an ionic, a free radical chain, or a concerted multicenter mechanism.

The free radical mechanism was rejected by experiments carried out in such solvents as chloroform, or acrylonitrile which traps or transfers free radicals. The radical-polymerization inhibitor, N-phenyl-α-naphthylamine, also had no effect on the rate or on the

stereochemistry. A stepwise ionic mechanism (see below) is also
rejected by following evidence.

1) The stereochemistry of addition is not influenced by polarity of
   solvents.
2) No deuterium is incorporated in the reaction in EtOD.
3) The preferential formation of Isomer I is not predictable by the
   ionic mechanism.
4) The trend in reactivity (Mo>W) does not reflect the trend in σ-
   basicity of the metal center or M-C bond stability (W>Mo).
5) The hydrogen-deuterium exchange of Cp_2MH_2 is slow even at room
   temperature in EtOD.

   Then, a concerted supra-antara addition, [σ2s + π2a], between
M-H σ-bond and C≡C π-bond remains as a possibility (cf. Figure 3).

Figure 3. Side-wise Overlap between M-H σ and Acetylene π[*] Orbitals
in the Skewed Disposition.

This mechanism can nicely explain our experimental observations.
Although similar supra-antara additions between two π-systems ([π2s
+ π2a] reaction) have been proposed[4] for dimerization or codimeriza-
tion of allenes, ketenes, and some olefins, the addition between a
σ- and a π-bond has never been proposed. In case of an interaction
between usual C-H and C≡C bonds, the overlap of the relevant orbi-

tals at the skewed fourcentered transition state (cf. Figure 3) may not become significant. However, when the bond length ($\sim$1.6Å) and the unique bonding scheme of transition metal hydrogen bonds are considered, the side-wise interaction between the electron-rich M-H bond and π-acidic orbitals of hexafluorobutyne may be such that it would lead to the thermal concerted reaction. Uniqueness of the hexafluorobutyne in concerted cycloaddition reactions is illustrated by its propensity to [π4s + π2s] reaction with various aromatic com-pounds.[5] The hexafluorobutyne is also unique in its π-acidic char-acter which is solely due to electronegativity of the fluorine atom. It is also a sterically demanding disubstituted acetylene resembling to di-t-butylacetylene in this respect. These two combined charac-teristics may prefer the skewed transition state over that involving a π-coordination of $CF_3C\equiv CCF_3$ to the excited molecule of Cp_2MoH_2 with parallel Cp rings[3](cf. Figure 1).

The postulated skewed transition state readily explains the preferential or exclusive formation of conformational isomer I from Cp_2MH_2 or $(CH_3C_5H_4)_2MH_2$ respectively in the kinetically-controlled low temperature reaction as illustrated in Figure 4. The attack

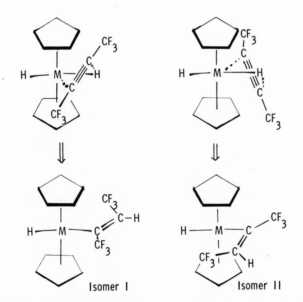

Figure 4. Pathways for the Kinetically Controlled Selective Forma-tion of Conformational Isomers.

from the back side will be sterically hindered in the dimethyl de-rivatives when the preferred conformations (cf. Figure 5) are taken into account. Then the kinetically controlled product in reaction with hexafluorobutyne is the isomer I.

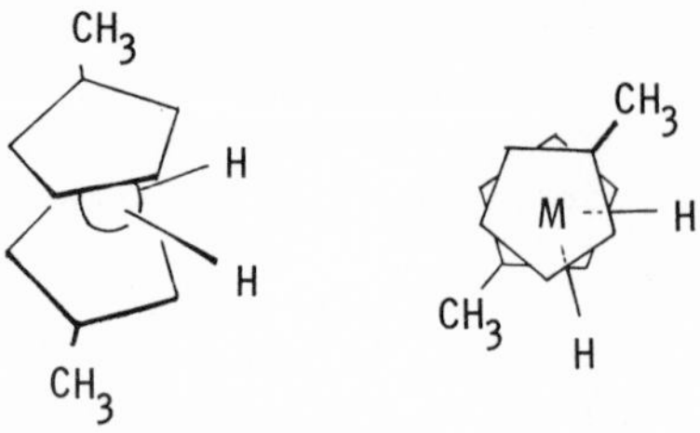

Figure 5. Preferred Conformation of the 1,1'-Dimethylcyclopenta-
dienyl Derivatives.

NATURE OF METAL-HYDROGEN BOND IN TRANSITION METAL HYDRIDES

Two types of transition metal hydride complexes are distin-
guishable, viz. coordinatively saturated and unsaturated ones. A-
mong unsaturated hydrides of low-valent metals, e.g. Cp_2MH_2, the
symmetry of highest-occupied M.O. (HOMO) determines its reactivity
towards unsaturated bonds, e.g. C=C, $-C\equiv C-$, or -N=N-. Thus, σ-basic
hydrides which have HOMO of σ-symmetry add to σ-acidic centers,
whereas π-basic ones form π-complexes with π-acidic unsaturations.

A comparison of the chemical behavior between σ-basic organic
amines and σ-basic transition metal hydrides, e.g. Cp_2MH_2(M=Mo, W)
revealed important differences. Hydrogen-deuterium exchange in
RNH_2 is very fast, but that in Cp_2WH_2 (pK_b=8.6 in aq. dioxane) is
slow even though the pK_b values are similar. No condensation occurs
between Cp_2MH_2 and ketones. In the reaction with a polar double
bond, e.g. acrylonitrile, amines give β-aminoethyl compounds (RNH-
CH_2CH_2CN) but Cp_2MoH_2 gives α-metalated products ($Cp_2MoH[CH(CN)CH_3]$).
Infrared and ^1H hmr data of protonated hydrides, $Cp_2ReH_2^+$ or $Cp_2MH_3^+$-
(M=Mo, W), indicate hydridic character despite the fact that these
species are acidic in solution. These aspects are explicable by invoking rehybridization at the metal upon an attack of a σ-acid (cf.
Figure 6).

Coordinatively saturated hydrides, e.g. $H_2Fe(dppe)_2$ (dppe=Ph_2-
$PCH_2CH_2PPh_2$) or $H_4Mo(dppe)_2$, are in general inert to insertion of
unsaturated bonds. Strongly π-acidic olefins give electron donor-
acceptor (EDA) complexes. Thus, a green CT complex forms from H_2-
$Fe(dppe)_2$ or $H_4Mo(dppe)_2$ and TCNE in benzene. These hydrides can
be made to react upon irradiation of visible light or upon an attack
of free radicals.

$$H_4Mo(dppe)_2 \xrightarrow{\;N_2,\; h\nu\;} (N_2)_2Mo(dppe)_2$$

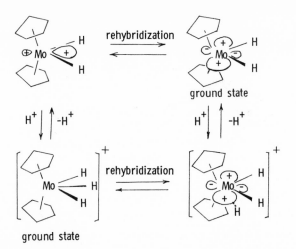

Figure 6.   Rehybridization at the Metal during the Protonation-De-protonation Equilibria.

In summary, energy level and symmetry of metal orbitals, especially those of frontier ones, are important in governing the nature of metal-hydrogen bond.

ACKNOWLEDGMENT

        The authors thank Mr. M. Aotaké for the experimental assistance in some part of this work.

REFERENCES

1. For previous papers of the series, see, a) A. Nakamura and S. Otsuka, J. Amer. Chem. Soc., 94, 1886 (1972); b) ibid., 95, 5091 (1973); c) ibid., 95, 7262 (1973); d) Tetrahedron Letters, 4529 (1973).
2. J. L. Thomas and H. H. Brintzinger, J. Amer. Chem. Soc., 94, 1386 (1972); J. L. Thomas, ibid., 95, 1838 (1973).
3. A. Nakamura, M. Aotaké, and S. Otsuka, J. Amer. Chem. Soc., in press.
4. P. R. Brook, J. M. Harrison, and K. Hunt, J. C. S. Chem. Comm., 733 (1973) and references therein.
5. C. G. Krespan, B. C. McKusick, and T. L. Cairns, J. Amer. Chem. Soc., 83, 3428 (1961).

DISCUSSION

DR. IBERS: In your reaction of substituted diazomethane with bis-cyclopentadienyl rare metals: Do you think that is carbon bound? If it were carbon bound, could you hit it with light to form a carbene?

DR. NAKAMURA: Yes, I think so.

DR. IBERS: Do you think you have formed a diazene complex? You do not know which way it has gone?

DR. NAKAMURA: I think it may be bound to two nitrogen atoms of the diazo compound.

DR. IBERS: Wouldn't that be an unusual method for making a diazene complex - to start with diazomethane and go on up?

DR. NAKAMURA: That is not a standard route for making what you might describe as diazonium complexes though I realize it isn't really diazonium. Substituted diazomethanes usually fall apart in the presence of metals. I just find it an amazing compound, that is all.

DR. OTSUKA: Actually, the carbon attached to the diazogroup is slightly anionic, so if you pour more electrons into the system, then you may weaken the carbon nitrogen bond, and release nitrogen.

DR. IBERS: Then you get carbenes.

DR. NAKAMURA: Yes.

DR. IBERS: You have not tried that?

DR. NAKAMURA: This is quite a stable compound.

KINETICS OF THE ADDITION OF TETRACYANOETHYLENE TO BIS(TERTIARY PHOSPHINE)BIS(ISOCYANIDE)RHODIUM(I) PERCHLORATES

Toshio Tanaka[*], Masa-aki Haga, and Katsuhiko Kawakami

Department of Applied Chemistry, Faculty of Engineering, Osaka University, Yamada-kami, Suita, Osaka 565, Japan

ABSTRACT

$[Rh(RNC)_2L_2]ClO_4$ (R=p-ClC$_6$H$_4$, p-CH$_3$OC$_6$H$_4$, $\underline{cyclo}$-C$_6$H$_{11}$, and L = PPh$_3$; R = p-CH$_3$OC$_6$H$_4$ and L = P(OPh)$_3$ react with tetracyanoethylene (TCNE) to give $[Rh(RNC)_2L_2(TCNE)ClO_4$. The IR and PMR spectra suggest that the adducts have a trigonal bipyramidal configuration with the TCNE coordinated rigidly, if TCNE is assumed to occupy one coordination site. The kinetic study of the reactions was carried out in various organic solvents by the use of a stopped-flow photometer; the reactions are first order with respect to both the Rh(I) substrates and TCNE, and the rate of reactions increases with increasing polarity of the solvents, and with increasing electron donating power of the R groups. The activation parameters of the reactions are discussed in terms of the reaction mechanism.

INTRODUCTION

Addition compounds formed between basic transition-metal complexes and various π-acid have widely been investigated, and most earlier works have dealt with the characterization and configuration of the adducts.

Attentions have recently been paid to mechanisms of such reactions. Here, we wish to discuss the kinetic study on the adduct formation between $[Rh(RNC)_2L_2]ClO_4$ (R= p-ClC$_6$H$_4$, p-CH$_3$OC$_6$H$_4$, $\underline{cyclo}$-C$_6$H$_{11}$ and L=PPh$_3$; R= p-CH$_3$OC$_6$H$_4$ and L=P(OPh)$_3$) and TCNE in various organic solvents.

191

STOICHIOMETRY

The Rh(I) substrates react easily with TCNE in dichloromethane to give 1 to 1 adducts, $[Rh(RNC)_2L_2(TCNE)]ClO_4$, in good yields. Elemental analysis of the adducts were in good agreement with the required values.

Infrared and PMR spectra indicated that the parent Rh(I) complexes take a square-planar trans-configuration, except for the p-chlorophenylisocyanide derivative, which was a mixture of cis and trans-isomers. On the other hand, all the TCNE adducts take a trigonal bipyramidal rigid structure with two isocyanides trans to each other, if TCNE is assumed to act as a monodentate manner.*

Table I shows electronic absorption maxima of the parent Rh(I) complexes in acetonitrile. The higher and lower frequency bands are assigned to d-d and charge-transfer bands, respectively.

Table I
Electronic spectra of $[Rh(RNC)_2L_2]ClO_4$

L	R	λ_{max}, nm	
PPh$_3$	p-ClC$_6$H$_4$	484	427
	p-CH$_3$OC$_6$H$_4$	478	419
	cyclo-C$_6$H$_{11}$	458	401
P(OPh)$_3$	p-CH$_3$OC$_6$H$_4$	455	398

The intensities of these bands decrease with increasing amounts of TCNE added to the solution, and an isosbestic point is observed around 415 nm. The mole-ratio method indicated the formation of 1 to 1 adducts in solution. With R= p-CH$_3$OC$_6$H$_4$ and L=PPh$_3$ the equilibrium constant, $[Adduct]/[Rh]\cdot[TCNE]$, in dichloromethane at 25°C was preliminary determined to be $3.3 \times 10^4 M^{-1}$. Thus, not only in other solvents used in kinetic runs but also for other systems, the reac-

---

*In the extreme case, this configuration may be replaced by an octahedron, in which TCNE is considered to act in a bidentate way.

tions have been assumed to proceed essentially to completion.

KINETICS AND MECHANISM

The reaction was followed by measuring the transmittance in the vicinity of 420 nm on a Union RA-1100 stopped-flow photometer with 2-mm quartz cell. Concentrations of the Rh(I) substrates in solution were varied between 1.2 to 2.0 x 10^{-4} M, in which region the Beer's law was shown to be valid in acetonitrile. This has been assumed to be the case in other solvents. All the reactions were carried out under pseudo-first order conditions, using at least a ten times excess of TCNE. The change in transmittance as a function of time were monitored by a memory scope. Five or six measurements were made for each run.

Plots of $Ln(A_t-A_\infty)$ vs time were found to be linear, and the p pseudo-first order rate constant , k_{obs}, was calculated by the method of least-squares, where A_t and A_∞ are absorbances at the time being "t" and "infinite" respectively. Again, plots of k_{obs} against the concentration of TCNE are linear with a zero intercept (Figure 1), which indicates that the reaction obeys the simple second order rate law: $-d[Rh]/dt=k_2[Rh][TCNE]$.

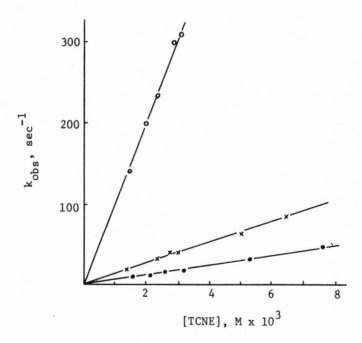

Figure 1. Plots of k_{obs} vs [TCNE]; [Rh(p-CH$_3$OC$_6$H$_4$NC)$_2$(PPh$_3$)$_2$]ClO$_4$ 1.6 x 10^{-4}M in CH$_3$CN(o), (CH$_3$)$_2$CO(×), and THF(•).

Second order rate constants (k_2) at 25°C obtained are summarized in Table II, which indicates the rate constant increases with increasing polarity of solvents.

Table II
Second order rate constants in the reaction
of $[Rh(RNC)_2L_2]ClO_4$ with TCNE at 25°C

L	R	$k_2 \times 10^{-4}$, $M^{-1}sec^{-1}$		
		CH_3CN	$(CH_3)_2CO$	THF
PPh$_3$	p-ClC$_6$H$_4$	2.33		
	p-CH$_3$OC$_6$H$_4$	9.23	1.40	0.59
	cyclo-C$_6$H$_{11}$	29.0		
P(OPh)$_3$	p-CH$_3$OC$_6$H$_4$	9.14	2.54	1.46

In the present work, the Rh(I) substrates are ionic, and therefore it might be possible that the difference of the reaction rate in various solvents is due to distinction of the ionic structure in these solvents. However, the addition of NaClO$_4$ up to 500 times the concentration of the Rh(I) complex in THF gave rise to only a slight change in the rate constant of the TCNE addition reaction. In addition, we observed a good second order kinetics in such a case. Thus, the solvent effect on the rate constant may not be interpreted only by a ion-pairing effect on the Rh(I) substrate in solution.

The activation parameters derived from a plot of log k_2 vs 1/T-(°K) and the Eyring's equation are listed in Table III.

There would be two different reaction mechanisms which have to be taken into account; (a) the reaction proceeds via an asymmetric three-center transition state involving a dipolar intermediate formed by a nucleophilic attack of the rhodium atom on TCNE, and (b) the reaction involves a symmetric concerted transition state, where is little charge separation. These possible pathways are depicted in Scheme I.

The mechanism (b) would predict that the $\Delta H^{\ddagger}$ value remains con-

Table III
Activation parameters for the reaction of
$[Rh(\underline{p}\text{-}CH_3OC_6H_4NC)_2L_2]ClO_4$ with TCNE

L	Solvent	$\Delta H^{\ddagger}$ Kcal/mol	$\Delta S^{\ddagger}$ e.u.
PPh$_3$	CH$_3$CN	2.2	-28
	(CH$_3$)$_2$CO	3.5	-27
	THF	7.6	-15
P(OPh)$_3$	CH$_3$CN	4.1	-22

$\Delta H^{\ddagger}$: $\pm$ 0.2 Kcal/mol, $\Delta S^{\ddagger}$: $\pm$ 3 e.u.

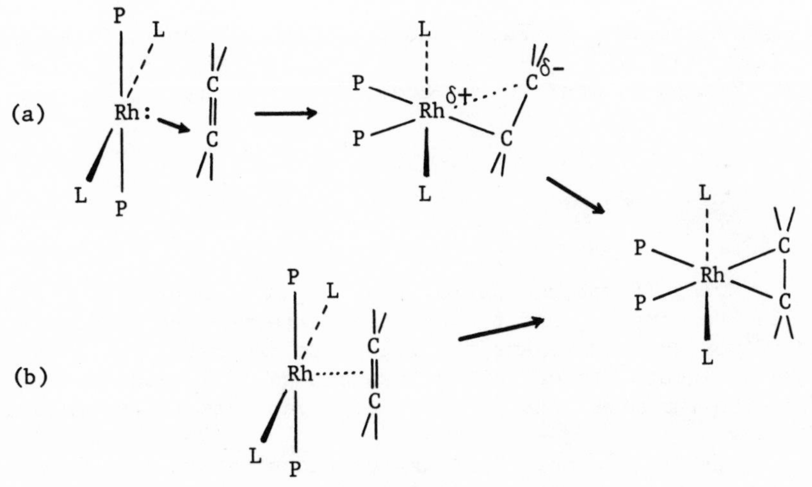

Scheme I

stant regardless of the solvents, because $\Delta H^{\ddagger}$ will be largely due to
a prerequisite deformation of the square-planar Rh(I) complex, as is

suggested for the reaction of $[Ir(Ph_2PCH_2CH_2PPh_2)_2]BPh_4$ with $(CH_3)_n$-$(C_2H_5O)_{3-n}SiH$.[3] On the other hand, marked solvent dependences of the $\Delta H^{\ddagger}$ and $\Delta S^{\ddagger}$ values observed here would explain the mechanism (a), in which the transition state is similar to that suggested recently for the reactions of trans-$IrCl(CO)(PPh_3)_2$ with trans-$CF_3(CN)C=C(CN)CF_3$[4] and of the related Ir(I) complexes with alkylhalides.[5]

Finally, it is to be noted that the rate constant increases with increasing electron donating power of the isocyanides. However, the substitution of $P(OPh)_3$ for PPh_3 resulted in a slight change in rate constant. This fact seems not to be explained only in terms of the electronic effect, because $\nu(C{\equiv}N)$ frequency is lower in the PPh_3 complex (2124 cm^{-1}) than in the $P(OPh)_3$ complex (2154 cm^{-1}). Thus, a steric effect may be more important than the electronic one in this case. One may assume that TCNE approaching to the rhodium atom is suffered a large steric hindrance from PPh_3 than from $P(OPh)_3$, according to the cone-angle of the phosphorus ligands.[6]

REFERENCES

1. A. J. Hart-Davis and W. G. A. Graham, Inorg. Chem., 9, 2658 (1970).
2. P. B. Chock and J. Halpern, J. Amer. Chem. Soc., 88, 3511 (1966).
3. J. F. Harrod and C. A. Smith, J. Amer. Chem. Soc., 92, 2699 (1970).
4. J. Ashley-Smith, M. Green, and D. C. Wood, J. Chem. Soc. (A), 1847 (1970).
5. R. Ugo, A. Pasini, A. Fusi, and S. Cenini, J. Amer. Chem. Soc., 94, 7364 (1972).
6. C. A. Tolman, J. Amer. Chem. Soc., 92, 2956 (1970).

DISCUSSION

DR. OTSUKA: If you use ultraviolet light to monitor the rate, you excite the complex, and the excited state may react very rapidly. So if you want to relate the ligand to the rates you must be very careful about this effect because change of ligand in tetra coordinated rhodium might change the energy gap between ground and excited states.

DR. COLLMAN: I do not think any of these things have long enough lived excited states to have any effect.

DR. TANAKA: If you use such low energy u.v. light, all the species may be in excited states.

DR. OSBORN: In fact, we have done some work on this, Jim. In the case of iridium, because we were interested in the fact that if our interpretation of the Pt(0) work is correct, then if you can produce a 17 electron system in iridium then you might be able to extend the reactivity of iridium(I).

DR. COLLMAN: You are dealing with a chain process?

DR. OSBORN: Well, we do not know what we are dealing with, but certainly if you take iridium(I) with trimethyl phosphine, or dimethylphenyl phosphine, and react it with butyl bromide you get huge accelerations in rate, even at something like 350 nanometers. In fact, a thermal reaction is negligible under the same conditions.

DR. COLLMAN: That is a different thing. You are generating a species that can carry a chain.

DR. OSBORN: The point has been made that probably where we do not know the mechanism, that ligand association is responsible, and one is picking up some reactive species which is photochemically dissociated. We do not know what the mechanism is in this case, but it could be that it would be possible in this case you would get acetylation.

DR. PETTIT: I am not sure I understand the impact of that statement. Surely that is not enough light intensity in the spectrophotometer. You could double your concentration and you should see an effect if it is an excited species. I would doubt you get enough light through the slits to execute many molecules.

DR. OTSUKA: Yes, we have such experiments, in the case of oxidative reaction of alkyl halides to rhodium complexes. If you carry it out in the dark, no addition takes place overnight, or over two days. But if you use u.v. light to monitor the rate, instantaneous oxidative addition takes place. You can immediately notice it by watching the cell.

DR. OSBORN: Are these the isonitrile compounds?

DR. OTSUKA: Yes, isocyanides.

CATALYTIC APPLICATIONS OF OXIDATIVE ADDITION OF ALKYL HALIDES TO TRANSITION METAL COMPLEXES

S. J. Lapporte[*] and V. P. Kurkov

Chevron Research Company, Richmond, California 94802

INTRODUCTION

The oxidative addition of alkyl halides is a reaction under-gone by coordinatively unsaturated transition metal complexes with concommitant increase in oxidation number and coordination number. In its simplist form, a metal in oxidation state (n) and coordination number (m) adds alkyl halides to give a new complex of oxidation number (n + 2) and coordination number (m + 2) (Equation 1).

$$L_mM^{(n)} + RX \longrightarrow L_mM^{(n + 2)} (R)(X) \qquad (1)$$

The kinetics[1,2,3] and stereochemistry[4,5,6] of this reaction have been intensively studied, particularly for square planar d^8 complexes and indicate metal reactivity increasing from right to left and from top to bottom in the periodic table. The mechanism is best interpreted as an S_N2 displacement[7] on carbon by nucleophilic metal with the expected $RI > RBr \gg RCl$ and primary > secondary > tertiary reactivity order for the alkyl halide. Coordination and subsequent insertion of carbon monoxide (Equation 2) or ethylene (Equation 3) to the organometallic halide followed by reductive elimination of acyl halide (Equation 4) or alkyl halide (Equation 5), respectively, provides a route to carboxylic acid derivatives or oligomeric alkyl halides.

$$L_mM^{(n + 2)} (R)(X) + CO \longrightarrow L_mM^{(n + 2)}(COR) X \qquad (2)$$

$$L_mM^{(n + 2)} (R)(X) + C_2H_4 \longrightarrow L_mM^{(n + 2)} (RCH_2CH_2) X \qquad (3)$$

$$L_mM^{(n+2)} (COR) X \longrightarrow L_mM^n + RCOX \longrightarrow products \qquad (4)$$

$$L_mM^{(n+2)} (RCH_2CH_2) X \longrightarrow L_mM^n + RCH_2CH_2X \qquad (5)$$

While the transition metal catalyzed carboxylation of olefins and alcohols had been known for years as a route to carboxylic acids, esters and anhydrides[8,9,10]; high temperatures, 250-300°C; and pressures, 500-800 atmospheres were often required. The carbonylation of olefins in the presence of HCl, catalyzed by rhodium salts, also required very high temperatures and pressures.[11] It was not until recently[12,13,14] that a proper mechanistic understanding of the role of the metal complex, alkyl structure, and leaving group in the oxidative addition, CO insertion, and reductive elimination reactions (Equations 1, 2, and 4) has allowed one to effect carbonylations at low temperatures and pressures. This understanding led in 1970 to the commercial introduction of a new acetic acid process based on the rhodium-iodide catalyzed carbonylation of methanol.[15,16,17] The rate controlling step in this reaction involves the oxidative addition of methyl iodide (formed from methanol and HI) to a coordinatively unsaturated RH(I) complex.

In this paper we describe the application of oxidative addition of alkyl halides to transition metal complexes to the catalytic synthesis of four compounds: methylene bispropionate, glycolic acid, acetylglycolic acid, and tetrahydrofuran.

Methylene Bispropionate

Interest in methylene bispropionate stems from its use as a grain preservative.[18] Many grains generally and corn specifically containing typically about 28-32% moisture are subject to fungal and bacterial attack. Such grains are routinely dried by mechanical means to a moisture content of ca. 15% or less where they become resistant to such attack. Needless to say, with increasing costs and decreasing availability of fuel, an alternative to mechanical drying becomes attractive. Such an alternative was introduced in Europe in the 1960's. Propionic acid at a dose rate of 4-6 oz/bushel effectively retarded the fungal spoilage of corn. In the U.S. propionic, propionic-acetic mixtures, and isobutyric acid have all been test marketed. The Chevron Chemical Company recently found that methylene bispropionate $(CH_3CH_2CO_2)_2CH_2$ (MBP) was fully equivalent to propionic acid in retarding fungal attack although the residual molds were often different. MBP was far superior to propionic acid in retarding growth of aerobes and lactobacillus. This is shown in Table I. Further, the chemical neutrality of MBP renders it less corrosive in shipping and storage.

Methylene bispropionate specifically and polyoxymethylene dicarboxylates generally have traditionally been prepared by the min-

Table I

Comparison of Bactericidal and Fungicidal Properties of
Propionic Acid (PA) and Methylene Bispropionate (MBP)

Sample: 22-24% High Moisture Field Corn
Application Date: December 8, 1972
Evaluation Date: May 29, 1973

Treatment	Dose, Wt %	Aerobes g	Lactobacillus g	Yeasts g	Molds g	Type
PA	0.23	5000	40	-	320	Penicillium, Pluer, Aspergillus flavus, glaucus
PA	0.33	650	15	-	1600	Penicillium
PA	0.44	370	25	-	1600	Penicillium
MBP	0.21	310	-	-	10	Penicillium, Paecilomycles
MBP	0.32	43	120	-	140	Diplodia
MBP	0.44	90	-	-	-	

eral acid catalyzed reaction of trioxane or paraformaldehyde and carboxylic anhydrous.[19]

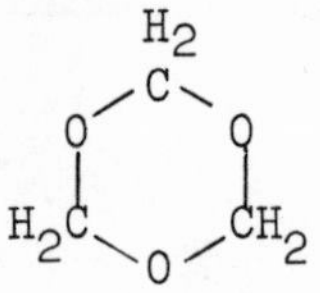

$$\text{(trioxane)} + (RCO)_2O \xrightarrow{HClO_4} RCO(OCH_2)_nOCOR \quad (6)$$

It occurred to us that the anhydrides required for this reaction could be prepared in situ by carbonylation of alkyl-rhodium halides in the presence of carboxylic acids:

$$C_2H_4 + HX \longrightarrow C_2H_5X \quad (7)$$

$$C_2H_5X + Rh^{(I)} \longrightarrow C_2H_5Rh^{(III)}X \quad (8)$$

$$C_2H_5Rh^{(III)}-X + CO \longrightarrow C_2H_5CORh^{(III)}-X \quad (9)$$

$$CH_3CH_2CORh^{(III)}X \longrightarrow CH_3CH_2COX + Rh^{(I)} \quad (10)$$

$$CH_3CH_2COX + CH_3CH_2CO_2H \longrightarrow (CH_3CH_2CO)_2O + HX \quad (11)$$

Propionic anhydride was indeed formed in 44% yield along with ethyl propionate by the reaction of ethylene, CO, and propionic acid catalyzed by 5 mole percent ethyl iodide and 0.1 mole percent rhodium as $[Rh(CO)_2Cl]_2$ after 15 hours at 150°C.

$$CH_2 = CH_2 + CO + CH_3CH_2CO_2H \xrightarrow[\substack{[Rh(CO)_2Cl]_2 \\ 150°C, 1500 \text{ psi}}]{C_2H_5I} (CH_3CH_2CO)_2O \quad (12)$$

44%

The same reaction carried out in the presence of trioxane (formaldehyde equimolar with propionic acid) gave 95 mole percent MBP based on the ethylene converted, Equation 13.

$$C_2H_4 + CO + 1/3 \text{ (trioxane)} \quad (13)$$

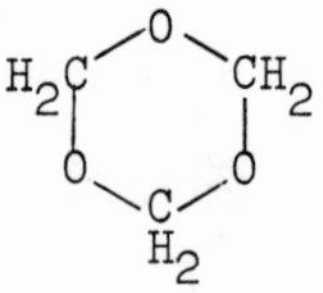

$$+ CH_3CH_2CO_2H \xrightarrow[[Rh(CO)_2Cl]_2]{C_2H_5I} (CH_3CH_2CO_2)_2CH_2$$

95%

Paraformaldehyde was fully equivalent to trioxane as a formaldehyde source. The synthesis of MBP as shown[20] in Equation 13 requires both a rhodium compound and a source of inorganic halide. While $[Rh(CO)_2Cl]_2$, $RhCl_3$, and $(P\phi_3)_2Rh(CO)Cl$ were comparable, $(P\phi_3)_2Ni(CO)_2$ and $(P\phi_3)_2PdCl_2$ failed. Yields of MBP varied with the presence and nature of the halide cocatalyst. This is indicated in Table II. In the absence of any halides neither MBP nor propionic anhydride were detected. Less than 5% of the propionic acid was converted to products of which only ethyl propionate was identified. Ethyl iodide cocatalyst gave an 80% conversion of propionic acid of which 84% was MBP along with small amounts of ethyl propionate. Anhydrous HCl gave approximately a 40% conversion of propionic acid, the uncertainty being due to interference of by-products with the infrared analyses of propionic acid. Only 26% MBP was detected along with 43% propionyl glycolic acid. The formation of this compound from formaldehyde, CO, and propionic acid will be discussed later.

The culmination of our studies was the direct synthesis of MBP by reaction of ethylene, CO, water, and formaldehyde in the stoichiometry shown by Equation 14.

$$2CH_2=CH_2 + 2CO + H_2O + CH_2O \xrightarrow[\text{125°C, 5 hours}]{[Rh(CO)_2Cl]_2} (CH_3CH_2CO_2)_2CH_2 \quad (14)$$

Yields of MBP and propionic acid were 84% and 16%, respectively, based on ethylene converted. This one-step synthesis of MBP from basic raw materials presumably proceeds via in situ formation of

Table II

Effect of Halide Cocatalyst on
Methylene Bispropionate Synthesis[1]

Halide Cocatalyst	None	C_2H_5I	HCl
Propionic Acid Conversion, %	<5	80	~40
Yield,[2] Mole %			
MBP	0	84	26
Propionic Anhydride	0	-	0
Propionylglycolic Acid	0	-	43

[1] 0.5 mole $CH_3CH_2CO_2H$, 0.5 mole CH_2O (as trioxane), 0.05 mole cocatalyst, 0.0005 mole $[Rh(CO)_2Cl]_2$, 1:1 C_2H_4/CO 1100 psi, 150°C, 16 hours.
[2] Based on propionic acid converted.

propionic acid and propionic anhydride followed by reaction with formaldehyde. While we have discussed the formation of propionate moe-
ities as involving oxidative addition of ethyl iodide to rhodium (I)
species, both independent and competitive studies on the RhCl$_3$-HI
catalyzed carbonylation of methanol and ethylene indicate that ethylene carbonylates much more readily than methanol. Since the rate
determining step in the carbonylation of methanol is oxidative addition of CH$_3$I to a rhodium (I) species,[16] it is not likely that
C$_2$H$_5$I would add more readily. On this basis, we suggest that carbonylation reactions involving ethylene involve intramolecular protonation of coordinated ethylene by a hydridorhodium species.

Glycolic and Acetylglycolic Acid Synthesis

The acid-catalyzed carbonylation of formaldehyde at 150-200°C
and ca. 10,000 psi pressure represents the commercial route to glycolic acid in ca. 95% yield.[21]

$$CH_2O + CO + H_2O \xrightarrow{\text{H}_2\text{SO}_4} HOCH_2CO_2H \tag{15}$$

The same reaction has been effected heterogeneously using nickel iodide on silica; however, temperatures of 200°C and pressures near
9000 psi were required to realize optimum yields of 42% glycolic
acid at 47% formaldehyde conversions.[22] In keeping with the theme
of oxidative addition of labile alkyl halides, it occurred to us that
the same reaction could possibly be effected homogeneously at lower
pressures by the oxidative addition of iodomethanol, a formaldehyde-
HI adduct, to a low valent second-row transition metal complex followed by carbonylation and reductive elimination, Equations 16-20.

$$CH_2 = 0 + HI \longrightarrow ICH_2OH \tag{16}$$

$$ICH_2OH + M \longrightarrow HOCH_2-M-I \tag{17}$$

$$HOCH_2-M-I + CO \longrightarrow HOCH_2CO-M-I \tag{18}$$

$$HOCH_2CO-M-I \longrightarrow HOCH_2-\overset{O}{C}-I + M \tag{19}$$

$$HOCH_2\overset{O}{\overset{\|}{C}}-I + H_2O \longrightarrow HOCH_2CO_2H + HI \tag{20}$$

Our results[23] are indicated in Table III. With HI alone at 150°C
and 1500 psi, 78% of the input formaldehyde is converted of which
84% are simple formaldehyde disproportionation products (methanol,
formic acid, and methyl formate). Less than 10% glycolic acid was
observed, along with only traces of acetic acid. With HI and several rhodium catalysts or with PdCl$_2$-HI formaldehyde disproportionation was also observed, but most of the methanol was carbonylated

to acetic acid. More than twice as much glycolic acid was observed with metal-HI systems than with HI alone. Further, the molar excesses of acetic acid over formic acid with rhodium catalysts are

Table III

Glycolic Acid Synthesis

1.0 Mole CH_2O (As Trioxane), 0.001 Mole Catalyst
0.05 Mole I^-, 3.0 Mole H_2O
150°C, 1000-1500 psi

Catalyst/Cocatalyst	Time, Hr	Observed, Mole %			Max. Mole %[1]
		HCO_2H	CH_3CO_2H	$HOCH_2CO_2H$	$HOCH_2CO_2H$
None/HI		27^2	2.3	~10	12
$RhCl_3/CH_2I_2{}^3$	5	0	14	86	-
$RhCl_3/CH_2I_2$	20	26	51	23	48
$RhCl_3/HI$	20	14	63	23	72
$(PO_3)_2Rh(CO)Cl/HI$	20	22	55	22	55
$PdCl_2/HI$	17	38	37	25	25

[1] $\Sigma(CH_3CO_2H-HCO_2H) + HOCH_2CO_2H$.
[2] Methyl formate (30%) and methanol (27%) were also formed.
[3] Control with glycolic acid instead of trioxane.

due to further reactions of glycolic acid leading to CO_2 and acetic acid. This was demonstrated by a control where after 5 hours at 150°C, glycolic acid gave a 14% conversion to acetic acid along with the requisite amount of CO_2. We assume these are formed by carbonylation-decarboxylation of glycolic and malonic acids.

$$HOCH_2CO_2H + HI \longrightarrow ICH_2CO_2H \qquad (21)$$

$$ICH_2CO_2H + CO \xrightarrow{RH} HO_2CCH_2CO-RhI \qquad (22)$$

$$HO_2CCH_2CO-Rh-I \longrightarrow Rh + HO_2CCH_2COI \xrightarrow{H_2O} CH_2(CO_2H)_2 \qquad (23)$$

$$CH_2(CO_2H)_2 \longrightarrow CH_3CO_2H + CO_2 \qquad (24)$$

On this basis, the maximum glycolic acid formed would be the sum of glycolic acid observed and the molar excess of acetic acid over formic acid observed. As indicated, maximum yields of glycolic acid as high as 72 mole % may have been produced. Indeed, these yields were realized by carrying out the reaction in acetic acid which trapped glycolic acid as acetylglycolic acid. It, in turn, was stable towards further carbonylation. The following mechanism was vi-

sualized for acetylglycolic acid formation; Equations 16-19 are the
same as indicated for glycolic acid. Equation 20, however, is replaced by Equations 25 and 26. Instead of water, acetic acid now

$$HOCH_2\overset{O}{\overset{\|}{C}}-I + CH_3CO_2H \longrightarrow HOCH_2\overset{O}{\overset{\|}{C}}-O\overset{O}{\overset{\|}{C}}CH_3 + HI \qquad (25)$$

$$HOCH_2\overset{O}{\overset{\|}{C}}O-\overset{O}{\overset{\|}{C}}CH_3 \overset{\lambda}{\longrightarrow} CH_3\overset{O}{\overset{\|}{C}}-OCH_2\overset{O}{\overset{\|}{C}}OH \qquad (26)$$

reacts with the acyl iodide to give a mixed anhydride. Intramolecular rearrangement of the mixed anhydride most reasonably accounts
for the acetylglycolic acid formation. Intermolecular acylation is
deemed less likely since it would lead to not only glycolic acid
but glycolide and open chain acetylated dimers as well, which were
not observed.

The sulfuric acid catalyzed synthesis of acetylglycolic acid
from formaldehyde, carbon monoxide, and acetic acid proceeds in
70% yield but requires 13,000 psi and 164°C.[24] In contrast, the
rhodium-catalyzed carbonylations proceed at good rates at 1000 psi
and 150°C. Results and conditions are shown in Table IV. $[Rh(CO)_2-Cl]_2$ with HI cocatalyst gave the highest yield of acetylglycolic
acid, 68 mole %. The only other volatile product was methylene bis-
acetate, 4%, most likely formed by the acid-catalyzed reaction of

Table IV

Acetylglycolic Acid Synthesis

0.5 Mole CH_2O (As Trioxane), 2 Moles
CH_3CO_2H, 0.05 Mole I^-, Catalyst
0.001 Mole, 150°C, 1000-1500 psi

Catalyst/Cocatalyst	Time, Hr	Product Yield, Mole %[1]	
		$CH_3CO_2CH_2CO_2H$	$CH_2(O_2CCH_3)_2$
$[Rh(CO)_2Cl]_2$/HI	5	68	4
$[Rh(CO)_2Cl]_2$/CH$_3$I	5	42	9
$[Rh(CO)_2Cl]_2$/CH$_3$I	5	47[2]	15[2]
None	5	0	6
None/CH$_3$I	5	0	6
None/HI	5	45	3
H$_2$SO$_4$	5	23	5[3]

[1]Based on formaldehyde charged.
[2]$CH_3CH_2CO_2CH_2CO_2H$, $CH_2(O_2CCH_2CH_3)_2$ with propionic acid
    solvent.
[3]Also contained 6% bis-acetoxydimethyl ether.

formaldehyde and acetic anhydride or a mixed anhydride.[19] Thus, in acetic acid solvent both side reactions, further carbonylation of glycolic, and disporportionation of formaldehyde were minimized. In addition to HI, organic iodides can be used as cocatalysts. For example, CH_3I-promoted reaction gave 42% yield of acetylglycolic acid and 9% of methylene bisacetate. The results were about the same using propionic acid instead of acetic acid. In this case the products were propionyl glycolic and methylene bispropionate.

In the absence of rhodium and iodide, no carbonylation occurred; only a small amount of methylene bisacetate was detected. Similarly, there was no carbonylation when CH_3I was used alone. However, HI in the absence of rhodium did catalyze carbonylation of formaldehyde and gave 45% yield of acetylglycolic acid. For comparison, sulfuric acid-catalyzed reaction gave only 23% yield. In the case of HI alone, either acid catalysis or trace metal catalysis are responsible for carbonylation.

In summary, clearly the rhodium-HI catalysis was superior and can be best accounted for by a scheme such as the one presented.

Tetrahydrofuran

Tetrahydrofuran (THF) is commercially made either from acetylene and formaldehyde via 1,4-butynediol or by decarbonylation-hydrogenation of furfural. An attractive alternative synthesis of THF would be cycloaddition of ethylene and ethylene oxide (Equation 27); however, such cycloadditions do not occur, except with very reactive

$$
\begin{array}{c}
CH_2 = CH_2 \\
+ \\
CH_2 - CH_2 \\
\diagdown \ \diagup \\
O
\end{array}
\longrightarrow
\begin{array}{c}
CH_2 - CH_2 \\
| \qquad | \\
CH_2 \quad CH_2 \\
\diagdown \quad \diagup \\
O
\end{array}
\qquad (27)
$$

epoxides such as tetracyanoethylene oxide[25] (Equation 28) It occurred to us that the cyclocooligomerization of ethylene and ethylene oxide could be effected by the sequence described by Equations 29-33.

$$
\begin{array}{c}
CH_2 = CH_2 \\
+ \\
(NC)_2C - C(CN)_2 \\
\diagdown \ \diagup \\
O
\end{array}
\xrightarrow{\quad 130\text{-}150° \quad}
\begin{array}{c}
CH_2 - CH_2 \\
NC \ | \qquad | \ CN \\
\quad C \qquad C \\
NC \diagup \ \diagdown \ O \diagup \ \diagdown \ CN
\end{array}
\qquad (28)
$$

87% yield

$$\underset{CH_2-CH_2}{\overset{O}{\triangle}} + HX \longrightarrow \underset{CH_2CH_2}{\overset{HO\ X}{|\ \ \ |}} \tag{29}$$

$$\underset{CH_2-CH_2}{\overset{HO\quad X}{|\quad\ \ |}} + M \longrightarrow HOCH_2CH_2\!-\!M\!-\!X \tag{30}$$

$$HOCH_2CH_2MX + C_2H_4 \longrightarrow HOCH_2CH_2CH_2CH_2MX \tag{31}$$

$$HOCH_2CH_2CH_2CH_2\!-\!M\!-\!X \longrightarrow HOCH_2CH_2CH_2CH_2X + M \tag{32}$$

$$HOCH_2CH_2CH_2CH_2X \longrightarrow \underset{O}{\Box} + HX \tag{33}$$

While ring opening of epoxides by anionic complexes such as $Co(CO)_4^-$ is well known,[26] activation of the carbon-oxygen bond in epoxides by neutral d^8 transition metal complexes is less common. On the other hand, acid-catlyzed ring opening (Equation 29) and oxidative addition of the β-hydroxyethyl halide to a coordinatively unsaturated low valent transition metal complex (Equation 30) is to be expected. Ethylene insertion into the transition metal alkyl, Equation 31, is also well known as in the $RhCl_3$-catalyzed dimerization of ethylene to butene,[27] the palladium-catalyzed arylation of ethylene by iodobenzene,[28] the cooligomerization of ethylene and alkyl iodides, catalyzed by copper and ruthenium complexes,[29,30] and indeed as the growth reaction in Ziegler polymerization of ethylene. THF would be formed by reductive elimination of 4-iodobutanol followed by ring closure with regeneration of HI (Equations 32 and 33) or possibly by direct metal-assisted cyclization. The realization of this scheme

$$\tag{34}$$

is shown in Table V. While THF was observed with several Group VIII complexes, $(P\phi_3)_2Ru(CO)_3$ and $(P\phi_3)_4Pd$, cocatalyzed by HI seemed best. Little THF was observed with $(\phi_3P)_4Pt$, $(\phi_3P)_2Fe(CO)_3$, $(\phi_3P)_2Ir(CO)-Cl$, $Cu(acac)_2$ and $[(\phi O)_3P]_4Ni$. Most of our studies were carried out with $(P\phi_3)_2Ru(CO)_3$. Both metal complex and HI are required for THF formation although 2-iodoethanol was fully equivalent to HI consistent with the proposed mechanism. Replacement of HI by HBr gave only a trace of THF. The major yield losses were due to formation of

homo- and copolymers of ethylene and ethylene oxide. In the absence
of ethylene, 75% of input ethylene oxide was converted to nonvolatile
products in five hours with $(P\phi_3)_2Ru(CO)_3$-HI. High molecular weight
polyethylene was formed by the same catalyst both in the presence and

Table V

Tetrahydrofuran Synthesis

0.5 Mole C_2H_4O, 1.4 Mole C_2H_4
0.0015 Mole Catalyst, 0.05 Mole HI
1.1 Mole H, 150°C, 600-1100 psi

Catalyst	Time, Hr	C_2H_4O Conv., %	THF, Mole %
None/HI	18	25	0
$(P\varphi_3)_2(CO)_3Ru$/None	18	30	0
$(P\varphi_3)_2(CO)_3Ru$/HI	3	14	49
$(P\varphi_3)_2(CO)_3Ru$/HI	18	65	32
$[Ru(CO)_4]_3$/HI	18	58	7
$[Ru(CO)_4]_3$-$P\varphi_3$/HI	18	33	5
Ru/HI	18	5	Trace
$(P\varphi_3)_3RuCl_2$/HI	17	87	5
$RuCl_3 \cdot XH_2O$/HI	18	34	12
$(P\varphi_3)_4Pd$/HI	5	13	30
$(P\varphi_3)_3RhCl$-$P\varphi_3$/HI	18	35	3

absence of ethylene oxide. Further, C_6, C_8, and C_{10} α,ω-iodohydrins
were also formed and identified as the primary alcohol after reduction by $LiAlH_4$. These were necessarily formed by further ethylene
insertion into the ω-hydroxyalkyl metal bond.

In conclusion, we hope we have demonstrated the synthetic utility of oxidative addition of alkyl halides to transition metal complexes. Undoubtedly, new catalytic reactions based on this principle will be discovered in the future.

REFERENCES

1. J. Halpern, Accounts Chem. Res., 3, 386 (1970).
2. J. P. Collman, ibid, 1, 136 (1968).
3. J. P. Collman and W. R. Roper, Advan. Organometal. Chem., 7, 53 (1968).
4. I. C. Douek and G. Wilkinson, J. Chem. Soc. (A), 2604 (1969).
5. G. M. Whitesides and D. J. Boschetto, J. Am. Chem. Soc., 93,

1529 (1971).

6. J. A. Labinger, R. J. Braus, D. Dolphin, and J. A. Osborne, Chem. Comm., 612 (1970).

7. J. A. Labinger, A. V. Kramer, and J. A. Osborne, J. Am. Chem. Soc., 95, 7908 (1973).

8. N. von Kutepow, W. Himmele, and H. Hohenschutz, Chem. Ing. Techn. 37, 383 (1965).

9. W. F. Gresham and R. E. Brooks, U.S. 2,497,304 (to DuPont Company) February 14, 1950.

10. W. Reppe, W. Schweckendick, and H. Kroeper, U.S. 2,658,075 (to BASF) November 3, 1953.

11. T. Alderson and V. A. Englehardt, U.S. 3,065,242 (to DuPont) November 20, 1962.

12. L. Cassar, G. P. Chiusoli, and F. Guerrieri, Synthesis, 509 (1973).

13. L. Cassar and M. Foa, J. Organometal. Chem., 51, 381 (1973).

14. C. W. Bird, "Transition Metal Intermediates in Organic Synthesis", Academic Press, New York, p. 149+ (1967).

15. F. E. Paulik and J. F. Roth, Chem. Comm., 1578 (1968).

16. J. F. Roth, J. H. Craddock, A. Hershman, and F. E. Paulik, Chemtech 1, 600 (1971).

17. H. D. Grove, Hydro. Processing 51, No. 11, 76 (1972).

18. D. L. Kensler, Jr., G. K. Kohn, and D. D. Walgenbach French 70/37,979 (to Chevron Research Company), July 12, 1971.

19. J. Tomiska and E. Spousta, Angew. Chem. Int'l Ed., 1, 211 (1962).

20. S. J. Lapporte and W. G. Toland, U.S. 3,720,706 (to Chevron Research Company), March 13, 1973.

21. A. T. Larson, U.S. 2,153,064 (to DuPont), April 4, 1939.

22. S. K. Bhattacharyya and D. Vir, "Advances in Catalysis", Vol. IX, Academic Press, New York, 1957, pp. 625-635.

23. S. J. Lapporte and W. G. Toland, U.S. 3,754,028 (to Chevron Research Company), August 21, 1973.

24. D. J. Loder and E. P. Bartlett, U.S. 2,211,624 (to DuPont), August 13, 1940.

25. W. J. Linn and R. E. Benson, J. Am. Chem. Soc., 87, 3657 (1965).

26. R. F. Heck, ibid, 85, 1460 (1963).

27. R. Cramer, ibid, 87, 4717 (1965).

28. K. Mori, T. Mizoroki, and A. Ozaki, Bull. Chem. Soc. Japan, 46, 1505 (1973).

29. W. W. Spooncer, U.S. 3,641,171 (to Shell Oil Company), February 8, 1972.

30. E. F. Magoon and L. H. Slaugh, U.S. 3,592,866 (to Shell Oil Company), July 13, 1971.

PHOSPHINE-NICKEL COMPLEXES AS CATALYSTS FOR CROSS-COUPLING REACTION

OF GRIGNARD REAGENTS WITH VINYLIC, AROMATIC AND HETEROAROMATIC HALIDES

Makoto Kumada

Department of Synthetic Chemistry, Kyoto University,
Kyoto, Japan

Recently, Corriu and Masse,[1] and we[2] independently reported the selective carbon-carbon bond formation by cross-coupling of Grignard reagents with olefinic and with aromatic halides in the presence of nickel complexes as catalysts. Here we briefly summarize our own work using various kinds of phosphine-nickel complexes, and consider possible effects of phosphine ligands on the catalytic activity of the nickel atom in both the cross-coupling and secondary reactions.

RESULTS AND DISCUSSION

Upon undertaking this series of studies, we designed the catalytic process of the cross-coupling reaction as pictured in Scheme 1 by combining two important observations on the chemistry in relation to σ-organonickel complexes bearing "supporting" ligand(s): (1) two organic groups on a nickel complex L_2NiR_2 are released by the action of, for example, a halobenzene $R'X$ to undergo coupling, along with the formation of $L_2Ni(R')(X)$, and (2) such a halogen-nickel bond is readily replaced by a Grignard reagent to form the corresponding organonickel bond, $L_2Ni(R')(R)$.

The reaction, indeed, could be achieved by the addition of a Grignard reagent to a vinylic, an aromatic, or a heteroaromatic halide in the presence of a catalytic amount of a dihalodiphosphine-nickel and the yields were generally very high. Phosphine ligands examined are shown with their abbreviations. Table 1 lists representative results obtained from reactions with some primary alkyl or aryl Grignard reagents.

A large body of data indicates that the catalytic activity of

211

$$L_2NiX_2 + 2RMgX_2 \longrightarrow L_2NiR_2 + 2MgX_2 \qquad (1)$$

$$L_2NiR_2 + R'X \longrightarrow L_2Ni(R')(X) + R-R \qquad (2)$$

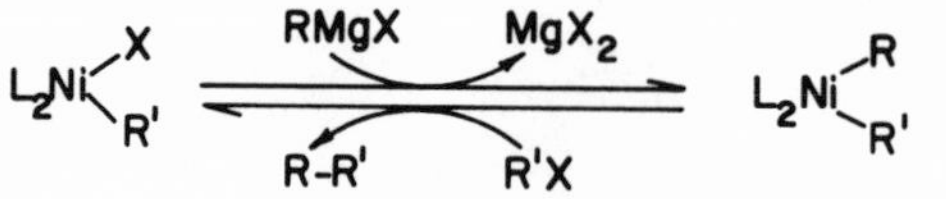

$$L_2Ni\begin{array}{c}X\\R'\end{array} \underset{R-R' \quad R'X}{\overset{RMgX \quad MgX_2}{\rightleftharpoons}} L_2Ni\begin{array}{c}R\\R'\end{array} \qquad (3)$$

Scheme 1

R_3P $Ph_2P(CH_2)_n PPH$ $Me_2P(CH_2)_2PMe_2$ cis-$Ph_2PCH=CHPPh_2$

(R = Et, Bu, Ph)   n = 2, dpe   dmpe   cis-dpen

                      n = 3, dpp

                      n = 4, dpb

dmpf

$H_{10}B_{10}$ C-PR$_2$ / C-PR   R = Me, dmpc   R = Ph, dppc

(-)-diop

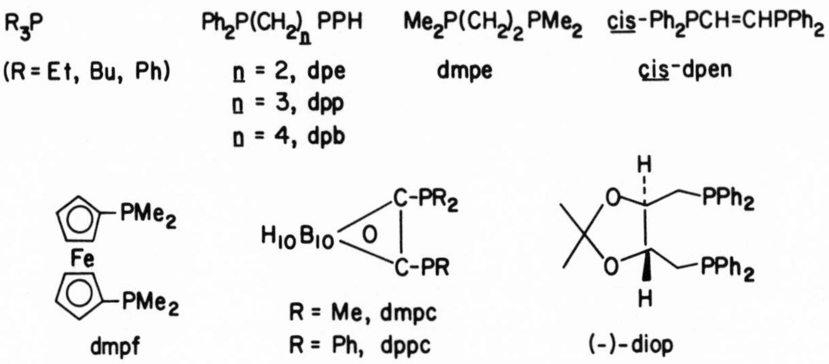

the nickel complexes depends considerably upon the nature of phosphine ligands; bidentate ligands, in general, exhibit the remarkable activity, whereas monodentate tertiary phosphines, especially those containing alkyl groups, are much less effective. Sometimes, a dramatic difference in catalytic activity arises from a seemingly slight change in structure of a phosphine ligand, suggesting profound importance of the geometry of the catalyst species (compare run 7 with run 8).

When a secondary alkyl Grignard reagent was used, the coupling reaction was accompanied by alkyl group isomerization from secondary to primary, its extent being strongly dependent upon the electronic nature of the phosphine ligand in the catalyst.[2b] The results of the reaction between i-PrMgCl and chlorobenzene are summarized in Table 2. Some of the main features observed are as follows:

(1) The catalysts which give rise to the preferential formation of the expected isopropylbenzene contain, as the ligand, dpe, dpp, cis-dpen and dppc. The complexes which induce the alkyl group isomerization to afford n-propylbenzene preferentially comprise dmpe, dmpf, dmpc, PEt$_3$, PBu$_3$ and PPh$_3$. This classification does not necessarily relate to the catalytic activity of the complex in relation to the coupling reaction.

(2) While the catalysts with bidentate phosphine ligands give

TABLE 1

Nickel-Phosphine Complex Catalyzed Cross-Coupling of RMgX with R'X[a]

Run	R'X	RMgX	L_2 in NiL_2Cl_2	Product	Yield, %
1	PhCl	EtMgBr	dpe	PhEt	98
2	(Cl-cyclohexenyl)	BuMgBr	dpe	(Bu-cyclohexenyl)	67
	Dichlorobenzene			Dibutylbenzene	
3	o-	BuMgBr	dpe	o-	89
4	m-	BuMgBr	dpe	m-	94
5	p-	BuMgBr	dpe	p-	95
6[b]	$CH_2=CHCl$	PhMgBr	dpe	$PhCH=CH_2$	89
7	PhCl	Me_3SiCH_2MgCl	dpe	Me_3SiCH_2Ph	trace
8	PhCl	Me_3SiCH_2MgCl	dpp	Me_3SiCH_2Ph	100
9[c]	(o-dichlorobenzene)	$BrMg(CH_2)_{10}MgBr$	dpp	(benzo-cyclophane, $(CH_2)_{10}$)	25
10[c]	(dichloropyridine)	$BrMg(CH_2)_{10}MgBr$	dpp	(pyrido-cyclophane, $(CH_2)_{10}$)	43
11	PhBr	$2,4,6-(CH_3)_3C_6H_2MgBr$	$2Ph_3P$	$2,4,6-(CH_3)_3C_6H_2Ph$	96

[a] The reaction was carried out in refluxing diethyl ether for 20 hr, unless otherwise noted; $NiL_2Cl_2/R'X = 10^{-2} - 10^{-3}$. [b] In a sealed glass tube at room temperature. [c] Solvent is THF.

Table 2

Products from the Reaction of i-PrMgCl with
Chlorobenzene in the Presence of NiL_2Cl_2

L_2 in NiL_2Cl_2	Total Yield (%)	Products distribution (%)		
$Ph_2PCH_2CH_2PPh_2$	74	96	4	0
$Me_2PCH_2CH_2PMe_2$	84	9	84	7
$Ph_2PCH_2CH_2CH_2PPh_2$	89	96	4	0
dmpf	48	8	74	18
dmpc	7	12	88	0
dppc	18	78	1	21
$Ph_2PCH=CHPPh_2$	8	92	8	0
$2PEt_3$	9	1	11	88
$2PBu_3$	8	2	16	82
$2PPh_3$	44	16	30	54

rise to the preferential formation of cross-coupling products, with
monodentate phosphine ligands the main product is reduction one, i.e.
benzene.

(3) Benzene is formed in the case where not isopropyl- but n-
propylbenzene is produced preferentially.

The results suggest that, as far as bidentate ligands are concerned, the more electron-donating the phosphine ligand on nickel,
the greater is the extent of the alkyl isomerization from secondary
to primary. Particularly, it should be noted that with the dmpe ligand the formation of n-propylbenzene is predominant, while with the
dpp ligand isopropylbenzene is formed preferentially.

The most extensive reduction occurring with monodentate phosphines may be understood if one reasonably assumes that the active
intermediate $NiP_2(Ar)(R)$ has trans configuration of the square planar structure like NiP_2Cl_2[3] and $NiP_2(Ar)(Cl)$,[4] and hence there will

be an increased probability for the reduction reaction to occur at the expense of the coupling between Ar and R groups remote from each other.

The similar electronic effects on the alkyl group isomerization during cross-coupling have also been observed for substituted halo-benzenes as reactants, although the substituent effects are smaller compared with the effects of phosphine ligands (Table 3).

Table 3

Products from the Reaction of i-PrMgCl with Various Halides in the Presence of $Ni(dmpe)Cl_2$[a]

Halide		Total Yield	Product distribution, %		
R	X	(%)	i-Pr-R	n-Pr-R	RH
C_6H_5	F	62	12	84	[b,c]
C_6H_5	Cl	88	12	78	6[c]
C_6H_5	Br[d]	83	10	69	8[e]
p-MeC_6H_4	Cl	24	5	84	11
p-$MeOC_6H_4$	Cl	45	6	65	29
m-$CF_3C_6H_4$	Cl	81	13	74	13
p-$CF_3C_6H_4$	Cl[f]	100	46	44	10
o-$CF_3C_6H_4$	Cl	40	51	33	16
α-$C_{10}H_7$	Br[f]	100	73	19	8

[a] Refluxed in diethyl ether for 40 hr, unless otherwise noted. [b] Benzene could not be determined, since the retention time on GLC coincided with that of fluorobenzene. [c] Biphenyl (4%) was also formed. [d] Refluxed for 20 h. [e] Biphenyl (13%) was also formed. [f] Refluxed for 13 h.

The alkyl group isomerization and reduction reactions can be best understood in terms of a mechanism involving σ-alkylnickel intermediates 1 and 3 and a hydrido-olefin nickel intermediate 2 (Scheme 2). The more favored carbanionic character of the primary alkyl group bonded to nickel (intermediate 3) than that of the secondary one (intermediate 1) must be responsible for the isomerization.

The mechanism for the catalytic cross-coupling process, alkyl

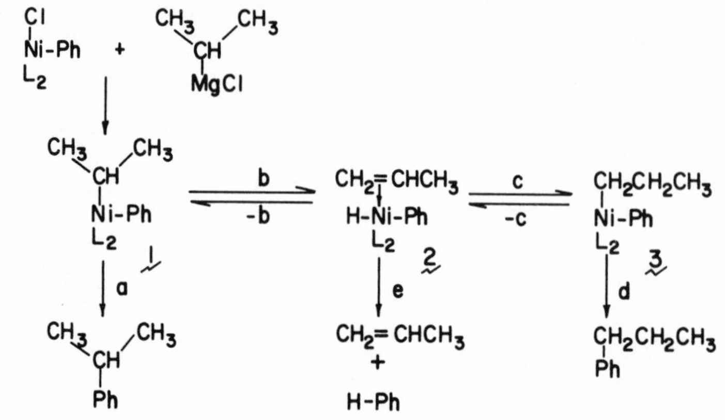

Scheme 2

group isomerization and reduction (Scheme 2) may be supported by the data of Table 4 on the reactions of preformed trans-bis(triphenyl-phosphine)(aryl)chloronickel(II)[4] (4a - 4e) with i-PrMgCl. Note-worthily, the presence of extra triphenylphosphine brought about im-

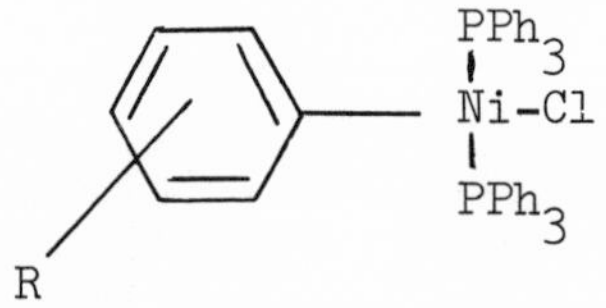

4a, R = H

4b, R = p-MeO

4c, R = p-Me

4d, R = p-CF$_3$

4e, R = o-CF$_3$

proved yields of the coupling products, and also considerably level-ed down the alkyl isomerization. This suggests that the added li-gand would block path b in Scheme 2 by forming Ni(Ar)(i-Pr)(PPh$_3$)$_3$, which might be expected to undergo reductive elimination more read-ily than otherwise.

Isomerization of the norbornyl group from endo to exo also oc-curred when 4a was treated with endo-norbornylmagnesium bromide in refluxing ether for 30 min. A mixture of two isomers of norbornyl-benzene (endo/exo = 27/73) was produced in 64% yield. Here again, the isomerization was decreased by the addition of extra triphenyl-phosphine.

trans-(PPh$_3$)$_2$Ni(Ph)Cl   +   (4)

4a

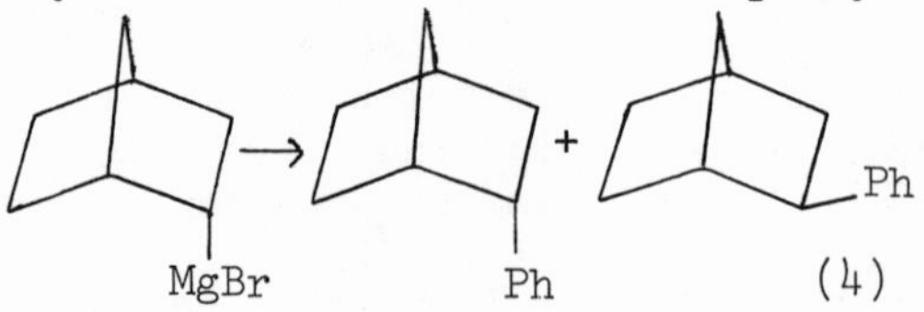

The nickel-phosphine catalyzed Grignard cross-coupling reaction with monohaloolefins took place stereospecifically with retention of

Table 4

Reaction of trans-bis(triphenylphosphine)(aryl)chloronickel(II)
with i-PrMgCl in the Presence (Figures in Parentheses) or
Absence of Added Triphenylphosphine[a]

Ar in Ni(PPh$_3$)$_2$(Ar)(Cl)	Total yield (%)	Product distribution		Ratio of n-PrAr / i-PrAr
		PrAr	: ArH	
p-MeOC$_6$H$_4$	97 (81)	32 (85)	: 68 (15)	12.7 (1.20)
p-MeC$_6$H$_4$	70 (71)	33 (78)	: 67 (22)	4.2 (0.62)
C$_6$H$_5$	100 (92)	33 (68)	: 67 (32)	3.7 (0.41)
p-CF$_3$C$_6$H$_4$	87 (83)	35 (76)	: 65 (24)	1.3 (0.74)
o-CF$_3$C$_6$H$_4$	43 (37)	0 (0)	: 100 (100)	—— ——

[a] At 20° for 30 min. Reactions with extra PPh$_3$ gave also Ni(PPh$_3$)$_4$ ($\sim$ 74% yield).

configuration, while that with cis- and trans-dihaloethylenes proceeded nonstereospecifically but stereoselectively to give cis-ole-
fins in many cases (Table 5).

The high cis-stereoselectivity observed with dihaloethylenes
strongly suggests that an elimination-addition mechanism[5] is operative on nickel. A plausible mechanism is illustrated in Scheme 3.
The evolution of some acetylene during the course of cross-coupling
reaction between an isomeric mixture of CHBr=CHBr with PhMgBr provides strong evidence for the mechanism.

A conceivable mechanism for the trans-selectivity observed with
Ni(dpp)Cl$_2$ as catalyst involves β-elimination of the intermediate 5
to form the corresponding acetylene complex before 5 reacts with a
Grignard reagent. In such a pentacoordinate complex the nickel center may well be blocked by the coordinating acetylene molecule from
the attack of a Grignard reagent. Consequently, the nucleophilic
attack of the Grignard reagent could occur on the acetylene carbon
atom from the outside, resulting in the formation of a trans-vinyl
complex 7 (Scheme 4).[2d]

Table 5

Yields and Isomer Ratios of Stilbene Formed from NiL_2Cl_2
Catalyzed Reaction of Haloolefins with PhMgBr

L_2 in NiL_2Cl_2	From trans-PhCH=CHBr[a] Yield (%) (cis/trans)	cis-PhCH=CHBr[b] Yield (%) (cis/trans)	trans-CHCl=CHCl Yield (%) (cis/trans)	cis-CHCl=CHCl Yield (%) (cis/trans)
dpp	88(7/93)	100(93/7)	93(26/74)	91(33/67)
dpe	74(10/90)	——	100(80/20)	92(90/10)
dmpe	80(8/92)	——	100(99/1)	95(100/0)
cis-dpen	——	——	98(74/26)	100(83/17)
2Ph$_3$P	87(8/92)	——	37(30/70)	19(84/16)

[a] The isomer purity was ca. trans/cis ~92.8. [b] The isomer purity was cis/trans = ~95/5.

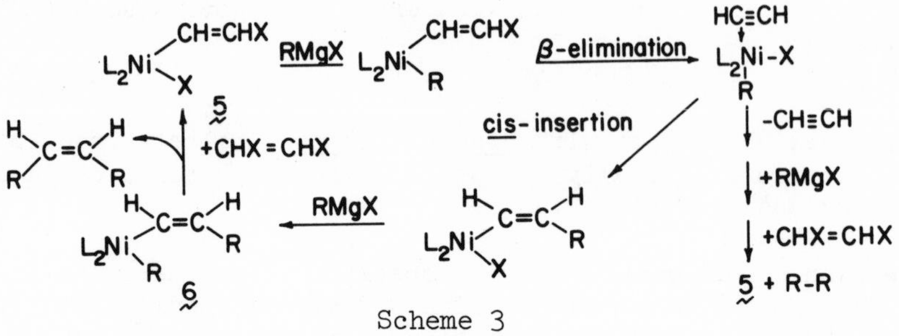

Scheme 3

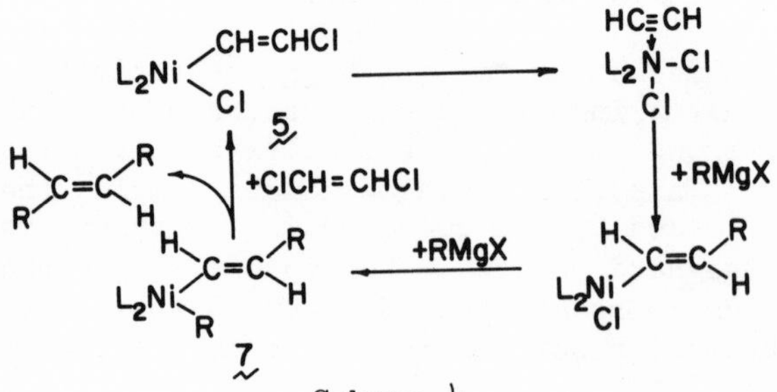

Scheme 4

Table 6

Asymmetric Cross-Coupling Reactions of sec-Alkyl Grignard Reagents
with Organic Halides in the Presence of Ni[(-)-diop]Cl$_2$[a]

No.	Grignard reagent[b]	Organic halide	Ratio of Grignard reagent / organic halide	Product[c]	Yield (%)[d]	$[\alpha]_D^{25}$ deg. of product[e]	Conf. (% ee)[f]
1	EtCHMeMgCl	PhF	2.0	EtC*HMePh	30	-3.74	R (13.6)
2	EtCHMeMgCl	PhCl	2.2	EtC*HMePh	31 (14)	-3.00	R (10.9)
3	EtCHMeMgCl	PhCl	1.0	EtC*HMePh	35 (35)	-2.81	R (10.2)
4	EtCHMeMgCl	PhBr	2.0	EtC*HMePh	20	-4.09	R (14.8)
5	EtCHMeMgCl	α-NpBr[g]	2.0	EtC*HMe-Np-α	44	+0.57	h
6	PhCHMeMgCl	H$_2$C=CHCl	3.85	PhC*HMeCH=CH$_2$	81 (21)	-0.83	R (13.0)[i]
7	PhCHMeMgCl	H$_2$C=CHCl	1.75	PhC*HMeCH=CH$_2$	90 (51)	-0.61	R (9.6)
8	PhCHMeMgCl	H$_2$C=CHCl	i	PhC*HMeCH=CH$_2$	i (63)	-0.50	R (7.8)

[a] Catalyst: Grignard reagent=2.5 x 10^{-3} for EtCHMeMgCl, 1 x 10^{-3} for PhCHMeMgCl. Solvent: diethyl ether. Reaction conditions: at 35° for 48-72 hr in the case of EtCHMeMgCl; at 0° for 10-60 min in the
case of PhCHMeMgCl. [b] Concentrations: EtCHMeMgCl 1.45 M, PhCHMeMgCl 0.70 M. [c] Only chiral products
are listed. [d] Based on the organic halide. Yields based on the Grignard reagent is described in parentheses. [e] Neat. [f] Configuration of the predominant isomer. ee = Enantiomeric excess. [g] Np = nap-
hthyl. [h] Maximum rotation is unknown. [i] (R)-(-)-3-phenyl-1-butene; $[\alpha]_D^{22}$ -6.39° (neat): D. J. Cram,
J. Amer. Chem. Soc., 74, 2141 (1952). [j] Not determined.

When a racemic secondary alkyl Grignard reagent having magnesium attached to a chiral carbon center was reacted with an aryl or vinyl halid in the presence of Ni[(-)-diop]Cl$_2$ as catalyst, an assymmetric cross-coupling reaction could be achieved (eq 5).[6]

$$HR^1R^2C-MgX + R^3X \xrightarrow{\quad Ni[(-)-diop]Cl_2 \quad} HR^1R^2C^*-R^3 \qquad (5)$$

Representative results observed are given in Table 6.

While the reaction of sec-butylmagnesium chloride with an aryl halide (35°, 48-72 hr) gave a mixture of products owing to the alkyl group isomerization, the reaction of α-phenylethylmagnesium chloride with vinyl chloride took place under much milder conditions (0°, 10 min vs. 35°, 3 days) to form 3-phenyl-1-butene as a sole product.

There are two cases to be considered. If the rate of cross-coupling reaction (r_c) is much smaller than that of inversion at the chiral carbon of the Grignard reagent (r_{inv}), the optical yield of the product should be independent of the amount of consumed Grignard reagent, while if $r_c > r_{inv}$ the smaller the amount of consumed Grignard reagent, the higher should be the optical yield. The reaction of sec-butylmagnesium chloride with chlorobuezene (Table 6, Nos. 2 and 3) may be the former case, while the reaction of α-phenylethyl-magnesium chloride with vinyl chloride (Nos. 6-8) may represent the latter case.

These results afford crucial evidence showing that the cross-coupling reaction occurs on a nickel center complexed with a phosphine ligand. A chiral carbon center is transfered from a Grignard reagent to a chiral nickel complex with the formation of diastereomeric intermediates containing a nickel-carbon σ-bond, from which an optically active cross-coupling product is released. In this respect, it is noteworthy that the optical yield depends upon the nature of the halide of halobenzenes (Nos. 1-4).

REFERENCES

1. R. J. P. Corriu and J. P. Masse, Chem. Comm., 144 (1972).
2. (a) K. Tamao, K. Sumitani and M. Kumada, J. Amer. Chem. Soc., 94, 4374 (1972); (b) K. Tamao, Y. Kiso, K. Sumitani and M. Kumada, ibid., 94, 9268 (1972); (c) Y. Kiso, K. Tamao and M. Kumada, J. Organometal. Chem., 50, C12 (1973); (d) K. Tamao, M. Zembayashi, Y. Kiso and M. Kumada, ibid., 55, C91 (1973).
3. P. M. Boorman and A. J. Carty, Inorg. Nucl. Chem. Letters, 4, 101 (1968).
4. M. Hidai, T. Kashiwagi, T. Ikeuchi and Y. Uchida, J. Organometal. Chem., 30, 279 (1971).
5. Z. Rappoport, Advan. Phys. Org. Chem., 7, 1 (1969).

6. (a) Y. Kiso, K. Tamao, N. Miyake, K. Yamamoto and M. Kumada,
   Tetrahedron Lett., 3 (1974); (b) G. Consiglio and C. Botteghi,
   Helv. Chim. Acta, 56, 460 (1973).

DISCUSSION

DR. TSUTSUI: During the 1930's and 40's Kharasch reported numerous papers about alkyl homo and cross coupling, and he checked all transition metal halides as catalysts, Co was the best one. George Casey's stoichiometric coupling reaction seems to be very clean. Have you tried any catalytic coupling reaction?

DR. KUMADA: Yes, I have done catalytic coupling reactions. I mentioned some stoichiometric reactions, but most are catalytic, with about 10^{-3} equivalents of catalyst.

DR. PETTIT: Can you get coupling with saturated halides?

DR. KUMADA: No, no reaction occurs at all. Only vinyl, aromatic, or SP^2 carbon halides.

DR. PETTIT: Your mechanism might suggest that you should get coupling.

DR. KUMADA: No. The diorgano nickel intermediate must first interact with aryl halide. As a result the cross coupling occurs to give us intermediate aryl or vinyl chloro nickel intermediates. The initial interaction between halide and the nickel catalyst is very important.

DR. PETTIT: It is conceivable that the Grignard reacts with the nickel chloride to reduce it to Ni(O), then you could have gotten oxidative addition with the Ni(O), to a saturated chloride, without going through the π-complex, but apparently it does not go.

DR. KUMADA: We tried, but we have never succeeded with saturated halides.

DR. BERGMAN: Your experiments with dichloroethylene suggest that perhaps one might be able to take diphorphine nickel dichloride and acetylene and phenyl magnesium halide and make stilbene. Does that work?

DR. KUMADA: No. I don't think so.

DR. BERGMAN: You cannot take the acetylene complex and add two phenyl groups?

DR. KUMADA: No. We carried out only the reaction between phenylmagnesium bromide and dichloroethylene complex. We have never gotten an acetylene complex intermediate, but its intermediacy is highly probable because sometimes during the cross-coupling reaction acetylene is evolved.

DR. BERGMAN: Apparently that is not a reversible reaction. You cannot make the acetylene complex readily from acetylene. Either that or the acetylene complex is not in fact an intermediate.

DR. KUMADA: Yes, it is a very tough problem. I have never gotten any direct evidence on it.

DR. COLLMAN: I notice that you seem to get much higher yields with the diphosphino propane than with the diphosphinoethane ligand. I was wondering if, in some stages of this complicated reactions one end of the phosphine is coming off. If that were true, it might be interesting to use a chelating phosphine in which one phosphorus was very sterically hindered, and one was not. Have you tried any such mixed chelating phosphines? One with one end a dimethylphosphine, and the other end a dicyclohexyl phosphine, for example.

DR. KUMADA: No, we have not.

DR. TSUTSUI: During this reaction with the Grignard, the metal might be reduced to zero-valent state. Might this isomerize your product?

DR. KUMADA: Yes, it might be, but we have not tried to check this as yet.

MECHANISMS OF THERMAL DECOMPOSITION OF TRANSITION METAL ORGANOMETALLIC COMPOUNDS

George M. Whitesides

Department of Chemistry, Massachusetts Institute of
Technology, Cambridge, Massachusetts 02139

Transition metal alkyls are intermediates in a variety of catalytic processes. These intermediate metal alkyls are often thermally unstable: in fact, their thermal instability -- that is, their propensity to transform into different compounds rapidly and in high yield at low temperature -- is frequently the characteristic that makes them catalytically important. Knowledge of the detailed mechanisms of these thermal decompositions is important both in the rational application of known catalytic reactions, and in the development of new ones.

A number of thermal decomposition mechanisms have been identified for transition metal alkyls; five are encountered commonly in processes that break the carbon-metal of the metal alkyl.[1] Careful mechanistic examination of particular instances of these reactions is slowly making obvious an important distinction between the mechanisms of typical organic reactions and these organometallic re-

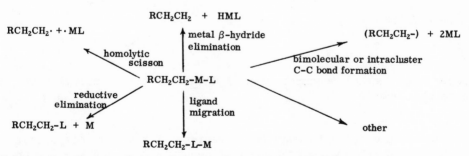

Scheme I. Common Paths for the Thermal Decomposition of Transition Metal Alkyls.

actions: viz, the reaction coordinates for organometallic reactions
are usually much more complex than those for organic reactions. If,
for example, one knows that reaction of an amine with a ketone yields
a Schiff base, it is usually possible without further information
(other than possibly the solution pH) to write a mechanism for the
transformation that has sufficient detail for most synthetic applications. By contrast, if one knows that thermal decomposition of
a transition metal alkyl generates the products expected of metal
β-hydride elimination, one presently knows almost nothing useful a-
bout the course of the reaction. The complexity behind apparently
straightforward thermal decomposition reactions is illustrated by
two examples from our recent work.

The thermal decomposition of di-n-butylbis(triphenylphosphine)-
platinum(II) (1) yields 1-butene, butane, and platinum(0) - containing species, by a kinetically first-order pathway involving elimination of a β-hydride from one butyl group, followed by reductive elimination of butane from a second.[2] Although this metal hydride elimination is the process that determines the products of the decomposition, the overall rate-determining step for the decomposition
is related: in solutions containing no added triphenylphosphine,
loss of triphenylphosphine is rate-limiting; in solutions containing
added triphenylphosphine, reductive elimination of butane is rate
limiting. Even with the information contained in Scheme II, it is
presently difficult to predict the mechanism of thermal decomposition
of related organoplatinum compounds; although comparitive studies

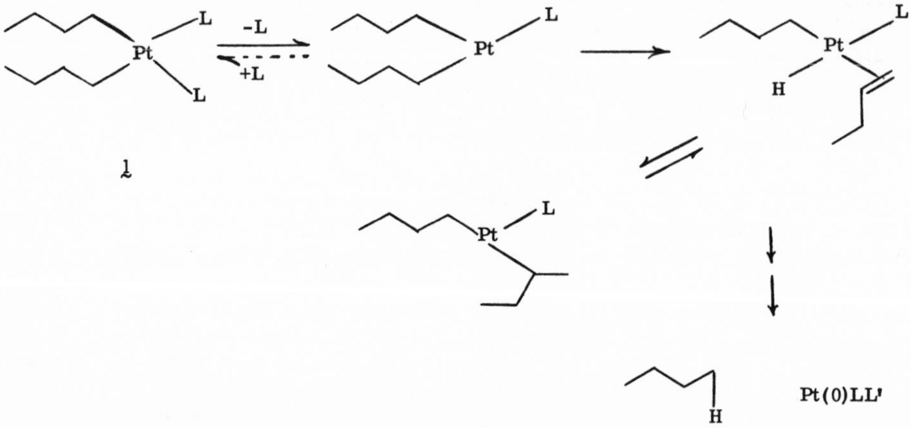

Scheme II. Thermal Decomposition of di-n-butylbis(triphenylphosphine)platinum(II).

have not been carried through for a variety of platinum compounds,
the tri-n-butylphosphine complexes of n-butyl-,[3] sec-butyl-,[3] 1-pro-
penyl,[4] and neophylsilver(I)[5] have each been established to decompose by separate mechanisms, and there is no reason to expect simpler

behavior for organoplatinum compounds. Thus, an apparently simple
β-hydride elimination mechanism from 1 is the end product of a much
more complex sequence of elementary steps, and even having established
at least some of the details of this sequence, there is no guarentee
that the derived mechanistic scheme describes any other decomposition.

The thermal decomposition of n-butyl(tri-n-butylphosphine)cop-
per(I) (2) also involves metal hydride elimination followed by reductive elimination, ultimately generating 1-butane and butane in
equal proportions. In this decomposition a different complexity occurs.[6] The decomposition of 2 shows a pronounced induction period,
which can be eliminated by addition of small quantities of Cu(I)H
or soluble "Cu(0)" species.[7] It is appealing to speculate, by partial analogy to the mechanism established for the decomposition of
1, that the Cu(0) in this reaction serves the function of providing
vacant coordination sites onto which hydride or olefin can be transferred during the β-hydride elimination step. This speculation seems

less convincing that it otherwise might in light of the observation
that the thermal decompositions of certain Cu(I) alkoxides -- reactions proceeding by an apparently unrelated homolytic scission of an
oxygen-copper(I) bond -- also show a similar induction period that
is reduced by addition of "Cu(0)" or Cu(I)H.[8] The influence of the
catalytic Cu(0) may thus be to change the structure of the organo-
copper clusters (in some presently unknown way) and thereby influence
the rate of the copper hydride elimination process.

Metal hydride elimination paths have been identified for a number of transition metal alkyls. In no case has the metal hydride e-
limination step been clearly established as rate-determining (although such reactions almost certainly exist), and in only a few instances is knowledge of the reaction coordinate less superficial than

speculation based almost entirely on the nature of the reaction products. The depth of our mechanistic ignorance concerning the other
thermal decomposition pathways listed in Scheme I is similar. Although patient accumulation of mechanistic detail will ultimately
make the classification of thermal decomposition pathways possible,
these mechanistic studies will be more difficult than those in conventional organic chemistry, and the predictive value of the currently available mechanistic information is small.

Since the prospects for successful manipulation of organometal-
lic-catlyzed reactions based on detailed mechanisms for the thermal
decomposition of the organometallic intermediates do not seem bright,
development of new reactions or modification of already existing ones
will probably have to proceed using less sophisticated approaches.
One attack we have taken to the discovery of new reactions with some
success is to reason along the following lines. Metal hydride eli-
minations, whatever their detailed mechanisms, dominate the chemistry
of many transition metal alkyls. To find new reactions, it might
help to design molecules in which metal hydride elimination is hindered. Since metal hydride elimination seems to occur most readily
from geometries in which the H-C-C-M dihedral angle is $0°$, construc-
tion and thermal decomposition of organometallic having H-C-C-M dihedral angles constrained to values far removed from $0°$ might re-
veal new reactions.

This reasoning led us to prepare dicyclopentadienyltitanium(IV)
1,4-tetramethylene (3).[9] Thermal decomposition of this substance

yields some of the butene expected for a metal hydride elimination-
reductive elimination pathway; it also yields significant quantities
of ethylene, the product resulting from β-cleavage of a carbon-carbon bond. Further studies have suggested both that the thermal stability of five-membered aliphatic metallocycles is appreciably higher than that of their acyclic analogs,[10] and that they may show unusual types of reactivity as a consequence of this enhanced resis-
tance to β-hydride elimination. Six-membered metallocycles also are
resistant to β-hydride elimination in at least one instance;[10] their
reactivity has not been tested.

Whether thermally stable metallocyclic organometallic compounds
will provide reagents for useful synthetic procedures remains to be
seen. It is pertinent that they might, in principle, be obtainable
by oxidative addition of olefins to metals, and that this route has

been successfully utilized in the preparation of 3.[9]

$$Cp_2TiN_2TiCp_2 \xrightarrow{CH_2=CH_2} 3 \xrightarrow{CO}$$

ACKNOWLEDGMENT

The research outlined in this paper was supported by the National Science Foundation. The individuals responsible for its successful prosecution are listed in the papers referenced.

REFERENCES

1. P. S. Braterman and R. J. Cross, Chem. Soc. Rev., 2, 271 (1973); M. C. Baird, J. Organometal. Chem., 64, 289 (1974).
2. G. M. Whitesides, J. F. Gaasch, and E. R. Stedronsky, J. Amer. Chem. Soc., 94, 5258 (1972).
3. G. M. Whitesides, D. Bergbreiter, and P. E. Kendall, J. Amer. Chem. Soc., 96, 2806 (1974); M. Tamura and J. Kochi, Bull. Chem. Soc. Japan, 45, 1120 (1972).
4. G. M. Whitesides, C. P. Casey, and J. K. Krieger, J. Amer. Chem. Soc., 93, 1379 (1971).
5. G. M. Whitesides, E. J. Panek, and E. R. Stedronsky, J. Amer. Chem. Soc., 94, 232 (1972).
6. G. M. Whitesides, E. R. Stedronsky, C. P. Casey, and J. San Filippo, Jr., J. Amer. Chem. Soc., 92, 1426 (1970); G. M. Whitesides, J. San Filippo, Jr., E. R. Stedronsky, and C. P. Casey, J. Amer. Chem. Soc., 91, 6542 (1969); A. E. Jukes, Advan. Organometal. Chem., 12, 215 (1974).
7. K. Wada, M. Tamura, and J. Kochi, J. Amer. Chem. Soc., 92, 6656 (1970).
8. G. M. Whitesides, J. S. Sadowski, and J. Lilburn, J. Amer. Chem. Soc., 96, 2829 (1974).
9. J. X. McDermott and G. M. Whitesides, J. Amer. Chem. Soc., 96, 947 (1974).
10. J. X. McDermott, J. F. White, and G. M. Whitesides, J. Amer. Chem. Soc., 95, 4451 (1973).

DISCUSSION

DR. MANGO: George, I was fascinated by that neopentyl platinum system. Did you get any β-methyl elimination?

DR. WHITESIDES: We looked a little bit, and did not see any.

There's a point concerning that that is interesting. Someone in
Sheffield mentioned that they had made the bis(neopentyl)bis(phenyl-
dimethylphosphine)platinum complexes and found them to be quite stable. This is obviously again one of those situations in which relatively small changes in the phosphine are making a large change in
whatever is going on. We have not looked at those other complexes.

DR. MANGO: Do you get any C_{10} products?

DR. WHITESIDES: We did not determine the products carefully.

DR. MARKS: I asked, because the β-methyl elimination is one way
to rationalize those kinds of differences in stability. It is just
the reverse of the Ziegler-Nata process.

DR. HALPERN: You described a case where, depending on whether
you had triphenyl- or tri-butyl phosphine, the metallocycle decomposed in different ways. Do you know whether either or both of
those was phosphine-inhibited?

DR. WHITESIDES: We do not know in that case, but we know that
the decomposition of this compound

$$\text{[cyclohexane ring]}\ PG \overset{PPh_3}{\underset{PPh_3}{\diagdown}}$$

is accelerated by phosphine, unlike the dibutyl compound. So it
looks like a 5-coordinate intermediate may occur here.

DR. HALPERN: This is a general point I wanted to introduce in
this discussion on insertions of olefins into platinum hydrides.
There seem to be at least two ways such reactions can go: one is
staying in the 4-coordinate regime, and going through a 3-coordinate
intermediate if you need to generate a vacant coordination site.
You have demonstrated examples of that, where you have to lose a
phosphine in order to get β-elimination. It is also possible, however, to go through a 5-coordinate intermediate. For example, the
insertion of olefins into the Chatt compound ($Pt(PEt_3)_2HCl$) is strongly catalyzed by $SnCl_2$, because if you replace Cl^- by $SnCl_3^-$ then an
olefin goes on very easily to form a 5-coordinate intermediate, i.e.
$Pt(PEt_3)_2(SnCl_3)(olefin)H$. That is an interesting situation, where
increasing the amount of $SnCl_3^-$ inhibits the reaction completely,
because a second $SnCl_3^-$ coordinates to form the five coordinate species, $Pt(PEt_3)_2(SnCl_3)_2H^-$, blocking the coordination of olefin.

DR. WHITESIDES: It is actually a little puzzling in the case of
the metallocycle, which we know decomposes by metal hydride elimination -- reductive elimination, it is not at all clear how that reaction can be accelerated by adding a phosphine to give a five-coordinate complex.

DR. HALPERN: In that connection, we have done some basic studies
that lead us to suspect that the way in which ethylene may insert into the Chatt complex is by displacement of a chloride, to form the
cationic trans-compound, $Pt(PEt_3)_2(C_2H_4)H^+$, (other ligands such as
CO do that fairly readily), and then because of the tendency of ethylene to stabilize five-coordinate platinum, a second ethylene goes
on much more easily than on to the original chloride complex [i.e.

via $Pt(PEt_3)_2(C_2H_4)_2H^+]$.

   DR. WHITESIDES: Of course that can not operate in our case, because it is certain you are not going to displace a carbanion.

   DR. HALPERN: An interesting implication of such a mechanism is that if you look at the reverse elimination of the ethylene from the ethyl compound, you can imagine a situation where that is actually catalyzed by ethylene!

   DR. ITOH: I was interested in the greater stability of the trimethylsilylmethyl platinum compound compared to the neopentyl compound. Do you think it possible that the silicon-carbon bond has some conjugation with the metal p-orbitals, and strengthens the metal-carbon bond.

   DR. WHITESIDES: I think that is certainly possible. Another possibility is simply that there is a significant difference in the electronegativity between silicon and carbon. Certainly trimethylsilylmethyl carbanion is much more stable than neopentyl anion, and that may have something to do with it.

NICKEL(0) CATALYZED REACTIONS INVOLVING STRAINED σ BONDS

R. Noyori

Department of Chemistry, Nagoya University, Chikusa,
Nagoya 464, Japan

Although extensive studies have been done in organic chemistry
with unsaturated carbon-carbon bonds, only little has been known
about reactions in which a carbon-carbon single bond is involved.
Certain strained σ bonds are activated under the influence of transition metal complexes and undergo characteristic transformations.
Examples illustrating the intriguing operations of nickel(0) complexes will hereinafter be described. One of the most important
characteristics of the Ni(0) catalysts is that the complexes, unlike other metal complexes, can promote the intermolecular coupling
reactions involving carbon-carbon σ bonds.

NICKEL(0) CATALYZED REACTION OF BICYCLO[2.1.0]PENTANE
AND ELECTRON-DEFICIENT OLEFINS[1,2]

Thermal and transition metal catalyzed reactions of bicyclo-
[2.1.0]pentane and an olefinic substrate display dramatically contrasting features. Thermal uncatalyzed reaction of bicyclopentane
and electron-deficient olefins takes place at high temperature via
a stepwise, diradical mechanism to afford bicyclo[2.2.1]heptane derivatives along with a variety of monocyclic adducts.[3] 1,2-Disub-
stituted olefins enter into the cycloaddition in a nonstereospeci-
fic manner. As shown by eq 1, the hydrocarbon undergoes the reaction with double inversion of stereochemistry at the C-1 and C-4
positions.

In the presence of bis(acrylonitrile)nickel(0), reaction of bicyclopentane with methyl acrylate or acrylonitrile proceeded smoothly under mild conditions to produce the [2+2] adduct effectively
(eq 2).

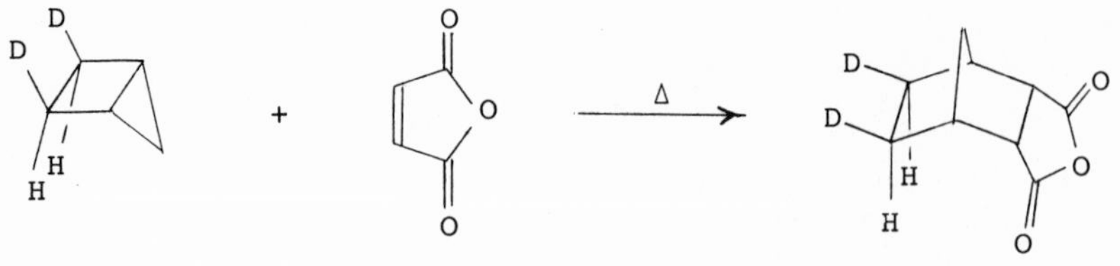

$$(1)$$

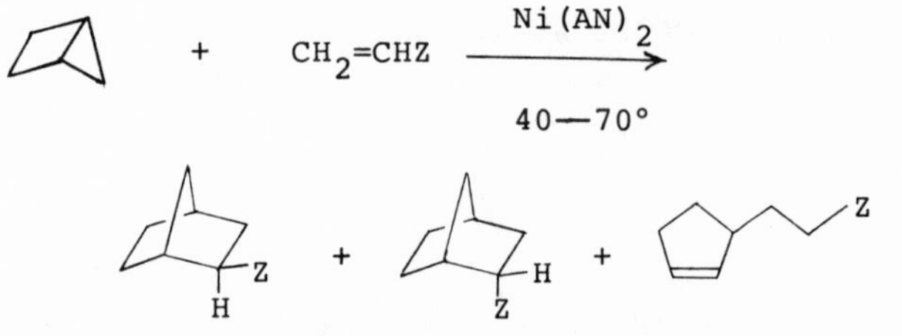

$$(2)$$

Z = COOCH$_3$	33%	33%	22%
Z = CN	47	28	16

Scheme I exhibits the stereochemical course of the reaction with respect to both olefin and bicyclopentane. Obviously the olefinic substrates enter into the cycloaddition in a highly stereospecific manner with retention of configuration. Stereochemical con-

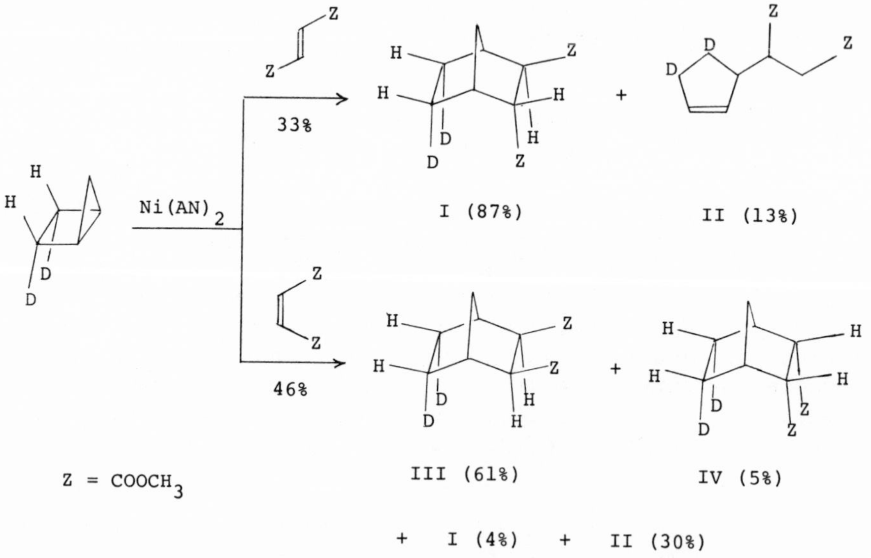

I (87%)   II (13%)

III (61%)   IV (5%)

+   I (4%)   +   II (30%)

Z = COOCH$_3$

Scheme I

sequence of bicyclopentane of the catalyzed reaction is virtually
the reverse of that encountered in the uncatalyzed, purely thermal
reaction; the hydrocarbon undergoes the reaction with retention of
original configuration.

An experiment performed with 5,5-dideuterated bicyclopentane
demonstrated that the formation of monocyclic adduct involves a specific deuterium shift during the reaction (eq 3).

$$(3)$$

These findings are fully in accord with the mechanism outlined
in Scheme II. The approach of Ni(0) atom to the σ bond of bicyclopentane from the exo side of the flap could be ascribed to prefer-
able [2+2] attraction between an occupied, antisymmetric d orbital
of the transition metal atom and a symmetric region of the bicyclopentane envelope. In the absence of Ni(0) complexes, olefins having symmetric character may approach to the σ bond from the antisymmetric endo side to obtain positive $[_\pi 2 + _\sigma 2]$-type interaction.[4]

NICKEL(0) CATALYZED REACTION OF BICYCLO[1.1.0]BUTANES AND ELECTRON-DEFICIENT OLEFINS[5-8]

Under the influence of a catalytic amount of Ni(0) complex such
as bis(acrylonitrile)nickel(0) or bis(1,5-cyclooctadiene)nickel(0),
bicyclo[1.1.0]butane experiences the C-1—C-2 and C-1—C-3 bond
cleavage and cycloadds to methyl acrylate or acrylonitrile to give

$$(4)$$

Z = COOCH₃ or CN

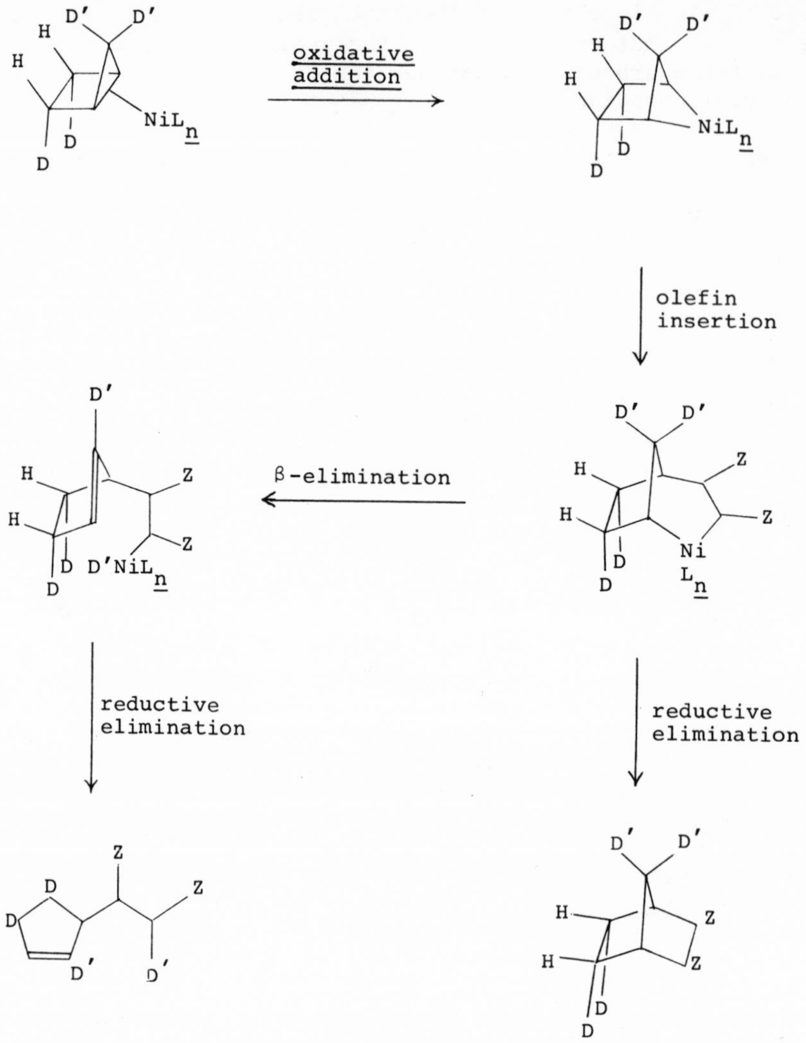

Scheme II

the allylcyclopropane derivatives (eq 4).

This type of two-bond cleavage of bicyclobutanes producing al-lylcarbene metal complex intermediates can be also achieved by transition metals such as Rh(I),[9] Pd(II),[10] Cu(I),[11] Ag(I),[12] etc. The regioselectivity of the intramolecular retro-carbene addition and reactivity of the carbene intermediates are markedly dependent on the nature of the transition metal catalysts employed. Our postulate on the reaction mechanism is as follows. In the VB description, the bicyclobutane-allylcarbene transformation is represented

generally as [VIa (formal oxidative addition product) ↔ VIb ↔ VIc]
→ [VIIa (ylene) ↔ VIIb (inverse ylide) ↔ VIIc (ylide)]; the polar
contributors VIb and VIIb correspond to the σ bonding component in
the MO treatment, whereas VIc and VIIc exhibit the π bonding charac-

VIa   VIb   VIc

VIIa   VIIb   VIIc

ter. Here the two-bond cleavage of an unsymmetrically substituted
bicyclobutane could be considered as occurring so as to lead to the
carbene complex most favorable from the viewpoint of resonance stabilization; the relative significance of the resonance structures is
primarily governed by the nature of the metal and substituents of
the carbenic moiety. Since Ni(0) atom has a great π-donating ability, its carbene complex is best represented by a resonance hybrid,
[VIIa ↔ VIIc] (M=Ni(0)), rather than that containing the form VIIb
as an important contributor.

Examination of regioselectivity attending the Ni(0)-catalyzed
reaction of bicyclobutanes which bear electron-donating and/or -with-
drawing group at the bridgehead position (eqs 5-8) indicates the va-
lidity of the hypothesis. In accord with the initial prediction, the
bridgehead substituents have proved to bias the reaction toward the
formation of the more stable Ni(0) carbenoids of type [VIIa ↔ VIIc];
the two-bond breakage reaction goes so as to have the more electronegative carbon ending up combined with Ni(0) atom. The related,
though oppositely operating, effect of alkyl groups have been observed for the Ag(I)-promoted rearrangement of bicyclobutanes (β
type process)[12] leading to the reactive species of type VIIb. Thus
contrasting electronic properties of Ni(0) atom and Ag(I) ion (strong
σ acceptor and weak π donor) are faithfully reflected in the regioselectivity of the retro-carbene addition of bicyclobutanes. Thus

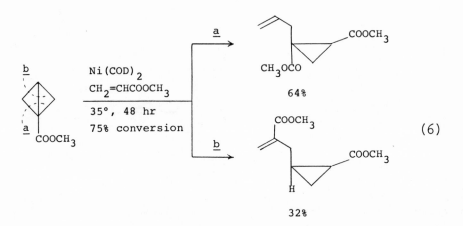

$$(5)$$

R = H or CH$_3$

R = H (90%)
R = CH$_3$ (87%)

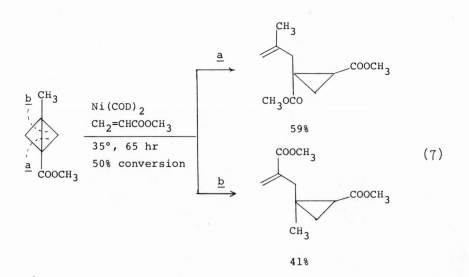

$$(6)$$

64%

32%

$$(7)$$

59%

41%

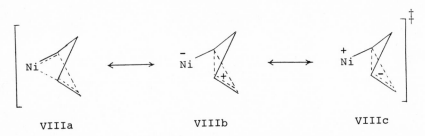

$$ \text{(8)} $$

$$ R = H \quad (62\%) $$
$$ R = CH_3 \quad (60\%) $$

this simple concept can account for most behaviors displayed by other transition metal complexes.

The present catalytic reaction follows the second-order kinetics dependent on the concentration of the Ni(0) catalyst and bicyclobutane substrate. Notably introduction of methyl group to bicyclobutane skeleton accelerates the two-bond cleavage reaction, while electron-withdrawing groups decelerate the reaction. The rate enhancement induced by methyl group could be associated with the growth of a positive charge in the σ bonding component of the transition state VIIIb in the conventional resonance treatment [VIIIa ↔ VIIIb ↔ VIIIc]. In conjunction with the regioselectivity, the dual pathways observed for the catalysis of the ester derivatives

(eqs 6 and 7) would indicate also the significance of π back-bonding from Ni(0) atom to bicyclobutane framework (VIIIc) throughout the two-bond fission; the somewhat higher degree of selectivity in the reaction of eq 6 (66:33 vs. 59:41) is probably related to this factor.

Another characteristic of the Ni(0) catalysis is that the carbenoid intermediates behave as a nucleophile. This could be most simply explained in terms of the low electronegativity (χ 1.1) of zerovalent d^{10} nickel atom.

Finally, it should be added that the carbene cyclopropanation reaction is taking place with an excellent stereospecificity as shown in Scheme III.

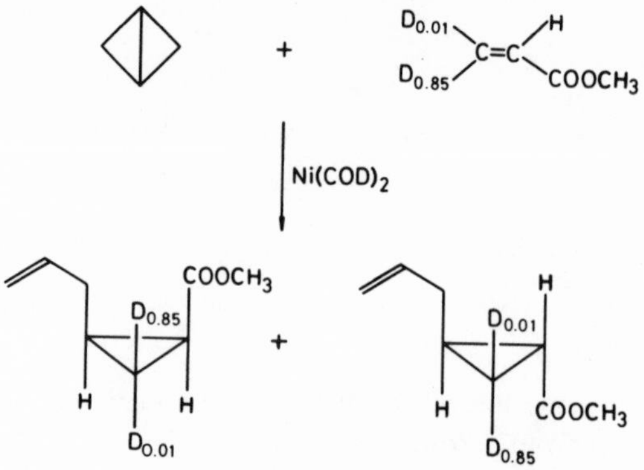

Scheme III

ACKNOWLEDGEMENT

    The author wishes to thank Dr. H. Takaya, Messrs. T. Suzuki,
Y. Kumagai, N. Hayashi, M. Hosoya, and H. Kawauchi for their valuable contribution to this work.

REFERENCES

1. R. Noyori, T. Suzuki, and H. Takaya, J. Amer. Chem. Soc., 93,
   5896 (1971).
2. R. Noyori, Y. Kumagai, and H. Takaya, J. Amer. Chem. Soc., 96,
   634 (1974).
3. P. G. Gassman, K. T. Mansfield, and T. J. Murphy, J. Amer. Chem.
   Soc., 91, 1684 (1969).
4. F. S. Collins, J. K. George, and C. Trindle, J. Amer. Chem. Soc.,
   94, 3732 (1972).
5. R. Noyori, T. Suzuki, Y. Kumagai, and H. Takaya, J. Amer. Chem.
   Soc., 93, 5894 (1971).
6. R. Noyori, Tetrahedron Lett., 1691 (1973).
7. R. Noyori, H. Kawauchi, and H. Takaya, Tetrahedron Lett., 1749
   (1974).
8. R. Noyori, M. Hosoya, N. Hayashi, and H. Takaya, to be published.
9. P. G. Gassman and R. R. Reitz, J. Organometal. Chem., 52, C51
   (1973), and references cited therein.
10. W. G. Dauben and A. J. Kielbania, Jr., J. Amer. Chem. Soc., 94,
    3669 (1972); S. Masamune, M. Sakai, and N. Darby, J. Chem. Soc.,
    Chem. Comm., 471 (1972).
11. P. G. Gassman, G. R. Meyer, and F. G. Williams, J. Amer. Chem.
    Soc., 94, 7741 (1972).
12. L. A. Paquette and G. Zon, J. Amer. Chem. Soc., 96, 203 (1974),
    and references cited therein.

DISCUSSION

DR. MANGO: I have a comment to make on your bicyclobutane work.
In doing work as a reviewer I had to read most of the literature in
this field. I cannot find one single body of chemistry so inconsis-
tent with any one theory. I got the impression from your talk you
were proposing a general intermediate - basically a resonance hybrid-
which would embrace the behavior of all the metals, and the observations of all those workers.

DR. NOYORI: I hope so.

DR. MANGO: My comment would be to go very carefully on that.
The impression I got from looking over that work is that the metals
were doing different things. The most impressive part of the work,
at least the most striking were papers by Paquette where they describe a silver catalyzed ring opening which was stereoselective and
they described it as 2a plus 2a (forbidden) and the selectivity was
around 96%. It was very puzzling - very difficult. You can draw
another mechanism for that, which is also concerted, as 2s plus 2s
plus 2s. It involves the central sigma bond, and you get identical
stereospecificity. The silver could very well simply be perturbing
the molecular framework of bicyclobutane sufficiently well so that
orbitals that previously did not overlap now begin to overlap, and
they simply undergo a thermally allowed rearrangement. That same
rearrangement, when you go to cubane and curiane also works beautifully. My feeling is that what these metals are doing, at least in
some cases is that kind of a thing, while other ones are doing different kinds of things. I think, it has been established in some
cases using a rhodium catalyst and methanol you are in fact generating peculiar acid functions, and you are getting straight protonic
catalysis. One further question involving the CNDO work that Trin-
del did, for he drew a conclusion that the underside of bicyclopentane, the symmetry of that side is antisymmetric and the upper side
was symmetric. I have to assume that he was talking about the occupied orbital and how it could have been anti-symmetric is difficult
to envision. Could you elaborate on what he was talking about?

DR. NOYORI: No, I do not know his exact meaning either. But I
think he has made a conclusion based on calculations involving all
the bonds.

DR. BERGMAN: He found an occupied MO, apparently, which is antisymmetric involving that bridge bond, on the under side. It does
not make any sense to me at all, because there is no such thing as
having one MO from the bottom and another from the top.

DR. NOYORI: No - I think the exo side has considerable bonding
character.

DR. MANGO: You do not have to do a CNDO calculation to figure
that out. You can just make a model of it, and see the bonds aimed
upwards.

DR. NOYORI: Yes, but he did.

DR. BERGMAN: It may be that you are getting some kind of a stepwise biradical mechanism coming in because it prefers to get the back

lobe coming down from under.

DR. NOYORI: Trindle proposed that the reaction of maleic anhydride and bicyclopentane proceeds in a concerted manner.

DR. BERGMAN: I think that what Trindle was concerned about was
not the structure of the transition states where either an olefin is
adding or something is coming out the alkylating agent. His conclusion, based on some extended calculations was that, in this reaction,
which eliminates nitrogen, which is in fact, the reverse of the addition of an olefin to this.

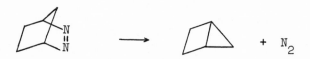

What he said was, his calculations indicated that there was some
bonding between the points on the other side of the molecule, if you
tried to close up the bond on the front, for reasons which are quite
obvious. Sure, there is an antisymmetric, antibonding orbital there,
but that, I think, is due to the presence of this group in the transition state, still interacting with the molecule. That is my understanding.

DR. WHITESIDES: Do compounds like cobalt phthalocyanine-things
that cannot undergo oxidative additions - catalyze these rearrangements?

DR. NOYORI: I do not know.

DR. WHITESIDES: Apparently it is a catalyst for the rearrangement of quadracyclane to norobornadiene. The kind of mechanism you
draw requires at some point or another carbon-metal bond formation.
It would be interesting to know if species that cannot do that can
also catalyze the rearrangements.

DR. NOYORI: You mean the electron transfer, or the oxidative
addition?

DR. WHITESIDES: I do not understand how cobalt phthalocyanine
does what it does, frankly.

DR. TSUTSUI: I think you have made an important point. Cobalt
phthalocyanine, you said it did not undergo oxidative addition, but
it may, in fact. Because you may have oxidative addition on one
side with alkylhalide.

DR. WHITESIDES: Right, on both sides I could believe.

DR. TSUTSUI: Yes, but on just one side too.

DR. WHITESIDES: It seems to me that would give you a rather
difficult dipolar intermediate.

DR. TSUTSUI: Yes, but a 7-coordinate intermediate has been proposed for the intermediate in the Vitamin B_{12} mechanism, so I do not
think you can exclude oxidative addition.

DR. WHITESIDES: That is possibly right, but it seems less attractive than with rhodium.

DR. HALPERN: I am not sure whether this is saying the same thing, but, we have also looked a little bit at Co (II) catalyzed rearrangements, particularly the quadricyclane system. The way we were led to this was thinking about the general problem of cleaving saturated molecules. We found that at least in our way of looking at these things, we could be guided pretty well by using alkyl halides or hydrogen as models, and rhodium adds these things by oxidative addition to give di-adducts. Silver typically cleaves hydrogen or alkyl halides heterolytically to give either a proton of carbonium ion, and can at least explain many of the silver catalyzed rearrangements. The way cobalt cleaves an organic halide is by pulling off a halogen atom and giving a radical and we found that if you had a sufficiently weak carbon-carbon bond, you could get homolytic cleavage of the bond with formation of an electron pair bond with the Co, which leaves a free radical at the other carbon center. Formally, that is a way of rationalizing the Cobalt (II) and of fitting it into the general scheme of a single bond cleavage.

DR. MARKS: Am I wrong or did I read in some place that weaker pi-acids like trinitrobenzene will cause some of these very strained molecules to pop open?

DR. HALPERN: Yes, to make one point on that, cyclopropane containing compounds can be cleaved by almost any acidic system. It is not true for cyclobutane. All of these reactions, transition metals acting on cyclobutane containing systems only, are specific. With cyclopropane you always have to worry whether you are looking at a general acid catalyzed process.

DR. PETTIT: Could I just make one comment on that? If one takes quadricyclane in methanol and adds sulfuric acid, you get an instant and quantitative yield of the methoxy compound. No norbornadiene whatsoever. If you use $AgBF_4$ you will get a 90% yield of the norbornadiene. It is difficult to understand what Paquette's data is, because he did not correct for the acidity of the solution in methanol. Can I ask you if your nickel catalyst isomerized quadricyclane to norbornadiene?

DR. NOYORI: Yes, it does.

DR. PETTIT: In methanol?

DR. NOYORI: No. We do not like methanol as solvents for Ni(0) catalysts. We use hydrocarbon solvents.

SOME SYNTHETIC REACTIONS BY MEANS OF NICKEL CARBONYL AND ORGANIC HALIDES

Membo Ryang[*], Ilsong Rhee, Shinji Murai, and Noboru Sonoda
Department of Petroleum Chemistry, Faculty of Engineering, Osaka University, Yamada-kami, Suita, Osaka, Japan

During the last decade, various types of reactions of nickel carbonyl with organic halides have been extensively studied and some of them have been established as synthetically useful reactions [1-4]. Herein one of the serious problems in applying the reactions to synthetic organic chemistry was that the reactions were limited to the special types of organic halides, such as allyl halides, aryl iodides, benzyl halides and α-haloketones. The high reactivity of these halides was considered to be due to the stabilization of the intermediate nickel complexes by the coordination of organic moieties as π-allyl or pseudo π-allyl type ligands to nickels and hence to a lowering of the activation energy of the reaction between nickel carbonyl and the organic halides. The typical example is the reaction of nickel carbonyl with allyl halides, and the synthetic application of the intermediate complexes, π-allylnickel halides, have been up to now most systematically investigated.[4]

As for the other halides, the reaction behaviors of iodobenzene and benzyl halides toward nickel carbonyl are suggestive of the intermediate complexes.[5] (Equations 1,2)

$$PhI + Ni(CO)_4 \xrightarrow{\quad C_6H_6 \text{ or } THF \quad} Ph\underset{O}{\overset{||}{C}}-\underset{O}{\overset{||}{C}}Ph \qquad (1)$$

$$PhCH_2X + Ni(CO)_4 \longrightarrow PhCH_2CH_2Ph + PhCH_2\underset{O}{\overset{||}{C}}CH_2Ph \qquad (2)$$

The reaction of benzyl halides with nickel carbonyl is remarkably affected by a solvent. When cyclohexane was used as the solvent,

bibenzyl was formed predominantly but, in the case of tetrahydrofuran (THF), nearly equal amounts of bibenzyl and dibenzyl ketone
were obtained. Furthermore, when N,N-dimethylformamide (DMF) was
used, the yield of dibenzyl ketone was very high (>90%). Therefore,
in nonpolar media the intermediate complexes are considered to be
benzoylnickel and benzylnickel complexes in dimeric forms, which
are stabilized by π-benzoyl- and π-benzyl type configuration and
the decompositions give bibenzyl and benzil, respectively. (Scheme
1).

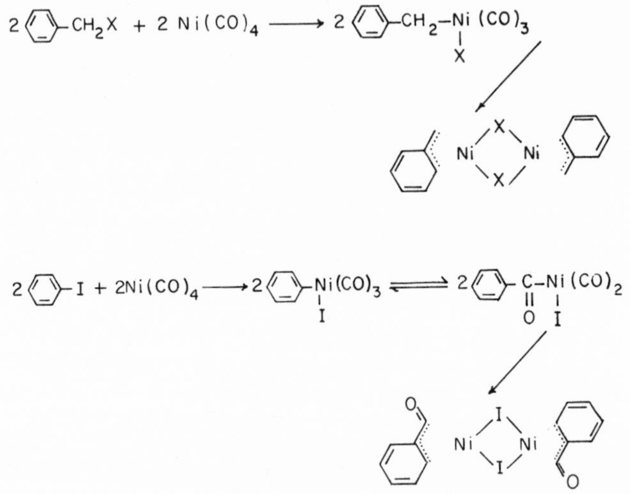

Scheme 1

On the other hand, when the more coordinating DMF was the solvent,
the intermediates transform into the monomeric configurations, of
which decompositions give a mixture of benzophenone (36%) and benzoyl iodide (64%) in the case of benzoylnickel complex and high
yield of dibenzyl ketone in the case of benzylnickel complexes. The
formation of bibenzyl and dibenzyl ketone from the reaction between
nickel carbonyl and phenylacetyl chloride would also support the
contribution of the energetically more stable π-benzyl type configuration. (Equation 3)

$$PhCH_2COCl \; + \; Ni(CO)_4 \xrightarrow[\text{n-hexane}]{} PhCH_2CH_2Ph \; + \; PhCH_2\overset{O}{\underset{}{C}}CH_2Ph \qquad (3)$$

                                              55 %   16 %

These results show that alkenyl halides, which are less susceptible
to nucleophilic substitution than alkyl halides, would be also reactive toward nickel carbonyl because of the coordination of alke-

nyl halides to nickel using π-electron of double bond and the stabilization of the intermediate alkenoylnickel complexes by pseudo
π-allyl type configuration. (Equation 4)

$$
R-CH = CHX \; \rightleftharpoons \;
\begin{array}{c}
R \quad H \\
\diagdown C \diagup \\
H-C \diagup \diagdown Ni(CO)_n \\
\diagdown C \diagup X \\
\parallel \\
O
\end{array}
\qquad (4)
$$

$$Ni(CO)_3$$

Indeed, recently, several types of reaction of alkenyl halides with
transition metal complexes have been reported,[6-10] and especially
the intermediate complexes from nickel carbonyl and alkenyl halides
would become versatile reagents to the introduction of alkenoyl
groups to unsaturated compounds and nucleophiles.

    Organotransition metal σ-complexes are known to add to unsa-
turated bonds in synthetically useful reactions. For example, when
benzyl halides or iodobenzene were used as the partners for the reaction of triiron dodecacarbonyl in the presence of olefins, benzylation or phenylation, respectively, of the olefins occurred.[11] Di-
sodium iron tetracarbonylate reacted with benzyl halides to give
dibenzyl ketone and, in the presence of olefins, to give benzylation
products of olefins, but the intermediate benzyliron complexes showed low reactivity towards alkynes. Similarly, benzoylnickel carbonyl complexes derived from iodobenzene and nickel carbonyl reacted
readily with olefins such as, styrene, acrylonitrile and ethylacry-
late, to give benzoyl adducts of the olefins and γ-lactones.[12]
(Equation 5)

The decomposition of phenyldiazonium tetrafluoroborate with nickel
carbonyl in the presence of acrylonitrile in acqueous acetone led
to the formation of 2-benzoyl-3-phenylpropionitrile, the coupling
product of phenyl and benzoyl groups to an intervening double bond.[13] (Equation 6)

$$PhN_2BF_4 \quad + \quad Ni(CO)_4 \xrightarrow{\quad CH_2=CHCN \quad} \quad \underset{O}{\overset{CN}{PhCH_2\overset{|}{C}H\text{-}\overset{\shortparallel}{C}Ph}} \tag{6}$$

On the other hand, the reaction of dicobalt octacarbonyl with benzyl bromide or iodobenzene gave the corresponding symmetrical ketones but, in the presence of olefins, did not give addition products of olefins.[14] These results show interesting differencies of
reactivities of organotransition metal complexes towards unsaturated
compounds.

In this paper, we wish to report our attempts to trap the acyl
groups in acylnickel carbonyl complexes derived from nickel carbonyl and several organic halides, especially alkenyl halides, by
the third components such as olefins, alkynes, imines, olefin oxides and organomercuric halides. (The study of the reactions described below is now continuing and any effort has not yet been made
to optimize the yields of the acylation products).

The Reaction of Nickel Carbonyl with Alkenyl Halides in the Presence of Olefins or Alkynes.

In general, acylnickel complexes are reactive towards olefins
and especially the alkenoylnickel complexes derived from nickel carbonyl and alkenyl halides in polar media would be a promising reagents for alkenoylation of various nucleophiles.

When DMF was used as a solvent, nickel carbonyl was found to
react easily with trans-β-bromostyrene, and the following products
were isolated; cinnamic acid (39%), cinnamic anhydride (7%) and N,N-
dimethylcinnamamide (26%). When the same reaction was carried out
in the presence of acrylonitrile, 3-(trans-cinnamoyl)-propionitrile
was obtained in moderate yields (28% in DMF, 41% in acetonitrile).
[15,16] (Equation 7) The results using n-propenyl- and iso-propenyl
bromide are summarized in Table 1.

$$PhCH=CHBr \quad + \quad Ni(CO)_4 \quad + \quad CH_2=CHCN \xrightarrow[\text{DMF or CH}_3\text{CN}]{60\sim65°C} \tag{7}$$

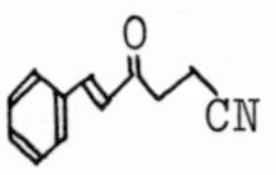

In connection with the above reaction, the reaction of potassium
hexacyanodinickelate was carried out and the styrylation products
of olefins were obtained.[15] (Equation 8)

$$PhCH=CHBr \quad + \quad K_4[Ni_2(CN)_6] \quad + \quad CH_2=CHY \xrightarrow[\text{aq.acetone}]{} \tag{8}$$

In the case of Y = CN; 56.5% yield.

Table 1

Formation of α,β-Unsaturated Ketones by the Reaction of Alkenyl Halides with Nickel Carbonyl in the Presence of Olefins. (Solvent; Acetonitrile, Temperature; 60-65°C)

Halide	Olefin	Product	Yield,%*
$CH_3CH=CHBr$	$CH_2=CHCO_2Et$	$CH_3CH=CHC-CH_2CH_2CO_2Et$ ‖ O	33
CH_3 | $CH_2=C-Br$	$CH_2=CHCN$	CH_3 | $CH_2=C-C-CH_2CH_2CN$ ‖ O	14
CH_3 | $CH_2=C-Br$	$CH_2=CHCO_2Et$	CH_3 | $CH_2=C-C-CH_2CH_2CO_2Et$ ‖ O	35
$PhCH=CHBr$	$CH_2=CHCN$	$PhCH=CHC-CH_2CH_2CN$ ‖ O	41

* Theoretical yields based on the alkenyl halides.

    The alkenoylnickel complex was found to be also reactive towards alkynes, forming dimers of γ-butenolactones.[17] For example, from nickel carbonyl and β-bromostyrene in the presence of 3-hexyne in DMF, the dimer of 2,3-diethyl-4-styryl-γ-but-2-enolactone was obtained in 65% yield. (Equation 9)

$$PhCH=CHBr + Ni(CO)_4 + EtC\equiv CEt \xrightarrow[DMF]{} \begin{bmatrix} Et\quad Et \\ C=C \\ O=C \quad C \\ \diagdown O \diagup \\ CH \\ \| \\ CH \\ | \\ Ph \end{bmatrix}_2 \qquad (9)$$

The Reaction of Aryl Iodide with Nickel Carbonyl
in the Presence of N-Benzylidene Alkylamine.

Although the carbonylation and ring closure of Schiff's bases

or aromatic ketoximes using dicobalt octacarbonyl had been established as a synthetic reaction of phthalimidine derivatives, the reaction of alkyl- or acylmetal carbonyl derivatives with imines which
contain carbon-nitrogen double bonds has not yet been reported.

When N-benzylidene alkylamine was used as an acyl trapping a-
gent, two types of novel and synthetically useful reactions occurred,
i.e., the acylation and cyclization of imines to indolinone derivative (in DMF) and the coupling reaction of two of acyl groups to an
intervening double bond (in benzene).[18] For example, iodobenzene reacted with nickel carbonyl in DMF at 75°C in the presence of N-benzylidene methylamine to give 1-methyl-2-phenylindolin-3-one (1) and
N,N'-dibenzoyl-N,N'-dimethyl-1,2-diphenylethylenediamine (2) in 28
and 13% yield, respectively. (Equation 10)

$$PhI + Ni(CO)_4 + PhCH=NCH_3 \xrightarrow{DMF}$$

1

2

$$(10)$$

It was suggested by the reaction using p-methyliodobenzene instead
of iodobenzene (product 4 in 60% yield) that, as for the formation
of 1, the attack of benzoyl group to the carbon site of the imine
double bond occurs first, followed by cyclization to give the indolinone derivative. (Equation 11)

$$p-CH_3C_6H_4I + PhCH=NCH_3 + Ni(CO)_4 \xrightarrow{DMF}$$

4

$$(11)$$

This reaction showed a remarkable solvent effect. When benzene was
used as a solvent instead of DMF, the main product was not the indolinone derivative but the coupling product of two of the benzoyl
groups to an intervening imine double bond, N-alkyl-N-(α-phenylphen-
acyl)benzamide (yields, R = CH_3; 53%, R = C_2H_5; 59%). (Equation 12)

$$PhI + PhCH=NR + Ni(CO)_4 \xrightarrow{C_6H_6} PhCN-CH-CPh$$

3

$$(12)$$

Various types of coupling reactions of alkyl, allyl or acyl
groups using transition metal complexes have been reported. More

recently synthetic interest on coupling reaction has been focused
on the reactions, which are formally considered to be coupling reactions of alkyl, acyl or alkoxy groups intervening the unsaturated
bond of the third component, such as carbon monoxide, olefins, alkynes and dienes. The above reactions present a new type of coupling reactions of acyl groups, in which the carbon-nitrogen double
bonds are the third components of the coupling.

To compare the reactivity of the corresponding anionic acylnickel complexes, the reaction of lithium p-toluoylnickel carbonylate with N-benzylidene methylamine was carried out at 50-60°C, and
the isolated product in a quantitative yield was found to be unexpected 4,4'-dimethylbenzylic acid.[19] (Equation 13)

$$\text{Li}[p\text{-}CH_3C_6H_4\underset{O}{C}Ni(CO)_3] \xrightarrow[\text{Et}_2O]{\text{PhCH=NR}} p\text{-}CH_3C_6H_4\text{-}\underset{\underset{COOH}{|}}{\overset{\overset{OH}{|}}{C}}\text{-}C_6H_4CH_3\text{-}p \qquad (13)$$

The similar reaction using phenyllithium in the place of p-tolyl-
lithium was also found to give benzylic acid in 26% yield. Although
N-benzylidene methylamine or its fragment was not incorporated into
the product, it was very interesting that in the absence of the i-
mine the reaction did not afford α-oxycarboxylic acid but gave only
the normal products such as ketones and/or an acyloin.

The Reaction of Organic Halides with Nickel Carbonyl in the Presence of Olefin Oxides.

Olefin oxides have been shown to have unique reactivities towards organotransition metal complexes and the following four types
of reactions have been reported: i.e. 1) carbonylation of olefin
oxides using dicobalt octacarbonyl and carbon monoxide; 2) isomerization of olefin oxides into ketones by the action of metal carbonyls; 3) dimerization of olefin oxides; 4) insertion of olefin
oxides into carbon-transition metal bond. The reactions of type 1
and 2 have been extensively studied, whereas the reactions of type
3 and 4 have not yet been clarified. Only one report concerning
with the reaction of type 3 dealt with the formation of α,β-diphe-
nyl γ-butyrolactone from styrene oxide by the action of lithium
aroylnickel carbonylate or dibenzyliron tetracarbonyl.[20] There are
two examples of type 4 reactions (with exception of the reactions
of lithium dialkylcuprate), i.e., the formation of 2-methyl-4-phenyl-
5-hydroxypentene-1 from the reaction of π-methallylnickel bromide
with styrene oxide[21] and, one ambiguous example, the formation of
p-tolylbenzyl ketone from the reaction between lithium p-toluylnic-
kel carbonylate and styrene oxide.[20]

Recently we found that the acylnickel complexes derived from nickel carbonyl and organic halides reacted with olefin oxides to give esters of β-halocalcohols, the products from the reaction of type 4.[22] For example, β-bromostyrene reacted with nickel carbonyl in DMF at 55°C in the presence of cyclohexene oxide to give β-bromo-cyclohexylcinnamate in a yield of 78%. (Equation 14,15)

$$ (14) $$

$$ (15) $$

In the case of propylene oxide, the isomeric mixture of the acylated products, β-bromo-n-propyl- and β-bromo-isopropylcinnamate, was obtained in a ratio of 1:2. (Equation 16)

$$ (16) $$

When iodobenzene was used instead of β-bromostyrene, the formation of β-iodo-isopropylbenzoate was predominant (n-propyl : isopropyl = 1 : 9). (Equation 17)

$$ PhI + CH_3CH-CH_2 \xrightarrow[DMF]{Ni(CO)_4} PhCOCH_2CHCH_3 + PhCOCHCH_2I $$

5%   51%   (17)

This reaction is the first case of the insertion of olefin oxide into acyl-nickel bond. In the hiterto known acylation of olefins, alkynes, and imines using acylnickel carbonyl halides, the acylated products did not bear halogen atom originated from organic halide, but, in the above acylation reaction of olefin oxides, the halogenated products were obtained. The reason is not yet clear. In contrast, when phenylmercuric chloride or bromide was used instead of organic halide, the acylated products were the esters of vinylalcohol and saturated alcohol and the halogenated product was not isolated. (Equation 18)

$$\text{PhHgCl} \quad + \quad \text{PhCH-CH}_2 \quad \xrightarrow[\text{DMF}]{\text{Ni(CO)}_4} \quad \text{PhCOCH=CHPh} \quad + \quad \text{PhCOCH}_2\text{CH}_2\text{Ph} \quad + \quad \text{PhCPh}$$

$$\underset{\text{O}}{} \qquad\qquad\qquad \underset{\text{O}}{20\%} \qquad\qquad \underset{\text{O}}{7.7\%} \qquad\qquad \underset{\text{O}}{72\%}$$

(18)

The Reaction of Nickel Carbonyl with Organic Halides in the Presence of Organomercuric Halides.

The irradiative carbonylation of diarylmercury with carbon monoxide is catalyzed by dicobalt octacarbonyl to give various symmetrical diaryl ketones.[23] This catalytic reaction is excellent for the synthesis of symmetrical diaryl ketones but does not appear to be applicable to the synthesis of dialkyl ketones from dialkylmercury compounds, most likely because of the well known photolability of the alkylmercury bond. (Equation 19)

$$\text{R-Hg-R} \quad + \quad \text{CO} \quad \xrightarrow[\text{THF}]{h\nu \ , \ \text{Co}_2(\text{CO})_8} \quad \text{R-}\underset{\text{O}}{\text{C}}\text{-R} \quad + \quad \text{Hg}$$

(19)

However, in the reaction of organomercuric halides with nickel carbonyl in DMF, the products were the symmetrical ketones quite free from any isomerized ketones.[24] (Equation 20)

$$2 \ \text{RHgX} \quad + \quad \text{Ni(CO)}_4 \quad \xrightarrow{\text{DMF}} \quad \text{R-}\underset{\text{O}}{\text{C}}\text{-R} \quad + \quad 2 \ \text{Hg} \quad + \quad \text{NiX}_2 + 3 \ \text{CO}$$

(20)

$$\text{R = Ph; } 97\%, \ \text{p-CH}_3\text{C}_6\text{H}_4; \ 100\%, \ \text{n-Bu; } 56\%, \ \text{iso-C}_5\text{H}_{11}; \ 59\%,$$

$$\text{n-C}_6\text{H}_{13}; \ 64\%.$$

The reaction seems proceed via the paths shown below (Scheme 2). If the reaction of 7 (R = Ph, X = Cl) with PhHgCl was slow enough, it is possible to trap the benzoyl moiety by a suitable reagent. The attempts using acrylonitrile or 3-hexyne were unsuccessful. Therefore, it was suggested that the reaction of 7 (R = Ph, X = Cl) with PhHgCl is too fast to admit the attack of the unsaturated compounds. Furthermore the reaction of organomercuric halides with nickel carbonyl in non-polar media did not give ketones. Hence, in non-polar media, organomercuric halide was considered to be a promising reagent for capture of aroyl or acyl groups in aroyl or acyl-nickel complexes, yielding unsymmetrical ketones. Indeed, the reaction of iodobenzene with nickel carbonyl in benzene in the pre-

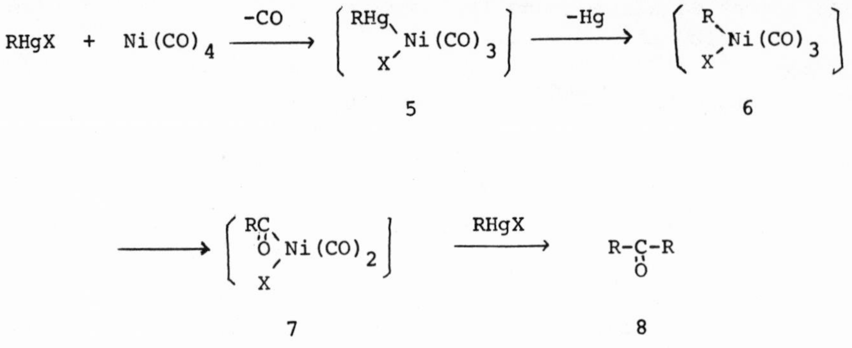

$$RHgX + Ni(CO)_4 \xrightarrow{-CO} \left[\begin{array}{c} RHg \\ X \end{array} Ni(CO)_3 \right] \xrightarrow{-Hg} \left[\begin{array}{c} R \\ X \end{array} Ni(CO)_3 \right]$$

$$\phantom{RHgX + Ni(CO)_4 \xrightarrow{-CO}} 5 6$$

$$\longrightarrow \left[\begin{array}{c} RC \\ \parallel \\ O \\ X \end{array} Ni(CO)_2 \right] \xrightarrow{RHgX} \underset{O}{R-\overset{\parallel}{C}-R}$$

$$ 7 8$$

Scheme 2

sence of p-tolylmercuric chloride led to the formation of p-tolyl-phenyl ketone in a yield of 88% as a result of rapid capture of benzoyl group in benzoylnickel complexes by p-tolylmercuric chloride.[25] (Equation 21, Table 2)

$$PhI + Ni(CO)_4 \xrightarrow{C_6H_6} \underset{O\ I}{Ph\overset{\parallel}{C}-Ni(CO)_2} \xrightarrow{p-CH_3C_6H_4HgCl} \underset{O}{Ph-\overset{\parallel}{C}-C_6H_4-CH_3-p}$$

$$\text{(21)}$$

Table 2

Formation of Unsymmetrical Ketones.

$$R-X + ArHgX + Ni(CO)_4 \longrightarrow \underset{O}{R-\overset{\parallel}{C}-Ar}$$

R-X	ArHgX	Solvent	Product	Yield,%
PhI	$p-CH_3C_6H_4HgCl$	C_6H_6	$Ph-\overset{\parallel}{\underset{O}{C}}-C_6H_4-CH_3-p$	88
PhI	$p-ClC_6H_4HgCl$	C_6H_6	$Ph-\overset{\parallel}{\underset{O}{C}}-C_6H_4-Cl-p$	90
$C_6H_5CH_2Br$	$p-CH_3-C_6H_4HgBr$	DMF	$PhCH_2-\overset{\parallel}{\underset{O}{C}}-C_6H_4CH_3-p$	24

REFERENCES

1. M. Ryang, Organometal. Chem. Rev., A 5, 67 (1970).
2. M. Ryang and S. Tsutsumi, Synthesis, 55 (1971).
3. G. P. Chiusoli, Accounts Chem. Res., 6, 422 (1973).
4. M. F. Semmelhack, Org. React., Vol. 19, (1972) 115.
5. I. Rhee, M. Ryang and S. Tsutsumi, unpublished results.
6. E. J. Corey and L. S. Hegedus, J. Amer. Chem. Soc., 91, 1233 (1967).
7. S. Fukuoka, M. Ryang, and S. Tsutsumi, J. Org. Chem., 36, 2721 (1971).
8. E. J. Corey and G. H. Posner, J. Amer. Chem. Soc., 89, 3911 (1967); ibid, 90, 5615 (1968).
9. M. Tamura and J. Kochi, J. Amer. Chem. Soc., 93, 1487 (1971).
10. M. F. Semmelhack, P. M. Helquist, and J. D. Gorzynski, ibid, 94, 9234 (1972).
11. I. Rhee, M. Ryang, and S. Tsutsumi, J. Organometal. Chem., 9, 361 (1967).
12. E. Yoshisato, M. Ryang, and S. Tsutsumi, J. Org. Chem., 34, 1500 (1968).
13. M. Ryang, N. Mizuta, N. Sonoda ,and S. Tsutsumi, unpublished results.
14. I. Rhee, M. Ryang, N. Sonoda, and S. Tsutsumi, unpublished results.
15. I. Hashimoto, M. Ryang, and S. Tsutsumi, Tetrahedron Lett., 4567 (1970).
16. M. Ryang, M. Takahashi, N. Sonoda, and S. Tsutsumi, unpublished results.
17. M. Ryang, Y. Sawa, S. M. Somasundaram, S. Murai, and S. Tsutsumi, J. Organometal. Chem., 46 (1972).
18. M. Ryang, Y. Toyoda, S. Murai, N. Sonoda, and S. Tsutsumi, J. Org. Chem., 38, 62 (1973).
19. M. Ryang, Y. Toyoda, N. Sonoda, and S. Tsutsumi, unpublished results.
20. S. Fukuoka, M. Ryang, and S. Tsutsumi, J. Org. Chem., 3184 (1970).
21. E. J. Corey and M. F. Semmelhack, J. Amer. Chem. Soc., 89, 2755 (1967).
22. M. Ryang, H. Hasegawa, N. Sonoda, and S. Tsutsumi, unpublished results.
23. D. Seyferth and R. J. Spohn, J. Amer. Chem. Soc., 91, 6192 (1969).
24. Y. Hirota, M. Ryang, and S. Tsutsumi, Tetrahedron Lett., 1531 (1971).
25. M. Ryang, T. Watanabe, N. Sonoda, and S. Tsutsumi, unpublished results.

SELECTIVE AND ASYMMETRIC REDUCTIONS OF CARBONYL COMPOUNDS USING

HYDROSILYLATION CATALYZED BY RHODIUM(I) COMPLEXES

Iwao Ojima

Sagami Chemical Research Center, Nishi-Ohnuma,
Sagamihara, Kanagawa 229, Japan

Metal hydrides such as $LiAlH_4$, $NaBH_4$, $NaBH_3CN$, $LiAl(O^tBu)_3H$ etc., have been widely used for the reduction of various functionalities. Hydrogenation catalyzed by transition metals and metal complexes has been also extensively studied. Besides these two famous reduction methods, we have developed in the last two years a new powerful method for the reduction of carbon-hetero atom double bonds using catalytic hydrosilylation.

Although organosilicon hydrides are very stable substances and virtually lack reducing ability, they can add to unsaturated bonds if a proper catalyst is added to the system. Since the resultant silicon-hetero atom bonds can be easily cleaved by the action of acids or bases, organosilicon hydrides can be used as reducing agents like usual metal hydrides, if the hydrosilylation of the compounds containing such a double bonds is effectively achieved. We found that a rhodium(I) complex such as $(Ph_3P)_3RhCl$ is extremely effective for the hydrosilylation of these compounds. This finding enabled us to establish a new powerful reduction method.

Selective Reduction of α,β-Unsaturated Carbonyl Compounds[1,2]

Calas,[3] Frainnet,[4] Petrov,[5] and their co-workers studied the catalytic hydrosilylation of carbonyl compounds using $ZnCl_2$, Ni and H_2PtCl_6 as a catalyst. However, the required conditions for these reactions were rather drastic and the formation of side products or the isomerization of both raw materials and products were often observed. Moreover, these catalysts have been used within the limit of monohydrosilanes. This limitation is mainly due to the fact that the disproportionation of polyhydrosilane is also catalyzed by these substances.[6] Thus, we cannot recommend to employ these catalysts.

We found that tris(triphenylphosphine)chlororhodium catalyzes the reaction under much milder conditions without any side reactions.[7,8]

The hydrosilylation of α,β-unsaturated carbonyl compounds using mono-hydrosilanes was found to proceed in a manner of 1,4-addition, while di-hydrosilanes undergo 1,2-addition to carbonyl functionalities specifically with little exception. Since either the resulting silyl ether or silyl enol ether can be easily hydrolyzed to a saturated ketone or an α,β-unsaturated alcohol, these reactions may furnish a convenient method for the selective reduction of α,β-unsaturated carbonyl compounds.[9] Thus, the efficiency and the selectivity of the reduction was investigated for a variety of α,β-unsaturated carbonyl compounds using various mono- and di-hydrosilanes. The results are summarized in Table 1.

Table 1

Selective Reduction of α,β-Unsaturated Ketones and Aldehydes
Using Hydrosilane-Rhodium(I) Complex Combinations

α,β-Unsaturated Carbonyl Compound	Hydrosilane	Amount of Cat. (mol %)	Conditions	Ratio of III/IV	Yield (%)
(structure)	Et_3SiH	0.1	50°, 2 hr	100/0 [a]	96
	Ph_2SiH_2	0.1	r.t., 30 min	0/100 [a]	97
(structure)	$EtMe_2SiH$	0.1	r.t., 3 hr	92/8 [a]	94
	Ph_2SiH_2	0.1	r.t., 30 min	0/100 [a]	98
(structure)	Et_3SiH	0.1	r.t., 1 hr	100/0 [a]	97
	Ph_2SiH_2	0.1	ice-cooled 30 min	0/100 [a]	97
(structure)	Et_3SiH	0.1	50°, 1 hr	99/1 [b]	95
	$PhSiH_3$	0.1	r.t., 1 hr	1/99 [b]	95
(structure)	$EtMe_2SiH$	0.1	r.t., 2 hr	70/30 [a]	97
	Ph_2SiH_2	0.1	r.t., 20 min	0/100 [a]	98
(structure)	$EtMe_2SiH$	0.1	45°, 4 hr	98/2 [b]	90
	Ph_2SiH_2	0.1	ice-cooled 30 min	1/99 [b]	97
(structure)	Et_3SiH	0.1	80°, 25 hr	94/6 [b]	95
	Et_2SiH_2	0.1	ice-cooled 30 min	3/97 [b]	98

[a] Products ratio was determined by nmr. [b] Products ratio was determined by glpc.

The α,β-unsaturated carbonyl compounds involving an isolated double bond in the same molecule were also chosen for the substrates. A selective hydrogenation of the double bond conjugated to carbonyl group can be easily achieved by the use of the rhodium (I) complex catalyzed hydrosilylation without any isomerization of the remaining double bond.

$$R^1 R^2 C=CHCR^3 \;(=O) \xrightarrow{[Rh]} \begin{cases} \xrightarrow{R_3SiH} R^1 R^2 CHCH=CR^3 (OSiR_3) \quad \text{I} \xrightarrow{\text{Hydrolysis}} R^1 R^2 CHCH_2 CR^3 (=O) \quad \text{III} \\[2mm] \xrightarrow{R_2SiH_2} R^1 R^2 C=CHCHR^3 (OSiR_2H) \quad \text{II} \xrightarrow{\text{Hydrolysis}} R^1 R^2 C=CHCHR^3 (OH) \quad \text{IV} \end{cases}$$

It was also demonstrated that the dihydrosilane-rhodium(I) complex combination in the reduction of α-ionone, pulegone, piperitone etc., displayed an exceedingly higher selectivity than that of metal hydrides such as lithium aluminum hydride and sodium borohydride. The comparison of the selectivities obtained by this method with those attained by usual metal hydrides are typically shown in Table 2.

Table 2
Reductions of Pulegone and Piperitone with Metal Hydrides [*]

Reducing Agent	Pulegone: A (=O)	B (OH)	C (=O)	D (OH)	Piperitone: E (=O)	F (OH)	G (OH)	H (=O)
Et_2SiH_2-[Rh]		100				100		
Ph_2SiH_2-[Rh]		100				100		
$LiAlH_4$	49	51				100		
$LiAl(O^tBu)_3H$	39	43	18		29	24		47
$LiBH_4$	7	93			23	48.5	28.5	
$NaBH_4$	18	36		46	60	22.5	18.5	

[*] Reductions were performed using 1.0–1.3 equivalent of metal hydride at room temperature for 24 hr.

Stereoselective Reduction of Terpene Ketones[10]

It has been known that metal hydrides can reduce ketones to alcohols stereoselectively and that each reducing agent has a characteristic stereoselectivity. We found that a stereoselective reduction of terpene ketones using hydrosilylation exhibited major differences in stereochemical selectivity from other reducing agents. Results of the reduction via hydrosilylation of camphor and menthone, representative terpene ketones, are listed in Table 3. Included also in the Table are the reported selectivities achieved by other known reducing agents.

Table 3
Stereoselectivities in the Reduction of Terpene Ketones

Reducing Agent	Camphor iso-Borneol/Borneol	Menthone Neomenthol/Menthol
PhSiH$_3$	90 / 10	90 / 10
Et$_2$SiH$_2$	91 / 9	83 / 17
PhMeSiH$_2$	75 / 25	87 / 13
Ph$_2$SiH$_2$	73 / 27	85 / 15
Et$_3$SiH	30 / 70	64 / 36
PhMe$_2$SiH	-------	0 /100
LiAlH$_4$	91 / 9	29 / 71
Al(OPr-i)$_3$		30 / 70
NaBH$_4$		51 / 49
B$_2$H$_6$	52 / 48	
Disiamylborane	65 / 35	
Dicyclohexylborane	93 / 7	
Diisopinocamphenylborane	100 / 0	
Ph$_2$SiH$_2$		52 / 48

As shown in the Table, the bulkiness of silanes exerts large influence in the stereochemical course of the reduction and a bulky hydrosilane favors the production of the more stable alcohols. This trend is quite unusual since it has been shown by Brown and Varma[11] that, in the reduction of monocyclic and bicyclic ketones, the bulkier hydroborane produces the larger percentage of less stable of the two possible alcohols. The peculiarity suggests that the transition state for the reaction cannot be accomodated with a simple four-centered type. However, a possible explanation for the observed trend can be advanced by taking into account the intermediacy of the organorhodium complex III as shown on the following page.

As the steric course of the reaction would be governed only by the size of the silyl moiety, it can be said that the bulkier the substituents on silicon, the more pronounced may be the formation of the complex IIIa which is a precursor of the more stable alcohol.

Asymmetric Reduction of Prochiral Ketones[12,13]

Asymmetric reduction of unsaturated compounds by the hydrogenation using homogeneous catalysts currently gathers much interest,
and recently, Scorrano and co-workers[14] reported the asymmetric hydrogenation of ketones by the use of a cationic rhodium(I) catalyst
with a chiral phosphine ligand even though in a low optical yield.
Asymmetric reduction of ketones via hydrosilylation was also achieved
using a platinum complex and a cationic rhodium complex by Yamamoto,
Kumada and co-workers,[15] and rhodium(I) complexes by Ojima,[12,13] Kagan[16] and co-workers with chiral phosphine ligands as a catalyst with
better optical yields.

Although discussions have been made on the mechanisms of asymmetric induction in the homogeneous asymmetric hydrogenation and hydroformylation, clear explanations are as yet unestablished. This
may be partly due to the failure of separating a steric effect or an
electronic effect from the many controlling factors in the asymmetric
induction. In our systems, however, only a steric effect plays a
key role in the induction of asymmetry, and an electronic one can be
excluded. Thus, we have succeeded in the effective asymmetric reduction of simple ketones via hydrosilylation and proposing the mechanism of the induction of asymmetry, which may be the first reasonable stereochemical interpretation on the asymmetric reductions
catalyzed by homogeneous transition metal complex.

The results of the asymmetric reduction of ketones using various
hydrosilanes are summarized in Table 4.

Table 4

Asymmetric Reduction of Prochiral Ketones via Hydrosilylation
Catalyzed by $[(PhCH_2)MePhP]_2Rh(S)Cl$ [S = Solvent]

Ketone	Hydrosilane	Yield[b] (%)	Carbinol[a] $[\alpha]_D^{21-28}$	Configu- ration	Optical[c] Yield (%)
Ligand = (−)-(S)-BMPP			(Optical purity[f] 62 %)		
PhCOCH$_3$	PhMe$_2$SiH	92	+ 14.32[d]	R	44
PhCOC$_2$H$_5$	PhMe$_2$SiH	96	+ 8.74	R	50
PhCOC$_3$H$_7$-i	PhMe$_2$SiH	95	+ 16.68[e]	R	56
PhCOC$_2$H$_5$	Et$_2$SiH$_2$	98	− 2.90	S	17
PhCOC$_3$H$_7$-i	Et$_2$SiH$_2$	98	− 6.73[e]	S	23
n−BuCOCH$_3$	Et$_2$SiH$_2$	93	+ 2.24	S	30
t−BuCOCH$_3$	Et$_2$SiH$_2$	95	+ 1.61	S	39
Ligand = (+)-(R)-BMPP			(Optical purity[f] 77 %)		
PhCOCH$_3$	Et$_2$SiH$_2$	97	+ 6.44[d]	R	16
PhCOC$_2$H$_5$	Ph$_2$SiH$_2$	98	+ 9.04	R	42
PhCOBu−t	EtMe$_2$SiH	97	+ 15.63[e]	R	56
PhCOBu−t	PhMe$_2$SiH	92	− 14.91[e]	S	54

a All optical rotations are for the neat liquid unless otherwise noted.
b Yield of a silyl ether (glpc analysis). The hydrolysis was accomplished
in almost quantitative yield.   c Optical yield is calculated from the
specific rotation of the pure enantiomer which reported in the literature,
and calibrated for the optical purity of the chiral phosphine employed.
d specific rotation in dichloromethane.   e specific rotation in ether.
f The optical purity was determined by quarternization using n-propyl
bromide.

Asymmetric Reduction of α-Keto Esters via Catalytic Hydrosilylation Using a Rhodium(I) Complex with Chiral Phosphine Ligands[17]

Asymmetric syntheses of α-hydroxy carboxylic acids and their derivatives have been extensively studied.[18] The addition of Grignard reagent to a chiral α-keto ester, e.g., menthyl pyruvate, or a reduction of a chiral α-keto ester, e.g., bornyl phenylglyoxylate, is a well known reaction[18] which proceeds in accordance with the Prelog's generalization.

However, no catalytic asymmetric hydrogenation of α-keto esters have been reported so far. This may be due to the fact that homogeneous catalysts such as $(Ph_3P)_3RhCl$ lack reactivity toward carbonyl compounds. These difficulties can be overcome since the hydrosilylation of α-keto esters is found to proceed in exceedingly high yield (87-100%).

The asymmetric catalytic reduction of α-keto esters has been successfully performed for the first time via hydrosilylation catalyzed by a rhodium(I) complex with chiral phosphine ligands. We used (+)-(R)-benzylmethylphenylphosphine [(+)-BMPP] or (+)-2,3-0-isopropylidene-2,3-dihydroxy-1,4-bis(diphenylphosphino)butane [(+)-DIOP] as a chiral ligand.

$$R^1COCOOR^2 \;+\; \equiv SiH \;\xrightarrow{[Rh]^*}\; \underset{OSi\equiv}{R^{1*}CHCOOR^2} \;\xrightarrow{TsOH-MeOH}\; \underset{OH}{R^{1*}CHCOOR^2}$$

$$CH_3COCOO\text{-}\bigcirc\!\!* \;+\; R_2SiH_2 \;\xrightarrow{[Rh]^*}\; \underset{OSiHR_2}{CH_3\overset{*}{C}HCOO\text{-}\bigcirc\!\!*}$$

$$\downarrow KOH-MeOH$$

$$\underset{OH}{CH_3\overset{*}{C}HCOOPr} \;\xleftarrow{PrOH/H^+}\; \underset{OH}{CH_3\overset{*}{C}HCOOH} \;\xleftarrow{H^+}\; \xrightarrow{Et_2O}\; HO\text{-}\bigcirc\!\!*$$

Results are summarized in Table 5. The optical yields realized for the asymmetric reduction of propyl pyruvate by this method are much higher than those attained by other methods, and especially, the optical yield obtained by the use of α-naphthylphenylsilane (81.5% e.e.) is the highest one ever known.

Table 5

Catalytic Asymmetric Reduction of Propyl Pyruvate,
Ethyl Phenylglyoxylate and Menthyl Pyruvate

α-Keto Ester	Hydrosilane	Chiral Ligand	Product	$[\alpha]_D^{20}$ [a]	Optical Yield (% e.e.) [b]
CH₃COCOOPr	Et₂SiH₂	(+)-BMPP	CH₃C̊HCOOPr OH	+ 3.67 [c] (R)	30.3
	PhMeSiH₂	(+)-BMPP		+ 6.05 [c] (R)	50.0
	Ph₂SiH₂	(+)-BMPP		+ 7.30 [c] (R)	60.3
	Ph₂SiH₂	(+)-DIOP		- 9.11 (S)	75.3
	α-NpPhSiH₂	(+)-DIOP		- 9.86 (S)	81.5
PhCOCOOEt	Et₂SiH₂	(+)-BMPP	PhC̊HCOOEt OH	+ 8.20 [c] (R)	6.4
	Ph₂SiH₂	(+)-BMPP		-13.24 [c] (S)	10.3
	Ph₂SiH₂	(+)-DIOP		+ 1.73 (R)	1.4
CH₃COCOO–	Et₂SiH₂	(+)-BMPP	CH₃C̊HCOOPr OH	+ 1.99 (R)	16.4
	Et₂SiH₂	(+)-DIOP		- 5.12 (S)	42.3
	Ph₂SiH₂	(+)-DIOP		- 7.55 (S)	62.4
	Ph₂SiH₂	(-)-DIOP		+ 7.96 (R)	65.8

a Optical rotations are for the neat liquid in the case of propyl lactate, and as for ethyl mandelate those are measured in chloroform. b Optical yields were calculated from the specific rotation of the pure enantiomer which is reported in the literature and confirmed on the basis of nmr spectra using a shift reagent, Eu(TFC)₃ [TFC: 3-trifluoromethylhydroxy-methylene-d-camphor]. c Optical rotation is calibrated by the purity of (+)-BMPP (77.7%).

In conclusion, it is demonstrated that the hydrosilylation catalyzed by a rhodium(I) complex with phosphine ligands furnishes a new powerful method for the reduction of carbonyl compounds. We have extended this method for the reduction of Schiff bases[19], isocyanates[20], and carbodiimides[21] and obtained good results. This method also provides an effective route to catalytic asymmetric reduction of these compounds.

ACKNOWLEDGEMENT

The author is grateful to Professor Yoichiro Nagai for his helpful discussions and encouragements. He is also indebted to Mr. Tetsuo Kogure and Mr. Mitsuru Nihonyanagi for their collaboration.

REFERENCES

1. I. Ojima, T. Kogure and Y. Nagai, Tetrahedron Lett., 5035 (1972).
2. I. Ojima, T. Kogure and Y. Nagai, to be published; T. Kogure, I Ojima and Y. Nagai, 21st Symposium on Organometal. Chem. (Japan), Abstr. No. 114 (1973 October).
3. R. Calas, E. Frainnet and J. Bonastre, Compt. rend., 251, 2987 (1960).
4. E. Frainnet, Pure. Appl. Chem., 19, 489 (1965).
5. S. I. Sadykh-Zade and A. D. Petrov, Zhur. Obshch. Khim., 29, 3194 (1959).
6. H. Gilman and D. H. Miles, J. Org. Chem., 23, 326 (1958).
7. I. Ojima, M. Nihonyanagi and Y. Nagai, JCS Chem. Comm., 938 (1972).
8. I. Ojima, T. Kogure, M. Nihonyanagi and Y. Nagai, Bull. Chem. Soc. Japan, 35, 3506 (1972).
9. E. Yoshii and M. Yamazaki, Chem. Pharm. Bull., 16, 1158 (1968).
10. I. Ojima, M. Nihonyanagi and Y. Nagai, Bull. Chem. Soc. Japan, 45, 3722 (1972).
11. H. C. Brown and V. Varma, J. Amer. Chem. Soc., 88, 2871 (1966).
12. I. Ojima, T. Kogure and Y. Nagai, Chem. Lett., 541 (1973).
13. I. Ojima and Y. Nagai, Chem. Lett., 223 (1974).
14. P. Bonvicini, A. Levi, G. Modena and G. Scorrano, JCS Chem. Comm., 1188 (1972).
15. K. Yamamoto, T. Hayashi and M. Kumada, J. Organometal. Chem., 46, C 65 (1972); K. Yamamoto, T. Hayashi and M. Kumada, ibid., 54, C 45 (1973).
16. J-C. Poulin, W. Dumont, T-P. Dang and H. B. Kagan, Compt. rend., 277, C 41 (1973); Idem, J. Amer. Chem. Soc., 95, 8295 (1973).
17. I. Ojima, T. Kogure and Y. Nagai, Tetrahedron Lett., 1889 (1974).
18. J. D. Morrison and H. S. Mosher, "Asymmetric Organic Reactions", Prentice-Hall, Inc., Englewood Cliffs, New Jersey, 1971.
19. I. Ojima, T. Kogure and Y. Nagai, Tetrahedron Lett., 2475 (1973).
20. I. Ojima, S. Inaba and Y. Nagai, ibid., 4363 (1973).
21. I. Ojima, S. Inaba and Y. Nagai, J. Organometal. Chem., in press.

DISCUSSION

DR. BERGMAN: You mentioned some spectacular yields in some of these reductions. Can you tell us if these are estimated yields, or isolated ones?

DR. OJIMA: Isolated yields.

DR. WHITESIDES: Are these reductions to any extent reversible? If you treat an alcohol with the catalyst system with the presence of a ketone, is there any oxidation of the alcohol? Like a Merwein-Pondorff process, sort of.

DR. OJIMA: I have not tried it. I think that the exchange does not occur.

DR. WHITESIDES: I asked because that last result of yours in which you get the same enantiomeric series with the menthyl ester regardless of whether you work with the (+)DIOP or (-)DIOP. That is a fairly remarkable result.

DR. OJIMA: No, you are mistaken - the sign of the product does change between (+) and (-)DIOP.

DR. TSUJI: I have two questions. You use dihydro and monohydro-alkylsilanes. If you use the very easily available trichlorosilane, what happens? Second, it has been known for a long time that a rhodium catalyst catalyzes hydrosilation of simple olefins, but still you said that ketones are selectively reduced in the presence of olefins. Does this mean that the reaction is very much different, because it is so much faster?

DR. OJIMA: Compared with the reaction of the carbonyl group, olefins react very slowly with hydrosilanes. And the first question, if we use trichlorosilane the oxidative adduct of the rhodium complex is very stable and does not react with carbonyl group.

DR. TSUJI: Do you have any idea of the difference between the hydrogens of the monohydrosilane and dihydrosilane? Is there a great difference in reactivity?

DR. OJIMA: Only the difference in reaction with α, β unsaturated ketones that I showed in my lecture.

CARBONYLATION REACTIONS OF ORGANIC HALIDES BY TETRAKIS(TRIPHENYL-PHOSPHINE)NICKEL (O) AND PALLADIUM (O) CARBONYL COMPLEXES UNDER MILD CONDITIONS

Masanobu Hidai[*] and Yasuzo Uchida
Ikuei Ogata
Department of Industrial Chemistry, University of Tokyo,
Hongo, Tokyo, Japan, and National Chemical Laboratory for
Industry, 1-Chome, Honmachi, Shibuya-ku, Tokyo, Japan

Synthesis of carboxylic acids and esters by carbonylation reactions at atmospheric pressure using transition metal catalysts is currently receiving considerable attention and has recently been reviewed by Cassar, Chiusoli, and Guerrieri.[1] Nickel-thiourea complexes,[2] nickel tetracarbonyl,[3] cobalt carbonyl anion,[4] and an iodide-promoted rhodium compound[5] are, for example, used for carbonylation reactions of organic halides under mild conditions. Coordination of carbon monoxide to metals (metal carbonyl complexes), oxidative additon of organic halides to low-valent transition metal complexes with the formation of carbon-metal bonds, and insertion of carbon monoxide into carbon-metal bonds with the formation of acyl complexes are important elementary processes involved in these reactions. We wish here to report the recent results of the study on carbonylation reactions of organic halides by tetrakis(triphenylphosphine)nickel (O) and palladium (O) carbonyl complexes under mild conditions.[6]

CARBONYLATION REACTIONS OF ARYL HALIDES BY TETRAKIS(TRIPHENYLPHOSPHINE)NICKEL(O) UNDER MILD CONDITIONS

We have reported in a previous paper[7] that aryl chlorides, bromides, and iodides are easily cleaved with tetrakis(triphenylphosphine)nickel(O) at room temperature to yield the arylnickel(II) complexes of the type $(PPh_3)_2NiX(Ar)$, where aryl is not only an _ortho_-substituted, but also phenyl or a _meta_- or _para_-substituted aromatic ligand.

Aromatic esters are almost quantitatively obtained by treating the arylnickel(II) complexes with methanol and triethylamine at room

$$Ni(PPh_3)_4 + ArX \longrightarrow (PPh_3)_2NiX(Ar) + 2PPh_3$$

$$(PPh_3)_2NiX(Ar) + CO + CH_3OH + NEt_3$$

$$\longrightarrow ArCOOCH_3 + Ni(CO)_2(PPh_3)_2 + NEt_3 \cdot HX$$

Ar = phenyl, o-tolyl, p-methoxyphenyl, 1-naphthyl

X = Cl, Br, I

temperature under carbon monoxide at atmospheric pressure. This means that all types of aryl halides are conveniently carbonylated with tetrakis(triphenylphosphine)nickel(0) to give aromatic esters at atmospheric pressure and room temperature, though Corey et al.[3b] reported that a mixture of nickel carbonyl (danger and toxic) and potassium t-butoxide in t-butyl alcohol effectively t-butoxycarbonylates not only trigonal halides but also alkyl iodides.

Several reaction pathways concerning the carbonylation reactions of organic halides using nickel carbonyl have been proposed, but no direct proof has been provided. In the present case, two reaction pathways are proposed. The one includes the insertion of carbon monoxide into the nickel-aryl bond to form the acyl complex $(PPh_3)_2NiCl(COAr)$ and the other contains the carbomethoxy complex $(PPh_3)_2Ni(Ar)(COOCH_3)$ as the intermediate, followed by the reductive elimination of aromatic esters.

An intermediate acylnickel compound $(PPh_3)_2NiCl(COC_6H_5)$ is isolated in a high yield as yellow crystals by the oxidative addition of benzoyl chloride to tetrakis(triphenylphosphine)nickel(0). Otsuka et al.[8] also studied the same reaction, but only the phenylnickel complex $(PPh_3)_2NiCl(C_6H_5)$, together with a small amount of a mixture of carbonyl-phosphine-nickel complexes, was obtained. The benzoylnickel complex isolated shows the infrared absorption bands at 1617 and 347 cm^{-1} assignable to the carbonyl stretching vibration and the Ni-Cl stretching vibration, respectively. The complex rapidly reacts with methanol in the presence of triethylamine to yield methyl benzoate in a high yield.

The infrared and NMR spectra of the reaction mixtures were taken to elucidate the reaction pathway. Although the carbonylation reaction in the presence of triethylamine is too rapid to be followed by an infrared spectrometer, the reaction without the amine is slow enough to be followed. The phenylnickel complex $(PPh_3)_2NiCl(C_6H_5)$ was dissolved in toluene and methanol, and the solution was circulated through an infrared spectrometer cell while carbon monoxide was bubbled into the solution. An absorption at 1620 cm^{-1} first appears, which is assignable to the benzoylnickel complex $(PPh_3)_2NiCl(COC_6H_5)$ (Figure 1). Although a carbomethoxy complex $(PPh_3)_2Ni(C_6H_5)(COOCH_3)$ is an unknown compound, its carbonyl stretching frequency may be higher than 1620 cm^{-1}. Then an absorption band at 1730

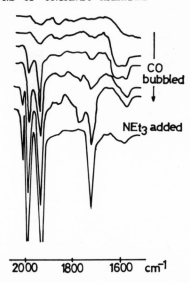

Figure 1. Time-dependent infrared spectra of a toluene-methanol of $(PPh_3)_2NiCl(C_6H_5)$.

cm^{-1} due to methyl benzoate and the bands at 1944 and 2000 cm^{-1} characteristic of $Ni(CO)_2(PPh_3)_2$ gradually appear. After further bubbling of carbon monoxide, the band at 1780 cm^{-1} assigned to benzoyl chloride also appears. Addition of triethylamine rapidly converts benzoyl chloride to methyl benzoate. The NMR spectrum of a

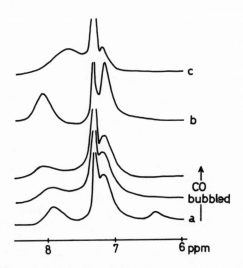

Figure 2. NMR spectra of a C_6D_6 solution of (a) $(PPh_3)_2NiCl(C_6H_5)$ (b) $(PPh_3)_2NiCl(COC_6H_5)$ or (c) $Ni(PPh_3)_4$.

C_6D_6 solution of the phenylnickel complex (Figure 2) shows a band
at 6.37 ppm due to <u>ortho</u>-phenyl protons of nickel-phenyl bond and a
band at 7.87 ppm assigned to <u>ortho</u>-phenyl protons of the coordinated
triphenylphosphine. After bubbling of carbon monoxide, the former
band disappears and the latter shifts to 8.01 ppm, which is assignable to <u>ortho</u>-phenyl protons of the coordinated triphenylphosphine
in $(PPh_3)_2NiCl(C_6H_5)$. These results indicate that the reaction proceeds through the insertion of carbon monoxide into the nickel-phenyl
bond. The insertion of carbon monoxide into nickel-carbon bonds is

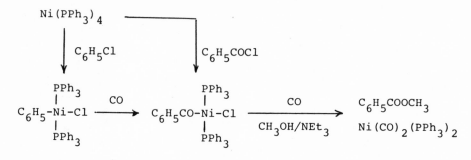

often postulated in many catalytic reactions, but no proof of such
a step has been provided except the recent reports on the preparation of acetylnickel complex $[P(CH_3)_3]_2NiCl(COCH_3)$ by the reaction
of the methylnickel complex $[P(CH_3)_3]_2NiCl(CH_3)$ with carbon monoxide[9]
and the isolation of $L_2NiCl(COC_6H_5)$ (L = tris(bornyl)phosphite) by
the carbonylation of the nickel-phenyl bond in $L_2NiCl(C_6H_5)$.[10]

The formation of methyl benzoate from the phenylnickel complex,
carbon monoxide, and methanol occurs only sluggishly in the absence
of triethylamine, even though NMR spectra suggest the stoichiometric
insertion of carbon monoxide into the nickel-phenyl bond. The effect of the base is not only to neutralize the acidity formed during
the reaction, i.e. to increase the reaction of benzoyl chloride with
methanol with the aid of the base, but probably also to enhance the
direct attack of methoxide anion on the $Ni\text{-}COC_6H_5$ moiety because the
acidity constant of methanol is 10^{-16} and the equilibrium constant
of $CH_3OH + NEt_3 \rightleftharpoons CH_3O^- + HN^+Et_3$ is about 10^{-6}.[11]

SYNTHESIS OF STABLE PALLADIUM(0) CARBONYL
COMPLEXES CONTAINING TRIPHENYLPHOSPHINE

As reported in previous papers,[12] we prepared the zerovalent
palladium carbonyl complexes stabilized with triphenylphosphine, i.e.
$Pd(CO)(PPh_3)_3$, $[Pd_3(CO)_3(PPh_3)_3]$, and $[Pd_3(CO)_3(PPh_3)_4]$, these being
the first stable carbonyl complexes of palladium reported. These
carbonyl complexes can be prepared by three different routes, (A):
the reduction of palladium(II) acetylacetonate with triethylaluminum
in the presence of triphenylphosphine under carbon monoxide, (B):

the reduction of $(PPh_3)_2PdCl_2$ with sodium borohydride in the presence of triphenylphosphine under carbon monoxide, and (C): the reaction of $(PPh_3)_2PdCl_2$ with carbon monoxide in methanol/amine systems, where the amine is a primary or secondary amine.

Method A

$$Pd(acac)_2 + 3PPh_3 + CO + AlEt_3 \xrightarrow[Toluene]{} Pd(CO)(PPh_3)_3$$

Method B

$$(PPh_3)_2PdCl_2 + PPh_3 + CO + NaBH_4 \xrightarrow[EtOH]{} Pd(CO)(PPh_3)_3$$

Method C

$$(PPh_3)_2PdCl_2 + CO + CH_3OH + R_1R_2NH$$
$$\xrightarrow{\hspace{2cm}} Pd(CO)(PPh_3)_3 + [Pd_3(CO)_3(PPh_3)_4]$$

$R_1 = C_2H_5$, $R_2 = C_2H_5$: $R_1 = n\text{-}C_4H_9$, $R_2 = H$

These palladium(0) carbonyl complexes are interconvertible under suitable conditions, as shown below.

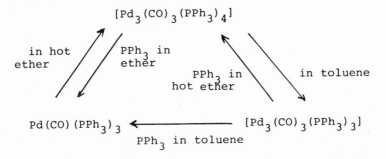

Use of a tertiary amine instead of a primary or secondary amine in the preparation method C yields a carbomethoxy complex $(PPh_3)_2$-$PdCl(COOCH_3)$ in a high yield. This gives a convenient route to the preparation of carboalkoxy complexes of nickel triad.[12b]

$$(PR_3)_2MCl_2 + CO + CH_3OH + NEt_3$$
$$\xrightarrow{\hspace{2cm}} (PR_3)_2MCl(COOCH_3) + NEt_3 \cdot HCl$$

$M = Ni$, Pd, or Pt: $R = C_6H_5$ or $\underline{n}\text{-}C_4H_9$

OXIDATIVE ADDITIONS OF ORGANIC HALIDES TO PALLADIUM(0) CARBONYL COMPLEX

So far the acyl complexes of palladium(II) of the type <u>trans</u>-

$(PR_3)_2PdX(COR')$ reported have been limited to acetyl or benzoyl complexes such as trans-$(PPh_3)_2PdCl(COMe)$ prepared by the oxidative addition of acetyl chloride to $Pd(PPh_3)_4$[13,8] and trans-$(PEt_3)_2PdX(COR)$ (R = Me, Ph) prepared by carbonylation of trans-$(PEt_3)_2PdX(R)$.[14] We found that vinyl chloride, allyl chloride, allyl bromide, methallyl chloride, benzyl bromide, methyl iodide, and iodobenzene react at room temperature with $Pd(CO)(PPh_3)_3$ to afford the corresponding acyl complexes of the type trans-$(PPh_3)_2PdX(COR)$.[15] The reactions proceed rapidly when the dissociation energy of carbon-halogen bonds in the saturated alkyl halides is smaller than about 60 kcal/mol. In the case of the unsaturated alkyl halides the more inert carbon-halogen bond in vinyl chloride or methallyl chloride can be cleaved to afford the corresponding acyl complexes. This may be explained by the initial coordination of the unsaturated compounds to palladium through a π-bonding, which would cause the activation of the carbon-halogen bonds, followed by isomerization to the acyl complexes. These oxidative addition reactions probably proceed according to the following scheme, since $Pd(CO)(PPh_3)_3$ dissociates to a coordinatively unsaturated species $[Pd(CO)(PPh_3)_2]$ in solution:

$$Pd(CO)(PPh_3)_3 \longrightarrow [Pd(CO)(PPh_3)_2] + PPh_3$$

$$\xrightarrow{RX} [PdX(R)(CO)(PPh_3)_2] \longrightarrow (PPh_3)_2PdX(COR)$$

CARBONYLATION REACTIONS OF ORGANIC HALIDES BY PALLADIUM(0) CARBONYL COMPLEXES

Iodobenzene is catalytically carbonylated with $Pd(CO)(PPh_3)_3$ at room temperature under carbon monoxide at atmospheric pressure to yield esters of benzoic acid, in contrast to the stoichiometric carbonylation reactions of aryl halides with $Ni(PPh_3)_4$ described above. The palladium(0) carbonyl complex $Pd(CO)(PPh_3)_3$ is easily prepared by the reaction of $(PPh_3)_2PdCl_2$ with carbon monoxide in diethylamine-methanol, and hence the latter complex can be used as the catalyst instead of the former carbonyl complex. For example, 19 mol of methyl benzoate per mol of $(PPh_3)_2PdCl_2$ was formed by treating iodobenzene (3 ml) with a suspension of $(PPh_3)_2PdCl_2$ (0.200 g) in methanol (5 ml) and diethylamine (3 ml) under carbon monoxide at atmospheric pressure for 6.3 hr at room temperature. The acyl complex $(PPh_3)_2PdI(COPh)$ was isolated from the reaction solution, which could also be used as the catalyst. The study on carbonylation reactions of other organic halides is now in progress.

REFERENCES

1. L. Cassar, G. P. Chiusoli, and F. Guerrieri, Synthesis, 509 (1973).

2. (a) G. P. Chiusoli, S. Merzoni, and G. Mondelli, Chimica e Industria, 47, 743 (1965); (b) G. P. Chiusoli, M. Dubini, M. Ferraris, F. Guerrieri, S. Merzoni, and G. Mondelli, J. Chem. Soc. C, 2889 (1968).
3. (a) L. Cassar and M. Foa, Inorg. Nucl. Chem. Letters, 6, 291 (1970); (b) E. J. Corey and L. S. Hegedus, J. Amer. Chem. Soc., 91, 1233 (1969); (c) M. Nakayama and T. Mizoroki, Bull. Chem. Soc. Japan, 44, 508 (1971); (d) L. Cassar and M. Foa, J. Organometal. Chem., 51, 381 (1973).
4. R. F. Heck in Organic Synthesis via Metal Carbonyls Vol. II, I Wender and P. Pino Ed., Interscience, New York, 1968.
5. J. F. Roth and J. H. Craddock, Abstracts of Papers, 162nd National Meeting of the American Chemical Society, Washington, 12-17, September, 1971.
6. M. Hidai, Y. Wada, S. Uzawa, Y. Uchida, I. Ogata, and K. Nobuchika, A part of this paper presented at the 30th Annual Meeting of the Chemical Society of Japan, April 1974, Osaka.
7. M. Hidai, T. Kashiwagi, T. Ikeuchi, and Y. Uchida, J. Organometal. Chem., 30, 279 (1971).
8. S. Otsuka, A. Nakamura, T. Yoshida, M. Naruto, and K. Ataka, J. Amer. Chem. Soc., 95, 3180 (1973).
9. H. F. Klein, Angew. Chem. Internat. Edit., 12, 402 (1973).
10. S. Otsuka and I. Kato, to be published.
11. J. B. Hendrickson, D. J. Cram, and G. S. Hammond, Organic Chemistry, McGraw-Hill, New York, 3rd ed., 304 (1970).
12. (a) K. Kudo, M. Hidai, and Y. Uchida, J. Organometal. Chem., 33, 393 (1971); (b) M. Hidai, M. Kokura, and Y. Uchida, ibid., 52, 431 (1973).
13. P. Fitton, M. P. Johnson, and J. E. McKeon, Chem. Comm., 6 (1968).
14. G. Booth and J. Chatt, J. Chem. Soc. A, 634 (1966).
15. K. Kudo, M. Sato, M. Hidai, and Y. Uchida, Bull. Chem. Soc. Japan, 46, 2820 (1973).

THE PALLADIUM CATALYZED REACTIONS FOR SYNTHESIS OF AMINES AND BENZOFURANS

I. Moritani, T. Hosokawa, and S.-I. Murahashi[*]

Department of Chemistry, Faculty of Engineering Science,
Osaka University, Machikaneyama, Toyonaka, Osaka, Japan,
560

Remarkable advances have recently been made in organic synthesis by using palladium complexes.[1] The catalytic reactions are of such variety that palladium promises to become as important in organic synthesis as the Grignard reagent. We report here nobel palladium catalyzed processes for synthesis of amines and benzofurans.

We have recently found[2] a new process for the synthesis of unsymmetrical secondary and tertiary amines from primary or secondary amines by dehydrogenation of amines by palladium black which is operationally simple, highly selective, and efficient, hence offers advantages over the previous methods.[3-5]

Treatment of primary amines bearing activated α-hydrogens with palladium at 25-80° afforded secondary amines and/or imines. Thus, allyl amines was converted into N-propylideneallylamine (95%), while benzylamine led to N-benzylidenebenzylamine (45%) and N,N-dibenzyl-amine (45%). These reactions will be elucidated by addition of the amine 1 to the intermediate imine 2 derived from dehydrogenation[5] of 1 to give 3 followed by elimination of ammonia (4, R^2=H) as shown in eq 1.

$$R^1CH_2NHR^2 \xrightarrow{\text{Pd}} \left[R^1CH=NR^2\right] \xrightarrow{1} \left[\begin{array}{c} R^1CH-NHR^2 \\ | \\ R^2CH_2NR^2 \end{array}\right] \xrightarrow[-R^2NH_2]{} R^1CH_2\overset{R^2}{N}CH_2R^1$$

$$\underset{1}{} \qquad\qquad \underset{2}{} \qquad\qquad \underset{3}{} \qquad\qquad \underset{4}{} \qquad\qquad \text{eq 1}$$

Analogous reactions were generally performed on secondary amines giving tertiary amines with two identical substituents. Rection (120°, 20 hr) of N-methylbenzylamine with palladium afforded N-methyl-

dibenzylamine in 85% yield in addition to methylamine. Treatment of
pyrrolidine (5) at 80° for 10 hr gave pyrrolidino-2Δ_1-pyrroline (6),
(65%), and 1-pyrrolidinebutylamine (7) (24%). When this reaction was
carried out at 150° for 5 hr, the diamine 7 was exclusively obtained
(85%). Further, under more severe condition (200°) α,δ-dipyrrolidi-
nobutane (8) (75%) was obtained. These results can be rationalized
in terms of either reductive cleavage or dehydrogenation of the intermediate (9)[7] as shown in eq 2.

eq 2

     In analogy with formation of tertiary amines in eq 1, amine ex-
change reactions might be expected by introducing a primary amine
(10,R^4=H) or a secondary amines (10) at the stage of addition of the
amine to the intermediate imine 2 followed by extruding R^2NH_2 as
shown in eq 3. Indeed, reaction of N-methylpropylamine with n-hexylamine (160°, 7 hr) afforded N-n-propylhexylamine (75%) by exchanging
the methyl group for the hexyl group. Further, addition of a secondary amine (10) to 2 to give 11 followed by extruding R^2NH_2 can
also be achieved selectively. Thus, reaction of N-methylbenzylamine
with N-methylbutylamine gave N-butyl-N-methylbenzylamine (80%), and
reaction of N-methylpropylamine with pyrrolidine led to N-propylpyr-
rolidine (98%).

eq 3

     The reaction can be envisioned as shown in eq 4. The key step
is the formation of a palladium π-complex of Schiff base bearing a
palladium-hydride bond (12) by dehydrogenation with palladium. Insertion of an amine (R^3R^4NH) into 12 leading to 13, followed by addition of the palladium-nitrogen across the nitrogen-carbon double
bond, would form 14 which subsequently cleaves to form products re-
ductively.[8]

$$R^1CH_2NHR^2 \xrightarrow{Pd^0} \begin{bmatrix} R^1CH{=}NR^2 \\ \vdots \\ H{-}Pd{-}H \end{bmatrix} \xrightarrow{R^3R^4NH} \begin{bmatrix} R^1CH{=}NR^2 \\ Pd{-}H \\ R^3N \overset{H\ H}{\diagup} \\ R^4 \end{bmatrix}$$

$$\underset{12}{} \qquad \qquad \underset{13}{}$$

$$\longrightarrow \begin{bmatrix} R^1CH{-}NR^2 \\ R^3{-}N \quad Pd{-}H \\ R^4 \quad H \end{bmatrix} \longrightarrow R^1CH_2 \;+\; R^2NH_2 \;+\; Pd^0$$

$$\underset{14}{} \qquad\qquad \underset{R^3 \quad R^4}{\overset{N}{\diagup}} \qquad\qquad\qquad eq\ 4$$

In relation to this amine exchange reaction, we have found an efficient process for synthesis of secondary or tertiary anines from the reaction of aryl or allyl alcohols with primary or secondary amines by palladium catalyst.[9] Treatment of a mixture of benzyl alcohol and n-hexylamine with palladium at 110° for 6 hr gave N-n-hexylbenzylamine exclusively (98% yield). Further, reaction of α-

$$PhCH_2OH \;+\; n\text{-}C_6H_{13}NH_2 \xrightarrow[110°,\ 6\ hr]{Pd} PhCH_2NHC_6H_{13}\text{-}n$$

methylbenzyl alcohol with pyrrolidine (110°, 26 hr) gave N-α-methyl-benzylpyrrolidine (83%). This method has advantages over the classical reductive amination of aldehydes and ketones where high pressure hydrogenation conditions are normally required and their yields are often low.[4]

These reactions can be rationalized by assuming the reaction pathway as shown in eq 5. Dehydrogenation of the alcohols (R^1R^2CHOH, 15) by palladium would lead to the intermediate 16. The amine (R^3R^4-NH) react with 16 to form Schiff base which is subsequently hydrogenated by the palladium-hydride species yielding the amines ($R^1R^2CHNHR^3$-R^4, 17) in addition to recovering palladium.

$$R^1R^2CHOH \xrightarrow{Pd^0} \begin{bmatrix} R^1R^2{=}O \\ \vdots \\ H{-}Pd{-}H \end{bmatrix} \xrightarrow[-H_2O \\ -Pd^0]{R^3R^4NH} R^1R^2CHNR^3R^4$$

$$\underset{15}{} \qquad\qquad \underset{16}{} \qquad\qquad \underset{17}{}$$

$$eq\ 5$$

It is noteworthy that the reaction of allyl alcohol with n-hexylamine afforded N-propylidenehexylamine in 87% yield. Reduction of
the carbon-carbon double bond of the intermediate Schiff base by palladium hydride species is superior to that of the carbon-nitrogen
bond. This result is parallel with that of the palladium catalyzed
reaction of allyl amine where propylideneallylamine is an exclusive
product (95%).

Application of this reaction provides extremely valuable method
for synthesis of N-substituted pyrroles.[9] A large variety of ways
such as condensations of primary amines with 2,5-diketones,[10] with
tetrabromobutanes[11] and with 1,4-diamino-1,3-butadiene[12] have been
established; however, these methods require 2,5-diketones or their
derivatives which are rather difficult to be prepared, and their
yields are generally low. The palladium catalyzed reaction of readily available buten-1,4-diol (18) with primary amines under mild conditions gave N-substituted pyrroles in highly selectively. Thus,
treatment of a mixture of n-hexylamine and 18 at 120° for 20 hr gave
N-n-hexylpyrrole in 93% yield. This reaction can be rationalized by
assuming that intramolecular cyclization of the intermediate secondary amine which can be derived from 18 via eq 5.

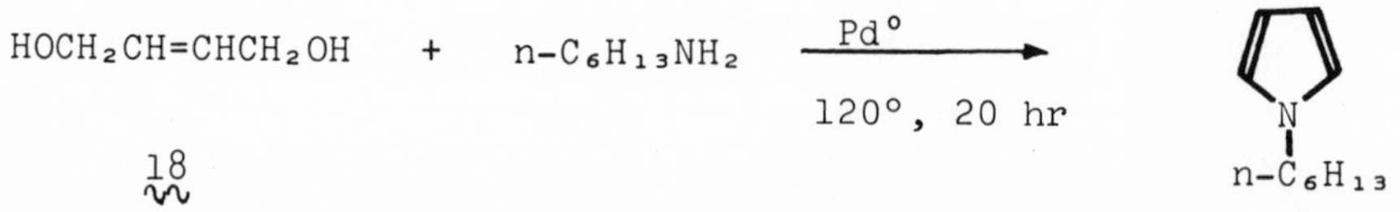

$$HOCH_2CH=CHCH_2OH \quad + \quad n\text{-}C_6H_{13}NH_2 \quad \xrightarrow[120°, \ 20 \ hr]{Pd°}$$

18

Recently, we have found the stoichiometric cyclization of allyl-
phenols with dichlorobis(benzonitrile)palladium to give 2-substituted
benzofurans.[13] This reaction can be carried out catalytically with
respect to palladium, and can be elucidated by the pathway involving
cis-oxypalladation step.

When the reaction of o-allylphenol (19) was performed by the
use of palladium acetate in the presence of cupric acetate and molecular oxygen as reoxidants, 2-methylbenzofuran (20) was obtained in
1960% yield based on used palladium acetate (19) was almost consumed
(98% conv), and the yield of 20 was 34% based on (19).

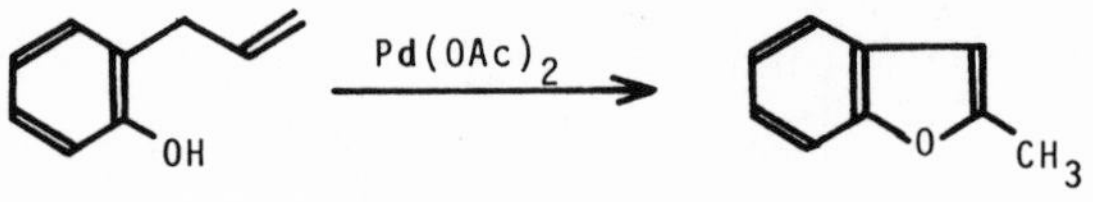

19   20

Figure 1

1-Allyl-2-naphthol afforded 2-methyl[2,1-b] naphthofuran in 786%
yield (53% conv, 16% yield). Similarly, 2-cinnamylphenol gave 2-
benzylbenzofuran in 1170% yield (63% conv, 23% yield). In contrast
to these results, indroduction of a methyl substituent to the γ-position of the allyl group led to vinyldihydrobenzofuran (21) along
with trace amount of benzofuran derivatives. Thus, a mixture of cis-
and trans γ-methylallylphenol(1:1.9) gave 21a (R=H) exclusively in
2720% yield (99% conv, 54% yield), while γ,γ-dimethylallylphenol
afforded 21b in 560% yield (60% conv, 11% yield) along with newly
formed 2,2-dimethylchromene (22) in 520% yield (10% yield). Furthermore, 2-cyclohexyl-2-enylphenol (23) gave dihydrobenzofuran 24 exclusively in 1630% yield (46% conv, 33% yield).

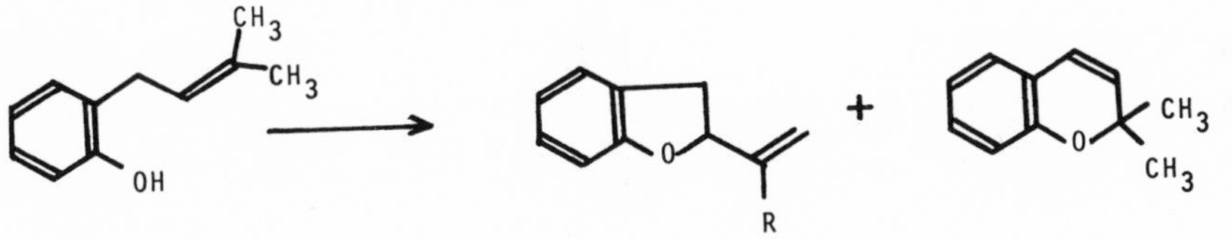

Figure 2

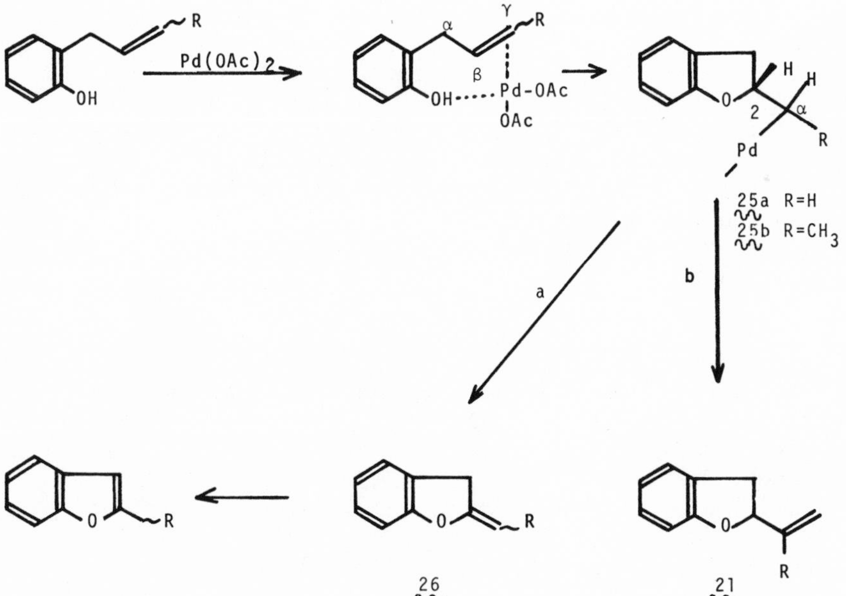

eq 6

These reactions can be reasonably explained by intramolecular oxypalladation followed by β-elimination of palladium hydride as shown in eq 6. In the reaction of o-allylphenol, the hydrogen on the C-2 carbon in the intermediate 25 (R=H) is an only β-hydrogen to be eliminated with palladium. Hence, the elimination gives the most thermodynamically stable 2-substituted benzofuran via exo-methylenebenzofuran (26) (path a). Introduction of a methyl group to the γ-position of the allyl group leads to two possibilities of β-elimination of palladium hydride in 25b.

eq 7

Evidently, exclusive formation of vinyldihydrobenzofuran 21 indicates that the β-elimination occurs preferentially from the methyl group (path b). Since cis-palladium hydride elimination is generally accepted,[14] the hydrogen on the C-2 must be considered to take the trans-configuration to the palladium, and hence the process is retarded. If the elimination is faster than the rotation about the carbon(C-2)-carbon(C-α) bond, it can be said that the reaction proceeds via a cis-oxypalladation step. Exclusive formation of 24 from 23, where the rotation is completely excluded, is best explained by the cis-oxypalladation followed by cis-elimination. cis-β-Elimination from the intermediate 27 which is derived from cis-oxypalladation of 23 and bears only one cis β-hydrogen to the palladium can give 24 exclusively as shown in eq 7.

REFERENCES

1. P. M. Maitlis, The Organic Chemistry of Palladium, Vol. 1 and 2, Academic Press, New York, 1971.
2. N. Yoshimura, I. Moritani, T. Shimamura, and S. -I. Murahashi, J. Amer. Chem. Soc., 95, 3038 (1973).
3. M. S. Gibson, The Chemistry of the Amino Group, S. Patai Ed., Interscience, New York, N. Y., 1968, p 37.
4. W. S. Emerson, Org. React., 4, 174 (1948).
5. A. A. Balandin and N. A. Vasyunia, Dokl. Akad. Nauk SSSR, 103, 831 (1955).
6. K. W. Rosenmund and G. Jordan, Chem. Ber., 58B, 51 (1925).

7. N. Yoshimura, I. Moritani, T. Shimamura, and S. -I. Murahashi,
    Chem. Comm., 308 (1973).

8. The footnote 6 in the reference 2.

9. S. -I. Murahashi, T. Shimamura, and I. Moritani, unpublished
    results.

10. R. Pips, Ch. Derappe, and Ng. Ph. Buu-Hoi, J. Org. Chem., 25,
    390 (1960); A. Kreutzberger and P. A. Kalter, ibid., 25, 554
    (1960).

11. A. Treibs and O. Hitzler, Chem. Ber., 90, 787 (1957).

12. M. F. Feyley, N. M. Bortnick, and L. H. McKeever, J. Amer. Chem.
    Soc., 79, 4144 (1957).

13. T. Hosokawa, K. Maeda, K. Koga, and I. Moritani, Tetrahedron
    Letters, 739 (1973).

14. R. F. Heck, J. Amer. Chem. Soc., 91, 6707 (1969), ibid., 93,
    6896 (1971); P. M. Henry, ibid., 94, 7305 (1972).

CHEMISTRY OF ALKYLNICKEL COMPLEXES. PREPARATION AND PROPERTIES OF ALKYLNICKELS WITH TERTIARY PHOSPHINE LIGANDS

Akio Yamamoto[*], Takakazu Yamamoto, Masakatsu Takamatsu,
Toshio Saruyama and Yoshiyuki Nakamura
Research Laboratory of Resources Utilization, Tokyo
Institute of Technology, Ookayama, Meguro, Tokyo 152,
Japan

INTRODUCTION

Nickel complexes with metal-to-carbon bond are generally regarded as the active species in various nickel-catalyzed organic reactions such as polymerization, oligomerization, hydrogenation, isomerization and isotopic hydrogen exchange of olefins.[1] Pertinent information regarding the mechanisms of these nickel-catalyzed reactions is expected to be gained by studying the behavior of the isolated alkylnickel complexes. Alkylnickel complexes without other
ligand may be obtained in special cases,[2] but the use of appropriate
stabilizing ligands often gives more stable alkylnickel complexes
which are suitable for studying the chemistry of these alkyls.[3-5]
Among various approaches for obtaining alkylnickel complexes the
alkylation of nickel acetylacetonate with dialkylaluminum monoethoxide in the presence of ligands has been found to provide the most
convenient route. Evidently the alkylation proceeds through stepwise
exchange reactions of the acetylacetonato ligands with the alkyl
group of the alkylaluminum compound with the formation of an intermediate having both alkyl and acetylacetonato ligands. In fact alkylnickel complexes of this type have been obtained by employment of
a suitable stabilizing ligand and by carrying out the alkylation under mild conditions,[6,7] and the formations of other alkyl(acetylacetonato) and hydrido(acetylacetonato) complexes have been reported.[8,9]
The alkyl(acetylacetonato) intermediates may be further alkylated
affording dialkylnickel complexes which are sometimes further reduced
to zero valent nickel complexes by splitting of the alkyl-nickel
bonds.

We report here the preparation of the dialkylnickel complexes
as well as the intermediate alkyl(acetylacetonato) complexes using

$$Ni(acac)_2 + nL \xrightarrow{\text{AlR}_2\text{OEt}} NiR(acac)L_n \xrightarrow{\text{AlR}_2\text{OEt}} NiR_2L_n$$

$$NiR_2L_n \longrightarrow NiL_n \quad + \quad R\text{-}R \quad \text{(or alkane + alkene)}$$

a variety of tertiary phosphine and phosphite ligands. Some of the
isolated alkylnickel complexes show an interesting fluxional behavior which is relevant to the discussion of the mechanism of the nickel-catalyzed reactions.

RESULTS AND DISCUSSION

Synthesis

Table 1 summarizes the nickel complexes prepared in this study.
The alkylation was carried out in ether under similar conditions
keeping the reaction temperature below -20°, but the products varied
as shown in the Table. The pronounced effects of specific ligands
are evident. Trialkylphosphines afforded dialkylnickel complexes of
the type NiR_2L_2, which are unstable at room temperature. Diphenyl-
phosphinoethane(dppe), a bidentate ligand, markedly enhances the
stability of the alkylnickel, and $NiMe_2$(dppe) and $NiEt_2$(dppe) can be
obtained as stable complexes at room temperature. Tertiary phosphines containing phenyl group(s) gave two types of complexes, NiR-
$(acac)L_n$ and NiL_4. The comparison of the products obtained using
PPh_2Me, PPh_2Et, PPh_2Me, reveals that a subtle difference in the electronic and steric effects of the tertiary phosphine exerts a pronounced influence on the course of the alkylation reaction and the
stability and reactivity of the alkylnickel complexes. Tricyclohexyl-
phosphine[6] and triphenylphosphine[7] give similar alkyl(acetylacetonato)
type complexes which differed in that triphenylphosphine afforded the
pentacoordinated complex $NiMe(acac)(PPh_3)_2$. The reaction of Ni(acac)$_2$, $AlEt_2OEt$ and PPh_3 at room temperature is known to give $Ni(PPh_3)_4$[10]
and $Ni(PPh_3)_2(CH_2=CH_2)$ depending on the experimental conditions.
Phosphite ligands did not give isolable alkylnickel compounds but
yielded only zero valent nickel complexes of the NiL_4 type. In some
cases (with PEt_3 and $P(OEt)_3$) the formation of binary complexes containing both nickel and aluminum components have been isolated. The
compositions of these complexes have not been determined.

These dialkyl, alkyl(acetylacetonato) and zero-valent nickel
complexes have been characterized by elemental analysis, ir and nmr
spectroscopy and chemical reactions. The following examples will illustrate the chemical properties of the alkylnickel complexes. Aci-
dolysis of $NiMe(acac)(PPh_3)_2$ and $NiEt(acac)(PPh_3)$ gave quantitative
amounts of methane and ethane and deuteriolysis afforded CH_3D and
C_2H_5D respectively. Thermolysis of $NiEt(acac)(PPh_3)$ (mp 93°) at 100°

Table 1
Nickel Compounds Isolated by Reactions of Nickel
Acetylacetonate, Dialkylaluminum Monoethoxide and
Tertiary Phosphines or Phosphites

	Ligand	AlR_2OEt employed	
		$AlMe_2OEt$	$AlEt_2OEt$
1	PEt_3	$NiMe_2(PEt_3)_2$	b
2	PBu_3	$NiMe_2(PBu_3)_2$	$NiEt_2(PBu_3)_2$
3	$(Ph_2PCH_2)_2$ (dppe)	$NiMe_2(dppe)_2$	$NiEt_2(dppe)$
4	PPh_3	$NiMe(acac)(PPh_3)_2$	$NiEt(acac)(PPh_3)$
5	PPh_2Et	$NiMe(acac)(PPh_2Et)$	$Ni(PPh_2Et)_n$
6	PPh_2Me	$Ni(PPh_2Me)_4$	——
7	$PPhMe_2$	——	$NiEt(acac)(PPh_2Me)$
8	$P(OPh)_3$	$Ni[(POPh)_3]_4$	$Ni[P(OPh)_3]_4$
9	$P(OEt)_3$	b	$Ni[P(OEt)_3]_n$
10	$P(C_6H_{11})_3$ [a] (PCy_3)	$NiMe(acac)(PCy_3)$	$NiEt(acac)(PCy_3)$

a.   Prepared by Jolly et al.[6]

b.   Binary complexes containing both nickel and aluminum components

     were obtained.

released ethylene, hydrogen, and a small amount of ethane, whereas
from NiMe(acac)(PPh₃) (mp 102-103°) methane and ethane were liberated
at 120° in a ratio of 1:2. Dialkylnickel complexes with basic phosphine ligands such as Pet₃ showed activity to initiate the polymeri-

zation of acrylonitrile and methacrylonitrile. In some cases the
reactions with CO_2 gave CO_2-coordinated complexes.

Nmr Spectra of Fluxional Alkylnickel Complexes

Examination of the nmr spectrum of $NiEt(acac)(PPh_3)$ measured in
various solvents and at various temperatures revealed a complicated
fluxional behavior of the complex.

Table 2 shows the nmr data of $NiCH_3(acac)(PPh_3)_2$ and NiC_2H_5-
$(acac)(PPh_3)$. The two methyl groups in the acetylacetonato ligand
of the methyl- and ethylnickel complexes are magnetically non-equivalent at room temperature in benzene, toluene, tetrahydrofuran and
acetone, whereas in pyridine they are equivalent at room temperature
and non-equivalent at $-36°$. The acetylacetonato ligand seems to be
labilized in pyridine solution and the partial dissociation of the
bidentate acetylacetonato ligand and its recoordination may be responsible as a mechanism to make the methyl protons of the acetylacetonato ligand equivalent. In fact the labilization of the acetylacetonato ligand in Lewis bases causes the disproportionation of
$NiC_2H_5(acac)(PPh_3)$ as follows:

$$2NiC_2H_5(acac)(PPh_3) \xrightarrow{2py} Ni(acac)_2 \cdot 2py + [(PPh_3)_2Ni(C_2H_5)_2]$$
$$(PPh_3)_2Ni(C_2H_4) + C_2H_6$$

Such a disproportionation in Lewis bases appears to be common among
$MR(acac)L_n$ type complexes.

As revealed by examination of the ^{31}P nmr spectrum of NiEt-
$(acac)(PPh_3)$, the triphenylphosphine ligand is extensively dissociated in pyridine and exchanges rapidly with the added triphenylphosphine even below $-40°$ where the methyl protons in the acetylacetonato
ligand are observed as nonequivalent. The degree of dissociation of
triphenylphosphine is small in toluene, benzene and acetone and the
exchange with the added triphenylphosphine is slower than in pyridine.

The outstanding feature of the ethylnickel complex is that the
proton nmr resonance of the ethyl group is observed as a singlet in
benzene, toluene, acetone, and tetrahydrofuran; whereas it is observed as a multiplet at 100 MH_z and as a pair of a triplet and a quartet at 220 MH_z in more basic solvents such as pyridine and triethylamine. There are two possible causes to be considered which make
the ethyl signal a singlet. One is the fortuitous coincidence of
the methyl and methylene protons in less basic solvents. The other
is the exchange of protons in the methyl and methylene groups bonded
with nickel. Although the former possibility can not be excluded,

Table 2

Nmr Data of the Methyl and Ethyl Nickel Complexes[a]

Complex	Solvent	temp, °C	δ, chemical shifts in ppm[b]		
CH$_3$Ni(acac)(PPh$_3$)$_2$			CH$_3$-Ni	CH$_3$(acac)	CH(acac)
	Benzene	25	0.09(3,s)	1.39(3,s)	5.30(1,s)
	Pyridine	25	0.04(3,s)	0.89(3,s) 1.78(6,s)	5.40(1,s)
CH$_3$CH$_2$Ni(acac)(PPh$_3$)			CH$_3$CH$_2$Ni	CH$_3$(acac)	CH(acac)
	Acetone-d$_6$	25	0.18(5,s)	1.42(3,s) 1.89(3,s)	5.37(1,s)
	Benzene	25	0.77(5,s)	1.43(3,s) 1.93(3,s)	5.30(1,s)
	Pyridine[c] (220 MHz)	64	0.60(3,t)[d] 0.89(2,q)[d]	1.76(6,s)	5.40(1,s)
		-36	0.6(3,br,s) 0.9(2,br,s)	1.62(3,br,s) 1.82(3,br,s)	

a   Chemical shifts are referenced to internal TMS.

b   Figures in parentheses mean peak intensity and the multiplicity:

s, singlet; t, triplet; q, quartet; br, broad.

c   Peaks due to ethane and the coordinated ethylene which were formed by decomposition are observed

at δ 0.72 and 2.5 - 2.8, respectively.

d   J = 7.4 Hz.

at the moment we favor the latter on the following basis. Despite
a large difference (0.6 ppm) in chemical shifts of the singlet ethyl
peak of NiEt(acac)(PPh$_3$) in benzene and acetone, the measurement of
the ^{1}H nmr spectrum of the complex in mixtures containing both solvents in various ratios revealed the appearance of the ethyl signal
as a singlet peak in the chemical shift region between δ 0.2 to 0.8.
When the complex was treated with D$_2$ in toluene at room temperature,
deuteriated ethanes were formed which contained considerable amounts
of C$_2$D$_6$, C$_2$D$_5$H, C$_2$D$_4$H$_2$, C$_2$D$_3$H$_3$ and C$_2$D$_2$H$_4$ in addition to C$_2$H$_5$D which
was expected as the sole product in the absence of H-D exchange re-
actions. On the other hand, no exchange of C$_2$D$_4$ with the ethyl complex was observed, the complex did not catalyze the isomerization
of butene-1, and no hydridic proton peak was observed in nmr of the
complex measured in toluene at low temperature (-100°). Therefore
the formation of a hydrido(ethylene) complex which is conceivable
as the intermediate in the proton exchange reaction can be excluded
or, if it should be present at all, its life time must be very short.

If the proton exchange should be actually taking place, this
complex may be regarded as an appropriate model relevant to metal-
catalyzed olefin catalysis such as isomerization, oligomerization
and isotopic exchange reactions. Examination of the nmr spectrum
of a specifically deuteriated ethyl complex such as Ni(CH$_2$CD$_3$)(acac)-
(PPh$_3$) would bring us unequivocal information regarding this question and the study on this line is now in progress.

Effects of Alkylaluminum Compounds on the Stability of Alkyltransition Metal Complexes

Many of these dialkyl- and alkyl(acetylacetonato)nickel complexes can be obtained as moderately stable complexes using dialkylaluminum monoethoxide, but further reaction of the isolated alkylnickel complexes with dialkylaluminum monoethoxide leads to decomposition with evolution of alkane. This apparent contradiction may
be understood if one takes into account the effect of alkylaluminum
compounds on the stability of alkyltransition metal complexes. We
have previously observed that addition of alkylaluminum compounds
to isolated alkyltransition metal complexes such as TiCH$_3$Cl$_3$, NiR$_2$-
(bipy), CrRCl$_2$Py$_3$, FeR$_2$(bipy)$_2$ and CuCH$_3$(PPh$_3$)$_3$(C$_6$H$_5$CH$_3$) leads to
destabilization of the transition metal-alkyl bonds and liberation
of alkanes.[11] The kinetic study revealed the presence of a maximum
decomposition rate at a certain concentration range of the alkylaluminum compound. The effect of the alkylaluminum compound was
accounted for as Lewis acid which withdraws electron from the transition metal and destabilizes the metal-alkyl bond. The decrease in
the decomposition rate at higher alkylaluminum concentration was
attributed to displacement of the complexed alkylaluminum component
by interaction with the added alkylaluminum compound.

Thus the stability of alkyltransition metal complexes is affected by the Lewis acidity and the concentration of the alkylaluminum compound. These results imply that the properties of alkyltransition metal complexes prepared in situ may be sometimes quite different from those of isolated alkyltransition metal complexes and extension of arguments based on the observation of the properties of alkyltransition metal complexes prepared in situ to those of pure alkyltransition metal complexes should be made with caution.

REFERENCES

1. For example, M. L. H. Green, Organometallic Compounds, Vol. 2, The Transition Elements, 1968, Methuen, London.
2. G. Wilke and H. Schott, Angew. Chem., 78, 592 (1966); T. Arakawa, Kogyo Kagaku Zasshi (J. Chem. Soc. Japan, Ind. Chem. Section), 70, 1738 (1967).
3. T. Saito, Y. Uchida, A. Misono, A. Yamamoto, K. Morifuji, and S. Ikeda, J. Amer. Chem. Soc., 88, 5198 (1966).
4. G. Wilke, G. Herrmann, Angew. Chem., 78, 591 (1966).
5. M. L. H. Green and M. J. Smith, J. Chem. Soc. A., 639 (1971).
6. P. W. Jolly, K. Jonas, C. Kruger, and Y. -H. Tsay, J. Organometal. Chem., 33, 109 (1971); B. L. Barnett and C. Kruger, J. Organometal. Chem., 42, 169 (1972).
7. A. Yamamoto, T. Yamamoto, T. Saruyama, and Y. Nakamura, J. Amer. Chem. Soc., 95, 4073 (1973).
8. Y. Kubo, A. Yamamoto, and S. Ikeda, J. Organometal. Chem., 46, C50 (1972).
9. T. Ito, T. Kokubo, T. Yamamoto, A. Yamamoto and S. Ikeda, J. Chem. Soc. Chem. Comm., 136 (1974).
10. G. Wilke and G. Herrmann, Angew. Chem., 74, 693 (1962).
11. T. Yamamoto and A. Yamamoto, J. Organometal. Chem., 57, 127 (1973).

DISCUSSION

DR. MARKS: Two questions: First of all, do you always prefer to make these alkyl compounds starting with the aluminum organometallic? Is there any reason for doing this rather than taking a nickel phosphinechloride plus Grignard or lithium reagent?

DR. YAMAMOTO: Sometimes we use Grignard reagent, but we have trouble in the isolation of the complex because you get magnesium halide which may be a problem to remove. In the case of alkyl aluminum compound the organoaluminum complex stays in the solution and you get crystals of product. Just by washing the crystals you can get the alkyl compound pure.

DR. MARKS: So it is not a matter of reduction, or anything like that, just a matter of purification.

DR. YAMAMOTO: Yes, but sometimes there are other reasons. For example, the organoaluminum compound used in the case of chromium compound can give the monoalkylchromiumdichloride but if you use alkyl lithium or alkyl magnesium compounds you may get the further alkylated product so it is somewhat specific in giving the monoalkyls.

DR. MARKS: Second question I have - These reactions with the Lewis acid activation of, say the methyl compound, give methane. Do you have any idea where the proton is coming from?

DR. YAMAMOTO: It seems at first thought to be coming from the solvent, but I am not quite sure. We did not examine this in detail.

DR. MARKS: Do you have the yields?

DR. YAMAMOTO: In the case of methyltitanium compound the yield is high.

DR. FALLER: I was curious about the NMR spectra that you have. There are possible problems with the mechanism. For instance, Ni-$(acac)_2py_2$ would be a paramagnetic compound with a relatively slow relaxation time. If reasonable amounts of it were produced you would lose the spectrum. I am particularly curious about some of the other intermediates that you have suggested for Ni(II). Once they become five-coordinate they would be paramagnetic, and any of them that went six-coordinate should be. Depending on conventions for counting electrons on the metal, there might be a problem, but no matter how you count it I think it comes out as Nickel (II). Once you have a five-coordinate compound, even if there is only a small amount there, I would anticipate a fast relaxation time and you might get slight contact shifts. Hence, there might be a lot more than a mere diamagnetic anisotropy that is causing the shift. I would think, in general, you might have to consider that seriously with some of the intermediates that you drew. As a second point, your proposals remind me of Schrauzer's addition product of fumaronitrile and cobaltdimethylgloxime-hydride. Apparent averaging in the proton nmr of the dicyanoethyl derivative was interpreted as fast hydrogen migration. The carbon resonances do not average, however, and hydrogen migration appears to occur primarily through dissociation to hydride and olefin. There may be some parallel with your work. Are there situations where you may actually have a large amount of hydride present? Have you looked above TMS to find the hydride peak?

DR. YAMAMOTO: Answering the second question first, I did look above TMS, and so far did not see any hydride peak appearing. Concerning your first question, the NMR spectrum was clean after it was left for some time in pyridine, and after we observed the appearance of ethane and ethylene. The nickel bisacetylacetonate might be formed eventually, but on this solution itself, it may still be one stage before giving the bisacetylacetonate.

DR. IBERS: These reactions of the acetylacetonate with nickel and chromium you obviously from past experience do under argon. If you try it under nitrogen, what happens?

DR. YAMAMOTO: So far, we did not get any nitrogen complex at all.

DR. IBERS: Cobalt is unique in that?

DR. YAMAMOTO: In the case of cobalt, yes.

DR. MARKS: Doesn't the tricyclohexyl nickel complex react with nitrogen?

DR. YAMAMOTO: Yes, but we did not try the tricyclohexyl phosphine. Also, with triethyl phosphine complex that Jolly prepared the formation of a nitrogen complex was reported. So we tried to get that complex but in this case, we failed.

DR. IBERS: With chromium, you mean?

DR. YAMAMOTO: With the triethylphosphine Ni(0) complex, and with the chromium complex.

DR. LAPPORTE: Did you ever look at the NMR spectrum of the nickel acetylacetone solution, and trialkyl aluminum or dialkyl aluminum in the absence of phosphines?

DR. YAMAMOTO: No, we did not.

DR. LAPPORTE: We did many years ago, and whereas with the ethyl nickel acac triphenyl phosphine you observe neither a nickel hydride nor a coordinated ethylene. In the absence of phosphine, we observed both coordinated ethylene and the metal hydride.

STUDIES OF HYDRIDO TRANSITION METAL CLUSTER COMPLEXES

Herbert D. Kaesz

Department of Chemistry, University of California,
Los Angeles, California 90024

ABSTRACT

The hydrido-metal cluster complexes comprise a small but grow-
ing group in the larger area of hydrido derivatives of the transition metals of which some 700 are now known. Metal hydrides have
come to be recognized for their role as catalysts or intermediates
in a number of reactions such as the hydroformylation and hydrogenation of olefins, the hydrogen-deuterium exchange and isomerization
of hydrocarbons, and synthetic nitrogen fixation. Only about 25
hydrido-metal clusters are now known.

We have found an example of a highly reactive electronically
unsaturated cluster, $H_4Re_4(CO)_{12}$, analogous in some ways to the coordinatively unsaturated complexes containing only one metal. Structural and spectroscopic studies on this and other hydrido cluster
complexes indicate that hydrogen atoms are located in positions
bridging between metal atoms. Magnetic resonance studies of the
substituted clusters $H_4Ru_4(CO)_{12-x}L_x$ (x = 1,2,3,4) and $H_2Ru_4(CO)_{13-y}L_y$ (y = 1,2) have shown that the hydrogens are involved in intramolecular tautomerism. These phenomena parallel the interstitial
nature and mobility established for hydrogen chemisorbed into metallic phases. The hydrido-metal clusters may thus provide an important link between the study of soluble metal complexes and the
problems of homogeneous catalysis on the one hand with that of solid
state and surface phenomena in connection with heterogeneous catalysis on the other.

Earlier[1] we had discovered the electronically unsaturated cluster $H_4Re_4(CO)_{12}$ containing 56 valence electrons as compared to a
total of 60 which can be accomodated in the full valence shell of a

closed tetrahedral metal cluster.[2] This is at present one of only
two such unsaturated carbonyl clusters known, the other being H_2Os_3-
$(CO)_{10}$,[3] with 46 valence electrons compared to a total of 48 which
can be accomodated in the completed valence shell of a closed triangular metal cluster.

The chemical reactions of $H_4Re_4(CO)_{12}$ illustrates its unsaturated nature. With $NaBH_4$ (Figure 1) two hydride anions are added to
give salts of $[H_6Re_4(CO)_{12}]^{2-}$ a saturated 60 electron complex previously isolated in the reduction of $Re_2(CO)_{10}$ with various reducing
agents.[4]

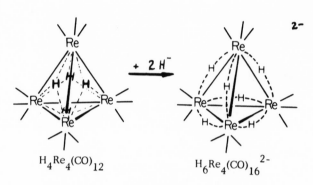

$$H_4Re_4(CO)_{12} \qquad\qquad H_6Re_4(CO)_{16}{}^{2-}$$

Figure 1

Contact between $H_4Re_4(CO)_{12}$ and other substances with well-
developed donor power including solvents such as ether and acetonitrile as well as the usual ligands such as CO or phosphines in hydrocarbon solution brings about reaction at room temperature, in
some cases, instantly. Most of these lead to degradation of the
tetrahedral cluster, as shown in Figure 2.

Of the several reactions in Figure 2, that with CO deserves
special note; at higher temperatures, we obtained $H_3Re_3(CO)_{12}$,
$Re_2(CO)_{10}$ and H_2. This suggested to us the possibility of a direct
synthesis of hydrido-metal clusters through the reverse of this
transformation, namely, the bubbling of hydrogen at atmospheric
pressure into solutions of the metal carbonyls at elevated temperatures. This synthesis is based on the displacement of CO by the
H_2 gas with subsequent formation of hydrido-metal and metal-metal
bonded species. As shown in Figure 3, we have demonstrated this
transformation for a number of carbonyl derivatives.[5] The products
in these reactions are obtained in higher purity than in previous
syntheses and in yields anywhere from 40 to 90% (representing an order of magnitude improvement in most cases).

NaBH4/25°
[H6Re4(CO)12]2-
H4Re4(CO)12
I2/25°
?
νCO: 2020s, 1984s cm-1
PPh3
25°
HRe(CO)4(PPh3)
+
HRe(CO)3(PPh3)2
+
[HRe(CO)3(PPh3)]2
Δ
Re2(CO)10
+
H3Re3(CO)12
+
Re
CO
(1 atm)
25°
H3Re3(CO)12
+
HRe(CO)5 [→ Re2(CO)10 + H2]
Br2
25°
Re(CO)5Br
+
Re2(CO)8HBr

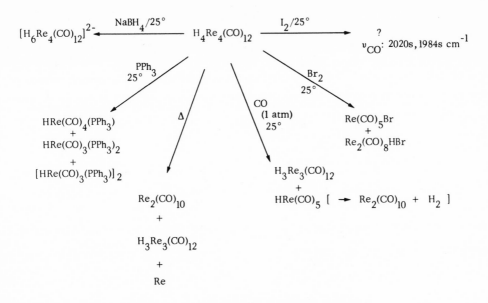

Figure 2. Reactions of $H_4Re_4(CO)_{12}$

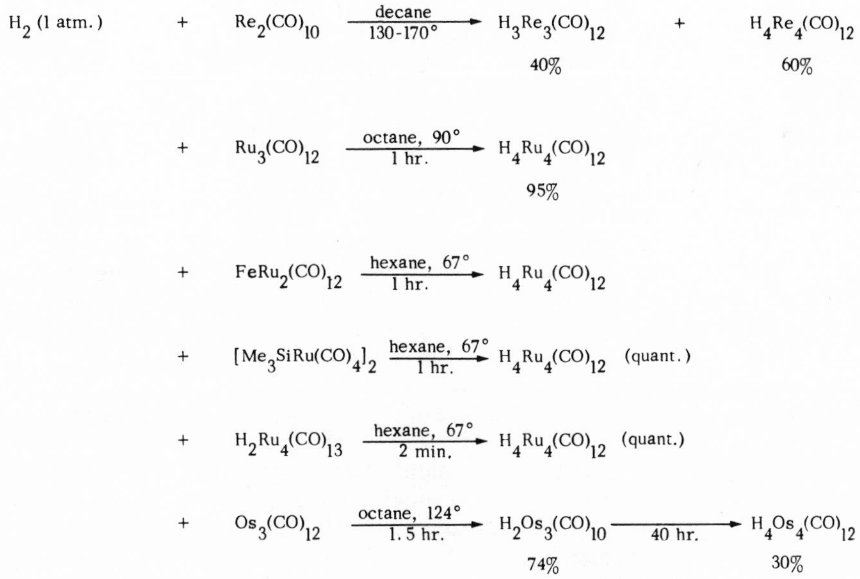

H2 (1 atm.)
+ Re2(CO)10
decane
130-170°
H3Re3(CO)12 40%
+ H4Re4(CO)12 60%
+ Ru3(CO)12
octane, 90°
1 hr.
H4Ru4(CO)12 95%
+ FeRu2(CO)12
hexane, 67°
1 hr.
H4Ru4(CO)12
+ [Me3SiRu(CO)4]2
hexane, 67°
1 hr.
H4Ru4(CO)12 (quant.)
+ H2Ru4(CO)13
hexane, 67°
2 min.
H4Ru4(CO)12 (quant.)
+ Os3(CO)12
octane, 124°
1.5 hr.
H2Os3(CO)10 74%
40 hr.
H4Os4(CO)12 30%

Figure 3. Reaction of hydrogen at low pressure with metal carbonyls.

The advantage of the direct synthesis is nowhere better illustrated than in the case of $H_4Ru_4(CO)_{12}$. This complex had previously been obtained in low yield and in the presence of numerous of the products under a variety of conditions, see Figure 4. $H_4Ru_4(CO)_{12}$ was first isolated in attempts to make $H_2Ru(CO)_4$ by the action of

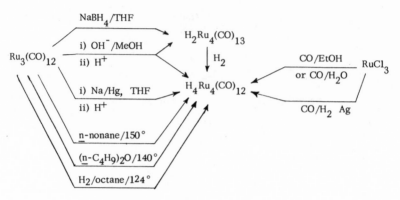

Figure 4

H_2, CO and silver powder on a carbonyl-containing chlororuthenium solution.[6a] This cluster is also obtained through paths involving the carbonyl anions,[6b] from $Ru_3(CO)_{12}$ in refluxing hydrocarbons[7a] or di-n-butyl ether[7b] and through reflux of $RuCl_3$ in alcohol[6b] or aqueous medium[8] accompanied by bubbling CO. The majority of these transformations involve abstraction of hydrogen from the solvent by coordinatively unsaturated species which are undoubtedly formed as intermediates. The formation of hydrido-metal clusters is no doubt facilitated by the presence of elemental hydrogen by which the success of our direct synthesis[5] may thus be understood.

With the pure samples of $H_4Ru_4(CO)_{12}$, we were able through its five principal carbonyl absorptions (Figure 5) to select the D_{2d} structure Figure 6., over the several others of lower symmetry which had been considered[6,7] owing to the presence of increased number of observed carbonyl absorptions arising from impurities.

By contrast the spectrum of $H_4Re_4(CO)_{12}$ contains only two principal carbonyl absorptions, in remarkable similarity to that of $Ir_4(CO)_{12}$ (Figure 7) indicating a much higher symmetry, T_d. An interesting contrast is thus noted between the structures of the two tetra-hydrido tetrametal dodecacarbonyl derivatives which differ in electron population namely $H_4Ru_4(CO)_{12}$, 60 e⁻, saturated, hydrogen most likely edge bridging as in Figure 6 and $H_4Re_4(CO)_{12}$, 56 e⁻, unsaturated, hydrogen most likely face bridging as in Figure 1.

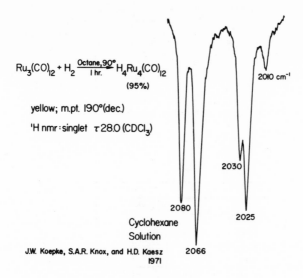

$Ru_3(CO)_{12} + H_2 \xrightarrow[\text{1 hr.}]{\text{Octane, 90°}} H_4Ru_4(CO)_{12}$

(95%)

yellow; m.pt. 190°(dec.)

^1H nmr: singlet τ 28.0 (CDCl$_3$)

2010 cm^{-1}

2030

2080

Cyclohexane
Solution

J.W. Koepke, S.A.R. Knox, and H.D. Kaesz
1971

2025

2066

Figure 5

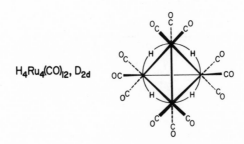

$H_4Ru_4(CO)_{12}$, D_{2d}

Figure 6

A portion of our studies have included Raman and infrared characterization of the hydrido-metal clusters. In all of those which we
we have studied, the hydrogen mode is shifted to lower energy ca.
1100 $\pm$ 300 cm^{-1} and considerably broadened, $\Delta\nu_{\frac{1}{2}} \approx$ 50 to 100 cm^{-1},
from what it is for terminally bonded hydrogen (ν_{M-H} = 1900 $\pm$ 300
cm^{-1}, $\Delta\nu_{\frac{1}{2}} \approx$ 10 to 30 cm^{-1}). This is illustrated in Figure 8 by the
Raman spectra of $H_4Ru_4(CO)_{12}$, $D_4Ru_4(CO)_{12}$, and a sample of $H_2D_2Ru_4$-
$(CO)_{12}$ obtained by direct treatment of $H_2Ru_4(CO)_{13}$ with D_2 (see Figure 3 above). We turned to Raman spectra because in many cases, the
broad absorptions due to hydrogen are difficult to see in the infrared. A position bridging between metals is thus indicated for
the hydrogen atoms which we have confirmed through structure studies
on a number of derivatives.[4,9]

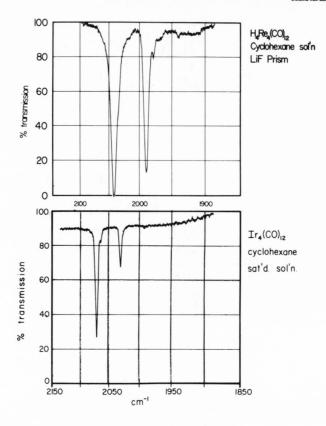

Figure 7

The bridging position for hydrogen results in a lengthening of the metal-metal bonds by about 15% in hydrido-metal clusters over comparable bonds in other metal clusters. This resembles the effects of hydrogen on the metal separations in the interstitial metal hydrides, as may be illustrated for binary compounds in which the metal sub-lattice retains the same symmetry of the free metal, namely (lattice parameter, Å): Ni = 3.52, NiH_x = 3.72 and Pd = 3.89, PdH_y = 4.03.[10]

It has also long been known that hydrogen displays mobility in the interstitial hydrides.[11] Based on the structural analogies mentioned above, this suggested to us that there might be other phenomenological similarities such as intramolecular tautomerism of hydogen in the cluster hydrides. We were therefore led to prepare phos-

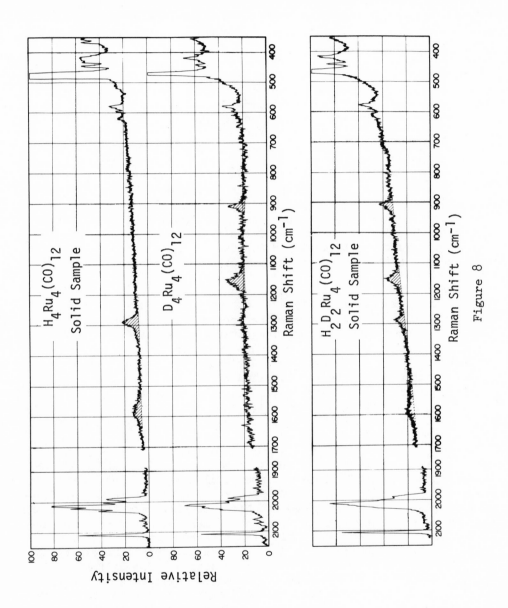

Figure 8

phine substituted derivatives of the hydrido-clusters in order to elucidate such fluxional behavior, it it existed, by means of spin coupling of ^{31}P with hydrogen. Similar conclusions could be sought in the ^{13}C coupling satellites of the unsubstituted clusters either in natural abundance or enriched in ^{13}C, however, solubility of the complexes presented a formidable obstacle. The substituted clusters proved to be more soluble and the nmr spectra of the various derivatives $H_4Ru_4(CO)_{12-x}L_x$ (L = P(OMe)$_3$; X = 1,2,3,4) and $H_2Ru_4(CO)_{13-y}L_y$ (L = P(OMe)$_3$, y = 1,2) were obtained. At rm.t°, these displayed single chemical shift for the metal-bonded hydrogen and averaged coupling to the phosphorus nuclei.[12] Upon lowering the temperature, however, these signals were observed to broaden and then to coalesce; an example of the temperature dependence and a limiting spectrum of the metal hydrogen resonances under optimum low temperature conditions is shown for $H_4Ru_4(CO)_{11}P(OMe)_3$, Figure 9. The activation energies for tautomerism lie in the range 8 to 10 kcal/mole in remarkable proximity to the barriers for diffusion of hydrogen in titanium (9.4 to 10.2 kcal/mole) or in palladium (4.6 to 17.8 kcal/mole) measured first by transport and then by broad line nmr techniques.[11] Two further examples of limiting spectra, those for two derivatives in the substituted $H_2Ru_4(CO)_{13}$ series and assignment of the chemical shifts and coupling patterns are shown in Figure 10.

Treatment of $H_4Ru_4(CO)_{12}$ with aqueous base leads to $[H_3Ru_4(CO)_{12}]^-$ an anion which also displays tautomerism, Figure 11. Under limiting conditions, we observe the spectra of two isomers, one showing a single resonance and the other two resonances with H-H coupling. The areas under the absorptions for the two isomers are in the ratio of 3 (isomer A) to 3.9 (isomer B); furthermore the isomer population can be shifted with polarity of solvent. Although symmetry assignments of C_{3v} and C_{2v}, respectively, are suggested by the limiting nmr spectra, it is not possible at present to choose between various models within each given symmetry, Figure 12.

We have also studied tautomerism and chemical transformations in substituted hydrido-rhenium carbonyl clusters.[13] The substituted derivatives $H_3Re_3(CO)_{12-x}L_x$ (L = Pϕ$_3$, x = 1,2,3; L = Pϕ$_2$Me, or P(OMe)$_3$, x = 1,2) have been prepared and geometrical isomers separated and characterized. Slow equilibration of isomers is observed, and experiments indicate an intra-molecular process. It is thus likely that structures containing bridging carbonyls might participate as high energy intermediates or transition states. The half-life of the tautomerism is on the order of days and represents a much higher barrier than observed in the tautomerism of the tetrahedral hydrido clusters discussed above or recently reported by Cotton et al.[14] for $Rh_4(CO)_{12}$. Experiments to determine the activation energy are under way.

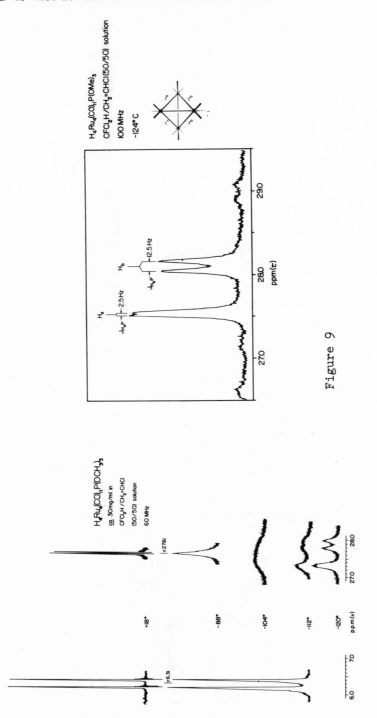

Figure 9

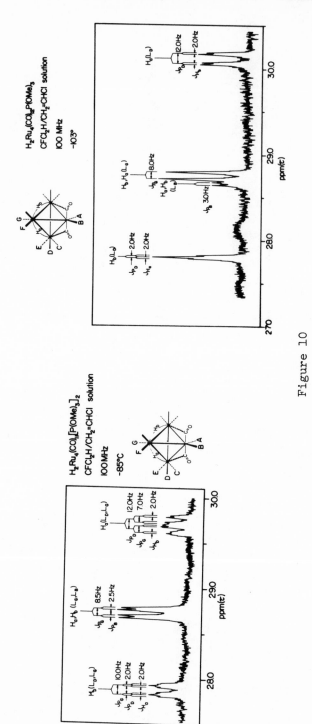

Figure 10

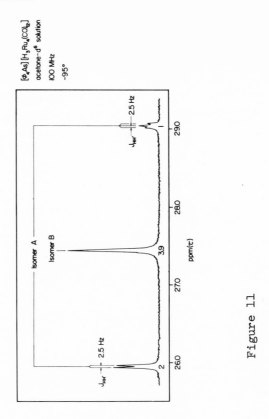

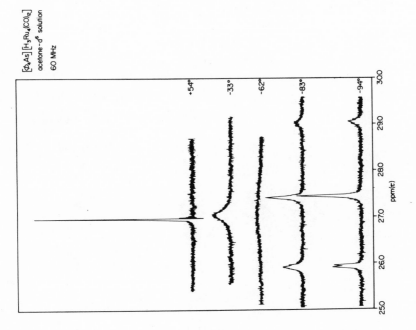

Figure 11

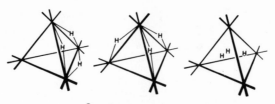

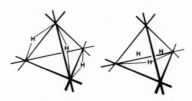

C_{3v} structures of $H_3Ru_4(CO)_{12}^-$

C_{2v} structures of $H_3Ru_4(CO)_{12}^-$

Figure 12

Under extensive heating in octane, $H_3Re_3(CO)_9(P\emptyset_3)_3$ is converted to $HRe_2(CO)_7(P\emptyset_2)(P\emptyset_3)$, a product also obtained in the treatment of $Re_2(CO)_8(P\emptyset_3)_2$ with H_2 at atmospheric pressure and 130° in xylene.[13] This is the first example of the cleavage of a ligand molecule on a hydrido-metal cluster. Among other transformations, a mole of ben-

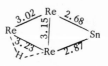

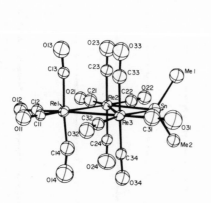

Figure 13

zene is eliminated through combination of a hydrogen of the cluster and a phenyl group of the ligand. Characterization of the dirhenium heptacarbonyl is under way; data indicates bridging both by hydrogen and a diphenylphosphido group between the two bonded rhenium atoms, as required by the diamagnetism of the compound.

The carbonyl anions are expected to be good nucleophiles and through this property provide pathways for the synthesis of new cluster complexes. We have investigated the reaction of $HRe_3(CO)_{12}^{2-}$ with $(CH_3)_2SnCl_2$ as a prototype for syntheses of a new class of planar, triangulated clusters of the $Re_4(CO)_{16}^{2-}$ family.[15] In this manner, an orange air-stable new hydride, $HRe_3(CO)_{12}Sn(Me)_2$ has been obtained[16] and its structure determined, Figure 13. The four metals are essentially planar; each rhenium atom is associated with four terminal carbonyl ligands, two of which can be considered axial and two equatorial. The non-methyl hydrogen atom, though not located, is believed to bridge the long Re-Re distance of 3.23 Å.

ACKNOWLEDGMENT

The following individuals have contributed to the work described in this lecture: Dr. R. B. Saillant, Dr. S. A. R. Knox, Dr. Carolyn Knobler, J. W. Koepke, M. A. Andrews, J. R. Blickensderfer, B. T. Huie and S. W. Kirtley. Work was supported under several grants of the National Science Foundation.

REFERENCES

1. R. Saillant, G. Barcelo and H. D. Kaesz, J. Amer. Chem. Soc., 92, 5739 (1970).
2. P. Chini, Inorg. Chim Acta Reviews, 2, 31 (1968).
3. B. F. G. Johnson, J. Lewis and P. A. Kilty, J. Chem. Soc. (A), 2859 (1968).
4. H. D. Kaesz, B. Fontal, R. Bau, S. W. Kirtley, and M. R. Churchill, J. Amer. Chem. Soc., 91, 1021 (1969).
5. H. D. Kaesz, S. A. R. Knox, J. W. Koepke and R. B. Saillant, Chem. Comm., 477 (1971).
6. (a) J. W. S. Jamieson, J. V. Kingston, and G. Wilkinson, Chem. Comm., 569 (1966);
   (b) B. F. G. Johnson, R. D. Johnston, J. Lewis, B. H. Robinson, and G. Wilkinson, J. Chem. Soc. (A), 2856 (1968).
7. (a) B. F. G. Johnson, R. D. Johnston, and J. Lewis, J. Chem. Soc. (A), 2865 (1968);
   (b) B. R. G. Johnson, J. Lewis, and I. G. Williams, ibid., 901 (1970).
8. B. R. James and B. L. Rempel, Chem. & Ind., 1036 (1971).
9. M. R. Churchill, P. H. Bird, H. D. Kaesz, R. Bau, and B. Fontal, J. Amer. Chem. Soc., 90, 7135 (1968).

10. cf. G. G. Libowitz, "The Solid State Chemistry of Binary Metal Hydrides", W. A. Benjamin (1965) and refs. cited therein.
11. (a) B. Stalinski, C. K. Coogan, and H. S. Gutowsky, J. Chem. Phys., 34, 1191 (1961);
    (b) see also Y. Ebisuzake and M. O'Keefe, Progress in Solid State Chemistry, H. Reiss, editor, 4, 187 (1968) and refs. cited therein.
12. "Intramolecular Tautomerism of Hydrogen in Hydrido Carbonylmetal Clusters", S. A. R. Knox and H. D. Kaesz, J. Amer. Chem. Soc., 93, 4594-4596 (1971).
13. "Tautomerism and Chemical Transformations in Substituted Hydrido-rhenium Carbonyl Cluster Complexes", R. B. Saillant, M. A. Andrews, J. R. Blickensderfer, and H. D. Kaesz, 165th Nat'l Meeting of the A.C.S., Dallas, Texas, April 8-13; paper INOR-33.
14. F. A. Cotton, L. Kruczynski, B. F. Shapiro, and L. F. Johnson, J. Amer. Chem. Soc., 94, 6191 (1972).
15. R. Bau, B. Fontal, H. D. Kaesz, and M. R. Churchill, J. Amer. Chem. Soc., 89, 6374 (1967).
16. "The Crystal and Molecular Structure of μ-(dimethylstannado)-μ-hydrido-dodecacarbonyl Trirhenium", B. T. Huie, C. B. Knobler, and H. D. Kaesz, paper in preparation.

DISCUSSION

DR. OGOSHI: Is it possible to study the structure of $H_4Re_4(CO)_{12}$ by far-infra-red spectroscopy? I think Re-Re stretching vibrations should occur at about 100 cm^{-1}.

DR. KAESZ: We have thus far had access only to Raman; that compound is dark red, and decomposes in the Raman beam, so we have not seen its low energy vibrations. We are collaborating with Tom Spiro who has methods of diluting colored compounds in NaCl and pelletizing them. What you say is correct - the low energy bands would be helpful in understanding its structure. This compound has not given single crystals suitable to study by crystallography yet.

A FACILE FORMATION OF π-ALLYLPLATINUM COMPLEXES BY DECARBOXYLATION
OF ALLYLOXYCARBONYLPLATINUM COMPLEXES: IMPORTANCE OF π-OLEFIN CO-
ORDINATION IN π-ALLYLPLATINUM FORMATION

Hideo Kurosawa[*] and Rokuro Okawara

Department of Applied Chemistry, Osaka University, Suita,
Osaka 565 Japan

SUMMARY

    Allyloxycarbonylplatinum(II) complexes readily afforded π-al-
lylplatinum(II) complexes by decarboxylation under several reaction
conditions. The importance of olefin coordination to a platinum a-
tom in the transition state is suggested. The other novel routes
to π-allylplatinum(II) complexes which apparently involve a similar
reaction intermediate are also described.

    π-Allylmetal complexes are often found to play an important
role in many transition metal catalyzed organic syntheses[1], where
the most extensively studied metal complexes include those of nickel and palladium. The way in which such π-allylmetallic complexes
can be formed spans many classes of reactions according to the nature of reactants, namely both metal complexes and organic substrates.
Typical examples are shown in eq 1-5. Eq 1 is the most widely applicable by the use of active allylmetal compounds such as the Grignard
reagents. Eq 2 represents oxidative addition or nucleophilic substitution, and eq 3 addition of M-X bond to conjugate or cumulative
double bonds. The product in eq 4 is often found as an intermediate
in metal catalyzed oligomerization and polymerization of butadiene.
A similar type product may be obtained from allene. Eq 5 shows the
conversion of π-olefin-metal complexes to π-allylmetal complexes on
treatment with a base. Although most of the known methods to prepare π-allylmetal complexes essentially parallel those for preparing
simple alkylmetal complexes (e.g., eq 1-3), there are several reactions which appear somewhat specific for the π-allylmetal formation
(e.g., eq 4 and 5). We report here a ready conversion of allyloxy-
carbonylplatinum(II) complexes into π-allylplatinum(II) complexes[2]

$$M-X + M'CH_2CH=CH_2 \longrightarrow \left\langle \!\!\! \begin{array}{c} \cdots M \end{array} \right. + \ M'-X \tag{1}$$

$$M^{(n)} + CH_2=CHCH_2X \longrightarrow \left\langle \!\!\! \begin{array}{c} \cdots M^{(n+2)} \end{array} \right. \tag{2}$$

(X= halogen, OH, OCOMe, OR etc.)

$$M-X + CH_2=C=CH_2 \longrightarrow X\!\!-\!\!\left\langle \!\!\! \begin{array}{c} \cdots M \end{array} \right.$$

$$\text{or } CH_2=CH-CH=CH_2 \qquad\qquad \text{or } \left\langle \!\!\! \begin{array}{c} \cdots M \\ CH_2X \end{array} \right. \tag{3}$$

(X= H, R, halogen, OR etc.)

$$M^{(n)} + 2CH_2=CH-CH=CH_2 \longrightarrow \left\langle \!\!\! \begin{array}{c} M^{(n+2)} \end{array} \right\rangle \tag{4}$$

$$\begin{array}{c} C' \\ \| \cdots M \\ C \\ \searrow CH \end{array} \xrightarrow{\text{Base}} \left\langle \!\!\! \begin{array}{c} \cdots M \end{array} \right. \tag{5}$$

where olefin coordination to Pt atom is thought to play an important role in the transition state. It may be noted that no alkylmetal complexes have been reported to be produced by decarboxylation of analogous alkoxycarbonyl complexes of metals. Also discussed in this lecture will be other novel routes to π-allylplatinum(II) complexes mechanistic aspects of which seem to be closely related to that of the decarboxylation above.

A synthetic scheme for the formation of π-allylplatinum(II) complexes by decarboxylation is shown in Scheme 1. As was reported previously[3], $[PtCl(PR_3)_2(CO)]^+$ are very susceptible to nucleophilic attack by alcohols to give allyloxycarbonylplatinum(II) complexes I. I could also be prepared by transesterification of $PtCl(COOMe)(PR_3)_2$. The infrared spectra of I showed two intense absorption bands at around 1640 cm^{-1} and 1050 cm^{-1}, characteristic of the PtCOOR moiety. We believe complexes I have trans square planar configuration without any interaction between the C=C bond and the Pt atom for the following reasons; 1) the olefinic proton signals in the NMR spectra of I in $CDCL_3$ showed no couplings with ^{195}Pt and ^{31}P nuclei, 2) two allylic methylene protons were observed magnetically equivalent, 3) the phosphine methyl proton signals in Ic and Id appeared as clear 1:2:1

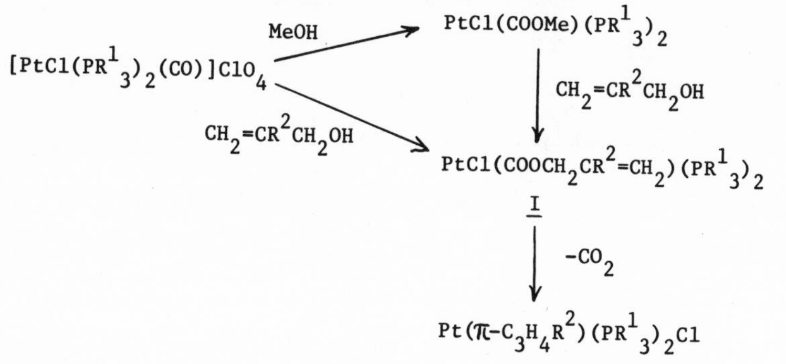

$$[\text{PtCl}(PR^1_3)_2(CO)]ClO_4 \xrightarrow{\text{MeOH}} \text{PtCl}(COOMe)(PR^1_3)_2$$

$$\downarrow CH_2=CR^2CH_2OH$$

$$[\text{PtCl}(PR^1_3)_2(CO)]ClO_4 \xrightarrow{CH_2=CR^2CH_2OH} \text{PtCl}(COOCH_2CR^2=CH_2)(PR^1_3)_2$$

$$\underline{I}$$

$$\downarrow -CO_2$$

$$\text{Pt}(\pi\text{-}C_3H_4R^2)(PR^1_3)_2Cl$$

(<u>Ia</u> PR^1_3= PPh$_3$, R^2= H; <u>Ib</u> PR^1_3= PPh$_3$, R^2= Me; <u>Ic</u> PR^1_3= PPh$_2$Me, R^2= H;
<u>Id</u> PR^1_3= PPh$_2$Me, R^2= Me)

Scheme 1

triplets accompanied by ^{195}Pt satellites, 4) the ν(Pt-Cl) value for
Ic (280 cm^{-1}) is much the same as that for trans-PtClMe(PPh$_2$Me)$_2$
(269 cm^{-1})[4].

A decarboxylation of $\underline{I}$ could be performed in several ways. Thus,
$\underline{Ia}$ and $\underline{Ic}$ were easily converted to Pt(π-C$_3$H$_5$)(PR$^1_3$)$_2$Cl in refluxing
benzene solution as shown in the scheme. [3] $\underline{Ia}$ and $\underline{Ic}$ also were found
to decompose slowly in CDCl$_3$ even at room temperature, but $\underline{Ib}$ and $\underline{Id}$
were more reluctant to undergo similar decarboxylation. Particularly
interesting is the fact that treatment of $\underline{Ia}$, $\underline{Ic}$ and $\underline{Id}$ with an equivalent amount of AgClO$_4$ in CH$_2$Cl$_2$/acetone gave the corresponding π-
allylplatinum(II) salts much more readily than the thermal process.

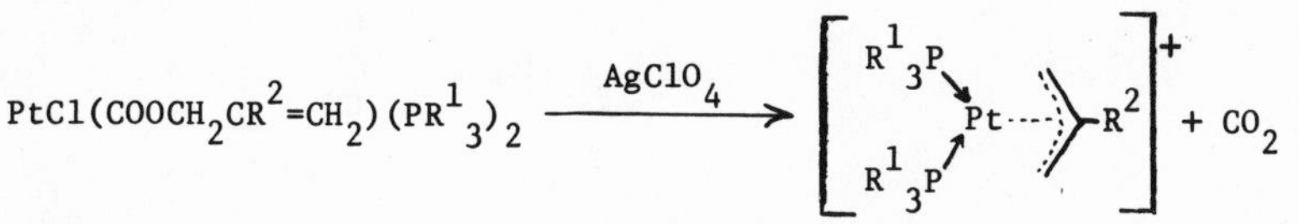

$$\text{PtCl}(COOCH_2CR^2=CH_2)(PR^1_3)_2 \xrightarrow{AgClO_4} \left[\begin{array}{c} R^1_3P \\ \\ R^1_3P \end{array} Pt \cdots \underset{}{\bigtriangledown} R^2 \right]^+ + CO_2 \qquad (6)$$

Furthermore, addition of a catalytic amount of SnCl$_2$ to a CH$_2$Cl$_2$ solution of $\underline{Ia}$ resulted in marked acceleration of decarboxylation.
Such enhanced reactivities of $\underline{I}$ by treatment with either AgClO$_4$ or
SnCl$_2$ may be compared with the ready addition of Pt-H bond of trans-
PtHCl(PR$_3$)$_2$ to a C=C bond of olefins when silver salts[5] or SnCl$_2$[6]
used as catalysts. The role of these catalysts in the olefin insertion was ascribed to the stabilization of an intermediate π-olefin-
platinum complex through removal of the chloride ligand by the former reagent or formation of SnCl$_3^-$ ligand with highly trans activating
ability in the latter case. Consequently, we assume that the decarboxylation described above proceeds through the transition state in-

volving intramolecular coordination of the C=C bond to the Pt atom.

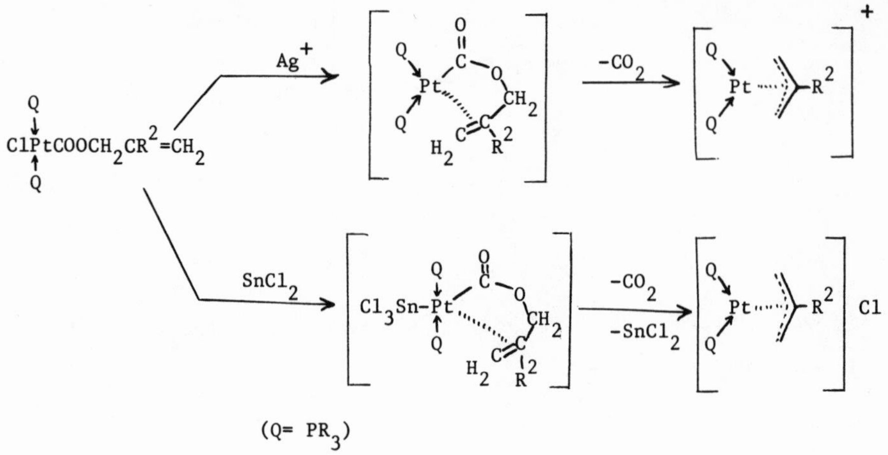

(Q= PR_3)

Scheme 2

It is of interest to note that occurrence of such cyclic intermediates has a precedent in the other way by which π-allylplatinum
(II) complexes are formed as shown in Scheme 3. Thus, a possible
sequence for the reaction of cationic platinum(II) hydrido complexes
with diallyl ethers was suggested[7] to involve an initial double-bond
migration from terminal to internal position followed by the forma-

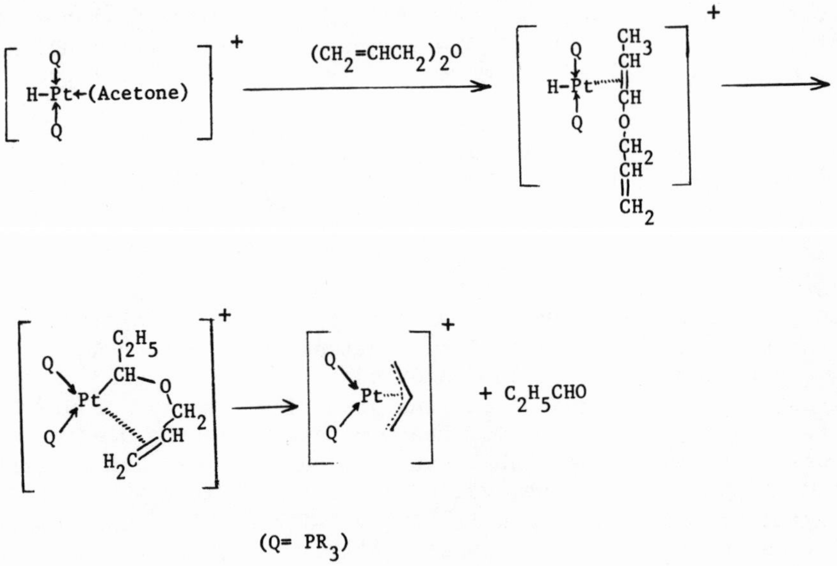

(Q= PR_3)

Scheme 3

tion of the cyclic organoplatinum(II) intermediate by olefin insertion. Elimination of an aldehyde moiety from this intermediate com-
pletes the reaction. Support for occurrence of such intramolecular
coordination comes from the fact that the reaction of cationic pla-
tinum(II) hydrido complexes with allyl acetates readily yielded stable organoplatinum(II) chelate compounds as shown in eq 7.

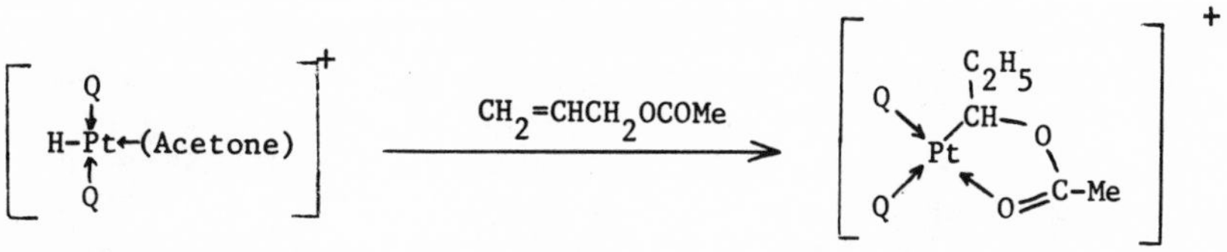

$$\left[H-\overset{Q}{\underset{Q}{Pt}}\leftarrow(Acetone) \right]^{+} \xrightarrow{CH_2=CHCH_2OCOMe} \left[\begin{array}{c} Q \diagdown \\ \\ Q \diagup \end{array} Pt \begin{array}{c} C_2H_5 \\ | \\ CH-O \\ \diagdown \\ O=C-Me \end{array} \right]^{+} \quad (7)$$

We could then suggest that the formation of π-olefin-metal intermediate is an important step not only in the specific preparation
of π-allylmetal complexes such as shown in eq 4 and 5, but in some
of other reactions to obtain π-allyl complexes such as oxidative addition or nucleophilic displacement. Although occurrence of olefin
coordination is apparently essential, such π-interaction may or may
not be followed by π→σ interconversion to form a σ-allylmetallic intermediate such as shown below for the present reactions.

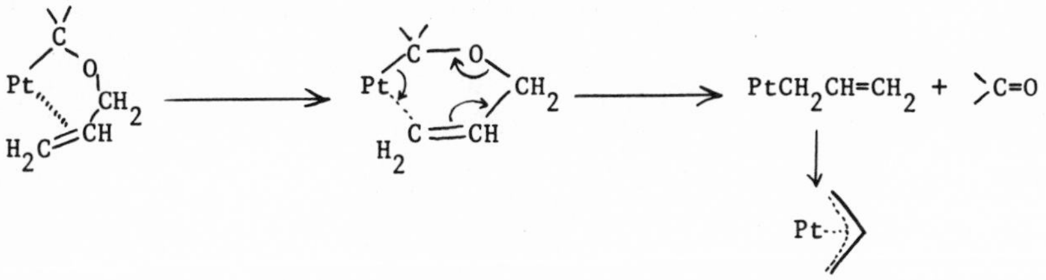

ACKNOWLEDGEMENTS

Thanks are due to Professor H. C. Clark, University of Western
Ontario for help in part of this work.

REFERENCES

1. See e.g., R. Baker, Chem. Rev., 73, 487 (1973).
2. Part of this work was reported; H. Kurosawa and R. Okawara, J.
   Organometal. Chem., 71, C35 (1974).
3. H. C. Clark and W. J. Jacobs, Inorg. Chem., 9, 1229 (1970) and
   references cited therein.
4. J. D. Ruddick and B. L. Shaw, J. Chem. Soc., (A) 2801 (1969).
5. H. C. Clark and H. Kurosawa, Inorg. Chem., 11, 1275 (1972).
6. R. Cramer and R. V. Lindsey, Jr., J. Amer. Chem. Soc., 88, 3534

(1966).
7. H. C. Clark and H. Kurosawa, Inorg. Chem., 12, 357 (1973).

DISCUSSION

DR. YAMAMOTO: Have you observed any CO_2 insertion reaction?
DR. KUROSAWA: No, not yet, not at all.
DR. HALPERN: A comment: You assumed in all these cases that
what the metal does is attack the olefinic bond, to pi-interact, but
there is another possibility. It is known that if you have an allyl
attached to a metal, then an electrophile, either a proton or another
metal ion can attack the allylic position, forming a sigma bond, and
shifting the double bond, which then comes away from the metal. With
propane, you will get propylene eliminated, or you get transfer to
mercuric ion. You can also have that kind of interaction internally,
the metal ion acting as electrophile and forming a sigma bond with
the terminal carbon atom, shifting the double bond, which can later
come back. If you make a cationic platinum complex, you would expect its electrophilic character to be enhanced. One could equally
well rationalize this chemistry in terms of sigma rather than pi interaction. It is a known type of reaction externally.

OPTICAL INVERSION AND SUBSTITUTION OF COORDINATED OLEFINS IN OPTI-
CALLY ACTIVE TRANS(N)-CHLORO-L-PROLINATOOLEFINPLATINUM(II) COMPLEXES
IN ORGANIC SOLVENTS

Kazuo Saito[*], Kazuo Konya, Junnosuke Fujita, Hiroaki
Kido, and Yoshiro Terai
Department of Chemistry, Faculty of Science, Tohoku Uni-
versity, Sendai, Japan 980

Optical isomerism due to asymmetric coordination of some olefins
such as propylene and trans-2-butene was first studied by Paiaro and
Panunzi.[1] We have synthesized diastereomers of the type trans(N-
olefin)[PtCl(L-prol)(olefin)] crystalline,[2] and discussed the absolute configuration of coordinated olefins on the basis of circular
dichroism (CD) spectroscopy.[3] The complexes with such olefins as
trans-2-butene and 2-methyl-2-butene gave no change in u.v. absorption and CD in organic solvents such as ethanol, chloroform and acetone. When the same olefin as the ligand was added, the CD changed,
the u.v. absorption remaining unchanged. (Figure 1) This fact suggests that only the inversion of the coordinated olefin takes place.
(Figure 2) The rate of decrease in CD strength obeyed the second order rate law, and no solvent path was observed. (Figure 3,4)

$$\text{Rate} = k_2[\text{complex}][\text{ligand}] \tag{1}$$

The results are summerized in Table 1 together with the activation
parameters.

The small E_a and the big absolute value of $\Delta S^{\ddagger}$ indicate that
inversion would take place through S_N2 mechanism. A probable reaction intermediate is shown in Figure 5. Proline has a large pyroli-
dine ring and the olefin may approach from only one side of the
square planar plane to results in big minus $\Delta S^{\ddagger}$ values. Acetone is
a very weak nucleophile and the contribution of the solvent path is
negligible.

311

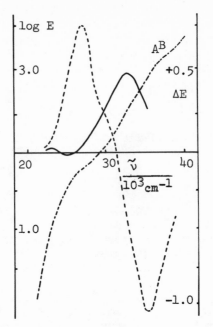

Figure 1. Ultraviolet absorption and circular dichroism spectra of $(+)_{380}^{\Delta\varepsilon}$-trans(N, olefin)-[PtCl(L-prol)(trans-2-butene)] in ethanol.

-.-.-. absorption,   ---- CD, fresh soln.,   _____ CD, after in-
                                                          finite time

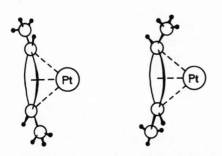

Figure 2. S,S and R,R configuration of coordinated trans-2-butene.

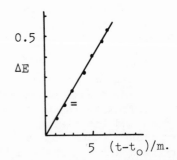

Figure 3. CD Change at 380 nm in
acetone. 0.0030 M [PtCl(L-prol)
(2-methyl-2-butene)] and 0.50 M
2-methyl-2-butene at 23.6°C.

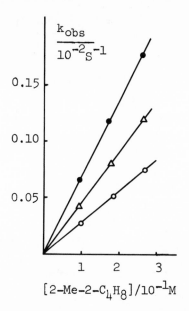

Figure 4. Relationship between k_{obs} and the concentration of added 2-methyl-2-butene in acetone.

——o—— 3.7 ± 0.3   ——△—— 13.2 ± 0.3   ——●—— 21.6 ± 0.3°C

Table 1

Second order rate constants and activation
parameters for the inversion of coordinated olefins in
[PtCl(L-prol)(olefin)] in acetone.

Ligand	t °C	k_2 $10^{-3}M^{-1}s^{-1}$	E_a kcal·mol^{-1}	$\Delta S^{\ddagger}$ e.u.
2-methyl-2-butene	13.2	4.44 ± 0.02	7.8	-44
trans-2-butene	17.0	6.3 ± 0.1	9.0	-39

Substitution of Olefins

When another kind of olefin was added to trans(N)[PtCl(L-prol)-(S,S-trans-2-butene)] in acetone, the change in u.v. absorption is very small, but the CD strength decreases with time. The change in u.v. spectrum was so small that the substitution rate was not followed from its change. However, the change in CD pattern is quite similar to that in Figure 1 and the final pattern is equal to that of trans(N)[PtCl(L-prol)(ethylene)] in which only the vicinal effect of coordinated L-proline is responsible. Hence the rate of substitution reaction was followed by measuring the decrease in CD strength at 380 nm.

$$[PtCl(L\text{-}prol)(S,S\text{-}trans\text{-}2\text{-}butene)] + olefin$$

$$\rightleftharpoons [PtCl(L\text{-}prol)(olefin)] + trans\text{-}2\text{-}butene \qquad (2)$$

With change in initial concentrations of the complex and the free olefin, similar diagrams as Figure 3 and 4 were obtained. The results are summarized in Table 2.

Table 2
Second order rate constants and activation parameters
for the substitution of various olefins for trans-2-butene
in [PtCl(L-prol)(S,S-trans-2-butene)] in acetone.

Added olefin	k_2 a) $10^{-3}M^{-1}s^{-1}$	E_a $kcal \cdot mol^{-1}$	$\Delta S^{\ddagger}$ $cal \cdot K^{-1}mol^{-1}$
Styrene	621	8	-34
cis- } 2-butene	880	10	-26
trans-)	9.7	9	-39
2-methyl-2-butene	7.3	9	-41
cis- } 1,4-dichloro-	310	8	-36
trans-) 2-butene	19.2	8	-40
cis-2-butene-1,4-diol	775	9	-29
trans-1,4-dibromo-2-butene	22	11	-31
cis- } 1,2-dichloro-	1.1	9	-44
trans-) ethylene	0.22	11	-40
maleic acid	47	10	-32
fumalic acid	ca.1 b)		
2,3-dimethyl-2-butene	0.22	9	-46

Table 2 continued

a) at 23.6°C measured by the decrease in CD strength due to the coordinated olefins.   b) at 9.0°C.

It is clearly seen that the cis-isomer of a given olefin gives greater second order rate constant that the trans-isomer by several tens times. When the same geometrical isomers are compared, it is evident that the substitution of chloro, bromo and hydroxy for the methyl protons of 2-butene gives only slight difference on the rate constant. Replacement of the methyl groups of 2-butene by carboxyl and chloro decreases the rate in this order, corresponding to the increase in electronegativity of the substituents. Further comparison suggests that the number of substituting groups on one side of –C=C– seems to give significant influence. Styrene, cis-isomers of the olefins and maleic acid give the largest k_2's. 2,3-Dimethyl-2-butene with two methyls on each side gives the smallest. Others with only one substituent on either side give intermediate values. The Arrhenius activation energies do not differ much from one another. Hence, in addition to the electronic effect mentioned above, steric influence in the rate determining step should play an important role. This is quite reasonable, since the substitution reaction proceeds only through the reagent path.

Table 3
Second order rate constants and activation parameters
of the substitution of 2-butene for coordinated S,S-trans-
2-butene of L-alaninato and L-prolinato complex of platinum(II).

Complex [a]	Added 2-butene	k_2 [b] $10^{-3}M^{-1}s^{-1}$	E_A kcal·mol^{-1}	$\Delta S^{\ddagger}$ cal·K^{-1}mol^{-1}
L-alaninato	trans	30.2	9.0	-35
L-prolinato	trans	4.1	9.2	-39
L-prolinato	cis	347	10.2	-26

a) [PtCl(L-am)(S,S-trans-2-butene)]   b) 8.0°C in acetone, measured by the loss of optical activity (decrease of CD strength).

Table 3 gives the second order rate constants and activation parameters of the substitution reaction of 2-butene for coordinated S,S-trans-2-butene in [PtCl(L-am)(S,S-trans-2-butene)], measured by

the decrease in CD strength of the complex. The alaninato complex
gives larger k_2 than the prolinato complex does with trans-2-butene
as nucleophile. This fact also supports the importance of steric
factor in the rate-determining step. L-prolinato anion with bulky
pyrrolidine ring on one side of the square plane may hinder the approach of the nucleophile.

Isotopic Exchange

Trans-2-Butene labelled with ^{3}H was prepared by the reduction
of trans-1,4-dibromo-2-butene with lithium aluminum triride LiAl^{3}H$_4$.
The complex [PtCl(L-prol)(S,S-trans-2-butene-^{3}H)] was prepared with
this labelled ligand and its substitution reaction with cis- and
trans-2-butene was examined in acetone. No net chemical change took
place during the kinetic runs, because the extinction at 280 nm remained unchanged. The results are shown in Table 4. The k_2(opt) and
the k_2 (iso) are equal within experimental error for the substitution
with cis-2-butene. Hence the decrease in CD strength should be exclusively due to the inter-molecular substitution, and there seems to
be no hydrogen exchange between the free and the coordinated olefin.

Table 4

Comparison of the second order rate constants of
the substitution of cis- and trans-2-butene for S,S-trans-
2-butene in [PtCl(L-prol)(S,S-trans-2-butene)] measured by
the loss of optical activity and by the isotopic exchange.

Added 2-butene	temp. °C	k_{opt} $10^{-3}M^{-1}s^{-1}$	k_{iso} $10^{-3}M^{-1}s^{-1}$
cis	8.0	347	———
	-20.0	70.9 ± 7.6	70.2 ± 4.1
trans	8.0	6.2	32.3
	-20.0	0.9	5.6

Rate constants k_{opt} and k_{iso} are second order rate constants, in
acetone.

Whenever the inversion of a coordinated olefin proceeds exclusively through an inter-molecular mechanism, the second order rate
constant measured by the decrease in CD strength should be the double of the second order rate constant for the inversion reaction,
k_2(opt) = $2k_2$(inv). On the other hand, the second order rate con-

stant for the isotopic exchange should be the sum of those for the inversion and retention process, $k_2(iso) = k_2(ret) + k_2(inv)$. Hence, if there were no stereoselective effect at the rate determining step $k_2(ret) = k_2(inv)$, the second order rate constants should be equal, $k_2(opt) = k_2(iso)$. Substitution with trans-2-butene shows significant stereo-selectivity. The retention is more feasible than the inversion so that the $k_2(ret)/k_2(inv)$ value is ca. 10.

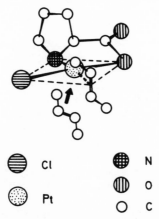

⊖	Cl	⊕	N
⊙	Pt	⊕	O
		○	C

Figure 5. Probable attack of the free olefin upon the complex.

Both the coordinated L-prolinate and the S,S-trans-2-butene could be responsible for the stereo-selectivity. However, as stated before, the final CD pattern, long time after the addition of free olefin, is almost identical with that of [PtCl(L-prol)(ethylene)] and no stereo-selectivity was appreciable at the equilbrium state. Hence L-prolinate ligand cannot be the source of the selectivity. Studies with molecular models suggest that there seems to be different steric interaction when trans-2-butene attacks the complex [PtCl(L-prol)-(S,S-trans-2-butene)], according to the orientation of the incoming ligand. Lewis and co-workers[4] studied the pmr spectrum of [PtX(acac)-(olefin)] (X = Cl, Br; olefin = methylated ethylenes) and observed coallescence of signals at 220 to 300K. The coordinated olefin seems to have free rotation around the Pt-olefin axis. Our optical active complex [PtCl(L-prol)(S,S-trans-2-butene)] gave only one methyl signal over -20°C but the free rotation was not verified. Nevertheless, the S,S-configuration of trans-2-butene must be responsible for the stereo-selectivity on the S_N2 attack.

REFERENCES

1. G. Piaro and A. Panunzi, Ric. Sci. Rend, Sez., A, 4, 601 (1964).
2. K. Konya, J. Fujita, and K. Nakamoto, Inorg. Chem., 10, 19 (1971).

3. K. Konya, J. Fujita, H. Kido and K. Saito, Bull. Chem. Soc. Japan, 45, 2161 (1972).
4. C. E. Holloway, G. Hulley, B. F. G. Johnson and J. Lewis, J. Chem. Soc., (A), 1969, 53; 1970, 1653.

DISCUSSION

DR. FALLER: I would just like to make one comment. I believe all of the mechanisms you have. As a general thing with relationship to overall aspects of adding olefins, and doing asymmetric synthesis, I think it shows up as a general rule, that anything that is four-coordinate and square planar will not show any sterospecificity itself. When you work with the aryl complex you never see any stereo selectivity with an optically active group on the olefin. Whenever you go to a 5-coordinate compound, that is the first chance you get of producing any stereoselectivity. As a point of reference, when one tries to design this kind of thing you almost have a mechanism which goes through at least 5-coordination, to get stereoselectivity.

DR. SAITO: The stereoselectivity should come from the S_n2 attack as the olefin approaches the metal to give effective 5-coordination.

DR. FALLER: My main question was, were you aware of any cases where anyone has actually seen stereoselectivity in four-coordinate compounds?

DR. SAITO: No, not so far as I know.

DR. PETTIT: In (Chloro)(L-prolinato)(Trans-2-buten)Platinum(II) is the top and the bottom of the complex the same?

DR. SAITO: No. When you change L-proline to alanine the rate increases. We have not studied the stereoselectivity of the alanine complexes yet, but the proline complexes have the smallest k_2 values among amino acids of the same type.

DR. PETTIT: My point was, that when you bring in the trans-2-butene you are making a new asymmetric center, and retention versus inversion might be associated with the rate at which you can form this new asymmetric center, not so much with the rotation of the olefin.

DR. SAITO: Actually, the presence of the pyrrolidine ring seems to allow the approach of the incoming trans-2-butene only from the rear side. I do not know which is the rear and which is the front, but from only one side of the olefin. Pyrrolidene comes this way so that it does not exclusively hinder the approach, but it will make the approach from that side more difficult than the approach from the other side. So preferentially, the trans-2-butene would come not from this side of the square plane, but from the other side, where the pyrrolidine does not cover the square planar complex. However, as a matter of fact, the olefin seems to be rotating because only one kind of methyl signal was observed in the pmr spectrum of the trans-2-butene complex at temperatures down to -20°. It would be easier for the incoming trans-2-butene to approach the apical site when the coordinated trans-2-butene assumes such an orientation.

DR. PETTIT: Even if it comes from the same side, it can come up either like this, or like this, to give either retention or inversion of the complex.

DR. SAITO: Yes, or inside out.

DR. PETTIT: And the rate at which that comes in would determine the retention or inversion of configuration.

DR. SAITO: Yes.

DR. PETTIT: Which is inducing the selectivity?

DR. SAITO: If the L-proline had been the source of the asymmetric coordination, the final pattern of the C.D. spectra would not be equal to that of, say, the ethylene complex, which has no other source of optical activity than coordinated L-proline, so that there is no selectivity at the equilibrated state, it is kinetic selectivity.

DR. BERGMAN: Yes, but which of the ligands induces the selectivity?

DR. SAITO: I would say that the S,S-trans-2-butene is the source. Otherwise, the final pattern of the "racemized" species will have some C.D. coming from asymmetric coordination of trans-2-butene.

CONFORMATIONAL EQUILIBRIA AND STEREOSPECIFIC SYNTHESIS

Jack Faller

Department of Chemistry, Yale University, 225 Prospect
Street, New Haven, Connecticut 06520

Pi allyl-complexes have been used as reactants or have been
postulated as intermediates in many of the reactions discussed at
this symposium. Our studies of rearrangement mechanisms and conformational equilibria often provide a rationale for the product
distributions and stereochemistry in these reactions. It appears
that steric factors tend to control stereochemistry. Kinetic and
thermodynamic parameters of some palladium complexes will be reviewed
in the perspective of their implications in rearrangements, stereospecific reactions, and racemization processes.

Nuclear magnetic resonance studies readily allow the identifi-
cation of isomers of the 1-3-π^3-allyl systems. Placement of a bulky
substituent at the central carbon atom causes a predominance of an
<u>anti</u> configuration at the terminal carbon atom for the other substi-
tuent; whereas for a nonbulky group only ∿5% is found in the <u>anti</u>
configuration.

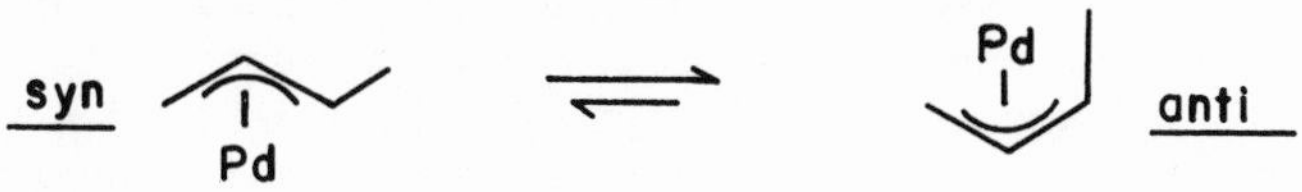

1-3-π^3-allyl Pd LX	% anti
1-methylallyl	5%
1-acetylallyl	5%
1,2-diphenylallyl	60%
1-acetyl-2-methylallyl	70%
1-acetyl-2-phenylallyl	100%

The mechanism of inconversion of these isomers has been clearly

321

shown to be pi→sigma→pi interconversions.[1] This is clearly illustrated in the temperature dependence of the proton magnetic resonance
spectrum of 1,1-dimethylallyl derivatives. These spectra show averaging of the <u>syn</u> and <u>anti</u> protons at 98°C, which corresponds to rapid
interchange of these protons on the nmr time scale (c. 10 sec^{-1}).
The methyl resonances, however, do not begin to average until 135°C
is reached.

The features of importance here are: 1) the rearrangement is quite
facile ($\Delta G^* = 15-24$ kcal/mole); 2) formation of a sigma bond at an unsubstituted end of the allyl is more favorable energetically; 3) formation of the 3-η allyl allows interchange of <u>syn</u> and <u>anti</u> substituents on the 3-carbon atom; whereas the substituents on the 1-carbon
atom retain their geometric relationship; 4) the chirality at carbon
atom #1 is inverted on 3-η formation (i.e., the metal is bonded on
the opposite side of the allyl carbon skeleton).

One can generally anticipate 1-3 kcal/mole increase in free energy of activation per substituent at the carbon atom to which the
σ-bond is formed. For example, free energies of activation vary from
18.9 to 20.9 to 23.7 kcal/mole for sigma-bond formations at unsubstituted, monomethyl substituted and dimethyl substituted terminal
carbon atoms.

Formation of a σ-bond at carbon-1 provides a pathway for the
interconversion of <u>syn</u> and <u>anti</u> isomers of 1,2-disubstituted-1-3-η^3-
allylpalladium complexes. In solution an equilibrium mixture of both
<u>syn</u> and <u>anti</u> forms are found; however, the interconversion is suffi-
ciently facile that upon crystallization only the more stable (in the
solid state) configuration is isolated. We have shown that the configuration isolated with certain crystalline π-allylpalladium chloride
dimers is dependent on both the bulkiness and the electronic nature

of the substituents.[2] Hence, formation of the amine derivative by
splitting of the dimer may allow isolation of the anti isomer in the
solid; whereas the original dimer may have had a syn configuration.

NMR studies of the (1,2-diphenylallyl)chloropalladium dimer and
several of its amine derivatives show a consistent preference for
the syn isomer in the solid phase, but approximately equal quantities
of syn and anti forms in solution. Consequently, one may prepare the
2,6-lutidine derivative of bis[(1,2-diphenylallyl)chloropalladium]
from trans-1,2-diphenylpropene in 95% yield.[2] NMR spectra show that
this complex crystallizes in the syn configuration, and that when it
is dissolved at temperatures below -10°C, it isomerizes very slowly
to the anti configuration. Treatment of the solid with 1.5 M sodium
methoxide produces cis-1,2-diphenylpropene (in an overall yield of
80%).[3]

The 1-3-η^3-(1-acetyl-2-methallyl)[(S)-α-phenethylamine]chloro-
palladium complex, which can be isolated in optically active form,
epimerizes rapidly in solution at ambient temperatures.[4] [The π-
allyl-palladium moiety is converted to its mirror image (an enantio-
merization); but the optically active amine retains its stereochemistry - hence the process is termed an epimerization]. This epimeriza-
tion occurs via a 3-η intermediate which inverts the chirality of
the allyl group relative to the metal. The only geometric changes

are the interchange of syn and anti protons.

If the 3-position of this π-allyl is also substituted, it becomes impossible for the allyl-metal moiety to invert its configuration via pi→sigma→pi processes. A 3-η process inverts the stereochemistry at the carbon-1, but isomerizes at carbon-3, thus producing an isomerization rather than a racemization of the complex.

Even a sequence of pi→sigma→pi rearrangements will not allow enantio-
merization of the π-allyl palladium fragment; hence, one isolated
enantiomer of a 1,3-disubstituted-π-allyl palladium dimer can isomerize or epimerize, but cannot racemize. The unfavorable thermodynamic situation which arises from placing the terminal substituents
in <u>anti</u> positions tends to keep certain enantiomers in high concentration. Thus, unlike the 1-acetyl-2-methallyl complex which loses
its optical activity rapidly at room temperature, the 1-acetyl-2,3-
dimethylallyl complex maintains optical activity for weeks.

Once having prepared a compound which maintains its chirality,
we are now in a position to use this derivative to study asymmetric
induction and mechanistic aspects of reactions of allyls. Although
we are unable at this point to ascribe an absolute configuration to
the resulting cobalt complex, we find that the allyl group can be
transferred from palladium to cobalt stereospecifically. Thus, the
(+) isomers of 1-acetyl-2,3-dimethylallylpalladium chloride dimer
give the (+) isomers of [1-acetyl-2,3-dimethylallyl]dicarbonyl(triphenylphosphine)cobalt upon reaction with sodium tetracarbonylcobal-
tate.[*] We anticipate that future studies of these complexes will
provide absolute stereochemistries and allow detailed aspects of
the mechanism of allyl transfer to be deduced.

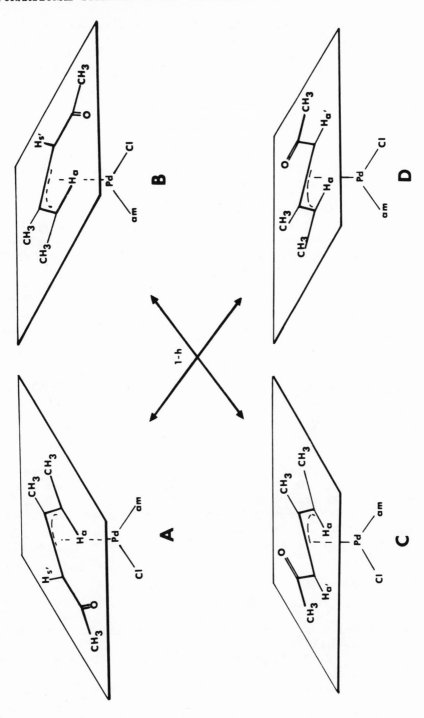

*We have shown that the only two isomers present at any significant
concentration in the palladium dimer are the (S)-anti-acetyl and (S)-
syn-acetyl isomers. These isomers are interconverted through a 1-η
σ-bonded intermediate and both rotate polarized light in the same
direction at 546 nm.

APPENDIX

Since the terminal carbon atoms of the allyl can be considered
to have four substituents (the syn and anti substituents, the central
allyl carbon, and the metal), the symbols (S) and (R) have been used
to denote chirality in the past. It should be noted that even though
(S) and (R) are adequate to describe the stereochemistry at a tetrahedral carbon atom, they do not provide sufficient information to
completely describe absolute configuration about the terminal carbon
atom in an allyl complex. In allyl and polyolefin complexes there
is an inherent chirality in the binding of the olefinic moiety to
the metal (absolute configurations at terminal carbons have been de-
noted by [+] and [-]). Hence even when the terminal carbon atoms of
a π-allyl complex are unsubstituted, the 1- and 3-carbon atoms are
chiral. Consequently, an unsubstituted or symmetrically substituted

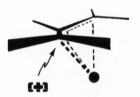
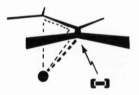

π-allyl may be considered to be a meso isomer. For a complete description of stereochemistry at a substituted terminal carbon atom,
one must define the traditional symbol with an additional modifier.
Since there appear to be no prior conventions we have found it convenient to describe the configuration at carbon-1 in the (S)-anti-
acetyl isomer as (S^+) and the configuration in the (S)-syn-acetyl
isomer as (S^-).

REFERENCES

1. J. W. Faller, M. E. Thomsen, and M. J. Mattina, J. Amer. Chem.
   Soc., 93, 2642 (1971).
2. J. W. Faller, M. T. Tully and K. J. Laffey, J. Organometal. Chem.,
   37, 193 (1972).
3. J. W. Faller and K. J. Laffey, Organometal. Chem. Syn., 1, 471
   (1972).
4. P. Ganis, G. Maglio, A. Musco, and A. L. Segre, Inorg. Chim. Acta., 3, 266 (1969).

5. J. W. Faller and M. T. Tully, J. Amer. Chem. Soc., 94, 2676
   (1972).

DISCUSSION

DR. PETTIT: What do you think the mechanism of transfer from
palladium to cobalt is, such that you still retain the asymmetry?
Is the absolute configuration maintained?

DR. FALLER: We have no way at present to correlate the absolute
configuration with the optical rotation in allyl cobalt complexes;
however, we are attempting to determine them now. For example, it
is feasible to transfer π-allyls derived from pinene, for which one
has a good idea of the absolute configuration based on steric con-
straints. We believe that the configuration I drew on the board
(and in the abstract) is the correct absolute configuration for palladium. One can work out quadrant rules for allyls on palladium that
really work. Now presumably, you can work out something similar for
cobalt. So by comparing the ord's of the pinene derivatives of palladium with those of cobalt we hope to get the absolute and relative
configuration. In some of these reactions, we have isolated intermediates which have a palladium-cobalt bond. Essentially, cobalt
carbonylate displaces a ligand on palladium (chloride, let us say)
and then remains there. My guess is that the carbon monoxide comes
off the cobalt and then when a σ-allyl is formed at palladium, the
double bond comes over and sits on the cobalt in place of the carbonyl. Then, you just have a σ-bond shift across to the cobalt.
So I do not think it will be difficult to come up with a stereochemistry, or to determine the factors which make the reaction stereospecific. It is an intramolecular transfer after you first form the
palladium-cobalt bond. Since you can even isolate that intermediate,
that aspect is quite clean. The thing that we believe you can get
out of this is whether a pi or a sigma bond forms first on cobalt.
I suspect that the olefin leaves first, binds to the cobalt and then
the sigma bond forms.

DR. MARKS: It has been some time since I have looked at some of
those papers of yours, Jack, but let us see. The pi-sigma equilibrium takes place with all these π-allyls. Do you have to add base
to induce that, or is this actually a 16 to 14 electron reaction?

DR. FALLER: As Jack Halpern noted in the rhodium system, there
are a number of equilibria in these systems. There are two or three
different mechanisms by which you can get interchange. There is the
Pd chloride dimer, which is analogous to the one Halpern has with
Rh, and you have similar bridge-splitting reactions of the dimer
which occur with the amines. All of these equilibrium reactions are
going on simultaneously, but you can forget all about them if you are
just interested in what is going on with the allyl group. Here also,
the reaction goes by several different paths. There is a first order

path where the double bond of the allyl group is detached (i.e., a
16 → 14 electron reaction), and there is a second order path where
you put in excess amine to drive it off faster. The reaction rate
for the first order path with the Pd chloride dimer is slowest. If
you have a monomeric amine complex, it is a little bit faster due to
the trans effect; with an excess of the amine you get a second order
reaction which begins to take over. The overall point is that once
you lock in a stereochemistry by putting a methyl group on the allyl
group, pi-sigma rearrangements produce only thermodynamically unfavorable isomers and you do not have to worry about all these other
equilibria.

SYNTHETIC MODELS FOR THE OXYGEN-BINDING HEMOPROTEINS

James P. Collman[*], Robert R. Gagné, and Christopher A.
Reed
Department of Chemistry, Stanford University, Stanford,
California 94305

The oxygen binding hemoproteins hemoglobin (Hb), myoglobin (Mb),
and cytochrome P-450 are important to the biological transport, storage, and metabolism of oxygen. Nevertheless the nature of the coordinate link between iron and dioxygen in these hemoproteins has
not been defined at the atomic level. Furthermore the way in which
the glogin proteinheme interaction directs reversible oxygen binding
has been obscure. My students have addressed and partially clarified these issues by preparing crystalline iron(II) porphyrin-dioxy-
gen complexes.[1-5] These remarkable Mb models which reversibly bind
oxygen in solution or in the solid state at ambient temperature have
been characterized by Mossbauer[1,5], ir spectra[4] and X-ray crystallographic analysis.[2]

The strategies which led to the synthesis of a stable iron porphyrin dioxygen complex were based on the structural characteristics
of Mb and consideration of the mechanisms for irreversible autoxidation of iron(II).

Myoglobin, the oxygen binding pigment found in muscle tissue
accounting for the red color of fresh beefsteak, is composed of a
153 residue peptide and an iron(II) protoporphyrin(IX) complex. The
ferrous protoporphyrin(IX) is held within a cleft in the globin much
as a silver dollar fits into a leather pouch. The only covalent interaction between the globin and the porphyrin arises from a coordinate bond between the so-called proximal imidazole of the histidine
residue F-8, and the Fe(II). In the deoxy form the five-coordinate
iron(II) is high spin with an ionic radius too large to fit in the
cavity within the porphyrin ligand.[6] A consequence of this is that
the high spin iron(II) ion projects out of the mean plane of the four
porphyrin nitrogens towards the proximal imidazole. The oxygen

329

binding site lies on the side of the porphyrin ring away from the proximal imidazole. Oxygenated myoglobin, MbO_2, is diamagnetic and the low spin iron is thought to lie in the porphyrin plane. During coordination with dioxygen the iron thus moves into the plane pulling along the proximal imidazole. Hoard[7] and Perutz[8] have proposed that this motion of the proximal imidazole is transferred through the four Mb like protein-porphyrin units of hemoglobin -- accounting for the cooperativity of oxygen binding between the four distant iron porphyrins in Hb.

Two structural models have been proposed for the iron-dioxygen group in HbO_2 and MbO_2: a sideways, π-type triangular bond $Fe\langle{}^O_O$ by Griffith,[9] and an angular ${}^{O-O}_{Fe}$ bond by Pauling.[10] The former has structural analogies in the diamagnetic dioxygen complex derived from coordinatively unsaturated d^8 and d^{10} complexes. However these are formally 6 or 4 coordinate and none involve a macrocyclic tetradentate ligand. Cobalt(II) forms angular dioxygen complexes, three of which have been structurally characterized.[11-13] However, these contain one more electron than the iron(II) dioxygen unit and esr studies indicate that this electron occupies an O_2 based π^* orbital.[14] The assignment of oxidation state to iron in MbO_2 has also been a matter of controversy but this question is less fundamental since oxidation states are not true physical properties, but rather subjective definitions.

Solutions of ferrous complexes and iron(II) porphyrins are irreversibly oxidized to iron(III) in the presence of oxygen. For example, six coordinate iron(II) tetraphenylporphyrin complexes, Fe-$(TPP)B_2$, are rapidly autoxidized to the corresponding μ-oxo ferric dimer (eq 1).

$$4\ Fe^{II}(TPP)B_2 + O_2 \longrightarrow 2\ Fe^{III}(TPP)-O-Fe^{III}(TPP) + 8\ B \quad (1)$$

B = axial base such as pyridine, imidazole, etc.

The rate of this reaction is strongly depressed by excess axial base suggesting a mechanism in which a five coordinate iron(II) complex reacts with oxygen followed by further reaction with another iron(II) complex which is probably also five coordinate (eqs 2, 3, 4) affording a very reactive complex of unknown constitution but

$$Fe(TPP)B_2 \rightleftharpoons Fe(TPP)B + B \quad (2)$$

$$Fe(TPP)B + O_2 \rightleftharpoons Fe(TPP)B(O_2) \quad (3)$$

$$Fe(TPP)B(O_2) + Fe(TPP)B \longrightarrow [BFe^{III}(TPP)-O-O-Fe^{III}(TPP)B] \quad (4)$$

$$\text{or } 2\ BFe^{IV}(TPP)=O$$

$$BFe^{IV}(TPP)O + Fe^{II}(TPP)B \longrightarrow (TPP)Fe^{III}-O-Fe^{III}(TPP) + 2\ B \quad (5)$$

which can be considered equivalent to two equivalents of iron(IV).
Further reaction with iron(II) gives the final observed product (eq
5). At present much of this scheme is speculative although consistent with Hammond's study[15] of the autoxidation of Fe^{2+} in aqueous
alcohol. It is clear that such a mechanism requires a bimolecular
reaction between two iron porphyrins to reach the observed product.

Consideration of this atom transfer redox mechanism suggests
the role of the protein in promoting reversible oxygenation of the
iron(II) porphyrin without irreversible oxidation to iron(III). By
encapsulating the porphyrin the globin prohibits reaction between
two iron centers. Furthermore in the deoxy form of Mb and Hb, the
globin imposes 5-coordination upon the iron by the agency of the
proximal imidazole. Solution equilibria involving free axial bases
favor the diamagnetic six coordinate form[16] which does not appear
to react with oxygen. Another probable role of the protein which is
simple to mimic is the hydrophobic nature of the oxygen binding environment and the shielding from acidic protons. Acids and other
electrophiles are known to react with the triangular $M\text{<}^O_O$ complexes
forming peroxides (eq 6). Such a reaction with a ferrous dioxygen

$$\begin{array}{c} Ph_3P \\ \diagdown \\ \diagup \\ Ph_3P \end{array} Pt \begin{array}{c} O \\ | \\ O \end{array} \quad + \quad HO\overset{O}{\overset{\|}{C}} - \overset{O}{\overset{\|}{C}}OH \longrightarrow H_2O_2 \qquad\qquad (6)$$

complex would probably initiate irreversible oxidation by forming
iron(III) and protonated superoxide (eq 7). Thus acids should catalyze the irreversible autoxidation of iron(II).

$$Fe(II)O_2 \quad + \quad H^+ \quad \longrightarrow \quad Fe(III) \quad + \quad HO_2 \qquad\qquad (7)$$

There are other potential roles which the protein may play that
are not well defined. A more distant imidazole (the "distal imidazole") from the histidine residue (E-7) projecting into the oxygen
binding pocket is not close enough to coordinate to iron but may
somehow stabilize coordinated dioxygen (as first suggested by Pauling[10]) by hydrogen bonding.

In summary kinetic stabilization of an iron dioxygen function
would seem to require at the minimum a pocket shielding coordinated
dioxygen from reaction with another iron porphyrin and providing a
non acidic environment.

We began our work in this area by developing a synthesis for a
ligand free tetraphenylporphyrin iron(II) which was isolated and
structurally characterized as a rigidly square planar intermediate
spin (S=1), 14 electron ferrous complex.[17] Addition of 2-methyl i-
midazole yielded a crystalline 5-coordinate high spin iron(II) complex which on the basis of its magnetic properties, Mossbauer[18]

(Lang) and X-ray crystallographic structure (Hoard)[18,19], seems to be an excellent structural model for the deoxy hemoproteins, Mb and Hb. Solutions of this five coordinate iron(II) porphyrin still oxidized at a very high rate, as the mechanisms discussed above would suggest.

The formation of a stable dioxygen complex derived from the so-called "picket fence concept" (Figure 1.) It was decided to con-

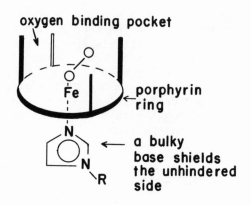

Figure 1. "Picket Fence Concept"

struct a porphyrin molecule in which one side was surrounded by bulky substituents (pickets) which would form a cavity for binding oxygen and thus shield coordinated dioxygen from intermolecular reactions with other iron porphyrins. The unencumbered side of the "picket fence porphyrins" was to be shielded by an axial imidazole. This axial base was to have a substituent sufficiently bulky as to disfavor coordination within the picket fence.

The synthesis of such a porphyrin is outlined in Figure 2 and relies upon the atropisomers which arise from tetra ortho-substituted phenyl porphyrins. Such atropisomers had been reported in the tetra-ortho-hydroxyphenyl case by Ullman.[20] Our preliminary experiments with the ortho-hydroxy complex were frustrated since the energy barrier to atropisomerism was sufficiently low and reactions of these sterically-hindered phenolic groups were sufficiently slow that attempts to synthesize a bulky picket fence porphyrin were frustrated by the more facile atropisomerism. Treatment of ortho-nitrobenzaldehyde with pyrrole afforded a modest yield of the tetra-ortho-nitrophenyl porphyrin as a mixture of atropisomers. This mixture was reduced in high yield to the tetra-ortho-amino derivative. Chromato-

graphic separation of the slowest moving atropisomer (the one we regarded as the tetra-α isomer—a hypothesis which was confirmed by
subsequent crystallographic analysis) was carried out and the remaining three other atropisomers were reequilibrated in boiling toluene. Again, the statistical mixture of four atropisomers was chromatographed affording the tetra-α isomer and reequilibration again
carried out so that ultimately all of this material was converted
into the desired tetra-α isomer, albeit, with modest material loss
from the repeated chromatographic separations. Fortunately, the or-
tho-amino atropisomers enjoyed a higher energy barrier to isomerization and at the same time the amino groups were much more reactive
than the phenolic groups had been. Subsequent elaboration to the
tetra pivalamide afforded the desired "picket fence porphyrins".
The energy barrier to atropisomerism was greatly increased in this
complex due to the steric bulk of the pivalamido substituents. Further the tertiary butyl groups afforded a welcomed higher degree of
solubility and provided a convenient pmr signal for chemical analysis.

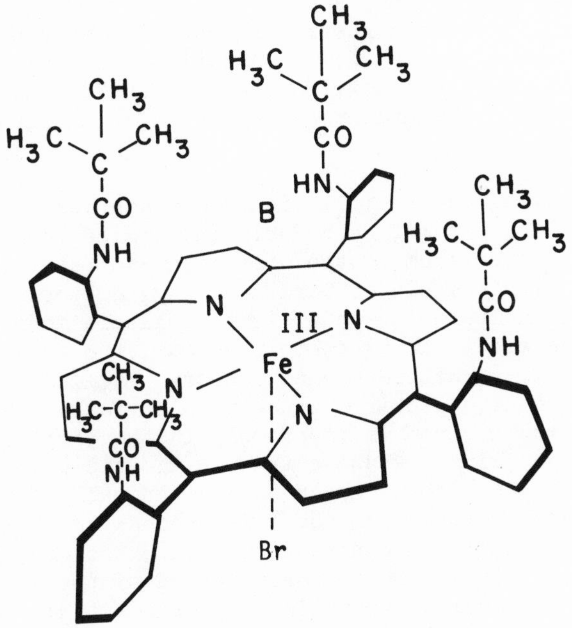

Figure 2. FeBr(α,α,α,αTpivPP)

Preparation of the iron derivative of this picket fence porphyrin
proved troublesome, but the ferric bromide complex was ultimately i-
solated in good yield (Figure 2). Subsequent reduction with chromi-
um(II) acetyl acetonate (the method developed for the model Mb stu-

dies outlined above) afforded an extremely air-sensitive four coordinate iron(II) picket fence porphyrin. Treatment of this intermediate with various axial imidazoles afforded a series of crystalline diamagnetic six coordinate iron(II) porphyrins[1], much to our surprise! This was even true with bulky imidazoles and thus the picket fence did not control five coordination to the degree we had anticipated. However, in all of these systems the product which is isolated depends more on the stability (solubility) of the crystals than on the nature of the solution equilibria.

Inspection of the electronic spectra of solutions of these six coordinate complexes in the presence of oxygen revealed reversible changes at ambient temperature when oxygen and nitrogen were alternately bubbled through the solution (Figure 3). Such changes were taken as evidence, albeit inconclusive, for the formation of some oxygen adduct. Others have relied more heavily on such spectral evidence in cases where the intermediate complex is not sufficiently stable to be isolated. Using N-methyl imidazole, we were able to isolate from such solutions a crystalline dioxygen complex.[1,5] This complex proved to be diamagnetic and careful gas evolution experiments demonstrated the Fe(II) to O_2 ratio to be 1:1. The Mossbauer spectra of this complex showed a chemical shift similar to that of HbO_2 and a similar temperature dependence for the quadrupole splitting (Table I).

Within a week of the isolation and analytical characterization of these single crystals, Ward Robinson carried them back to New Zealand and mounted them on the diffractometer at the University of Canterbury. With the aid of Gordon Rodley, Robinson obtained a crystallographic analysis for this first myoglobin model.[2] Although there are severe crystallographic problems, the semi quantitative features of this structure are now clear. Pauling was right.[10] Oxygen is coordinated to iron in an angular fashion with the Fe-O-O angle of $136 \pm 4°$. Iron is rigorously in the porphyrin plane with the first oxygen directly above iron in an axis normal to the porphyrin plane. The second oxygen is statistically four-fold disordered, the O-O vector bisecting the N-Fe-N bonds in the porphyrin plane. The imidazole occupies a constant position, two way statistically disordered with respect to the orientation of the N-methyl group (Figure 4). This arrangement gives rise to two different types of coordinated dioxygen. However, the structural parameters for these two types of dioxygen are crystallographically indistinguishable. The structure shows the oxygen to be within the picket fence and that the tertiary butyl groups are turned inward. The amide N-H group is also oriented in that direction; but this hydrogen is not situated in a position which would enable hydrogen bonding. Other structural features of interest are the Fe-O distance $1.75 \pm .02$ Å which is about 0.1 Å shorter than the calculated summation of covalent radii.[6] The Fe(II) axial imidazole nitrogen distance is $2.07 \pm .02$ Å ---

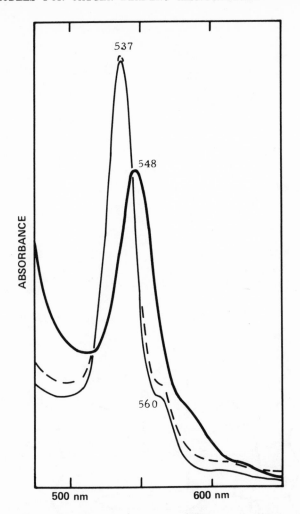

Figure 3. Visible spectrum of 3 x 10^{-5} M Fe(1-Me-imid)$_2$ ($\alpha,\alpha,\alpha,\alpha$-TpivPP) with 10^{-4} M 1-Me-imid in benzene: _____ under nitrogen; _____ under oxygen (1 atm); - - - - under nitrogen after two oxygenation/deoxygenation cycles.

somewhat elongated from the anticipated covalent radii. Thus it appears that dioxygen is asserting a mild trans effect on the axial base and that the iron oxygen distance is suggestive of modest π bonding. The O-O separation is 1.24 $\pm$ .08 Å, in the superoxide range, but the figures are too imprecise to have meaning.

Recently we have synthesized the iron complex of the $\alpha,\alpha,\alpha,\beta$-

Table I
Mossbauer Data

	T($^\circ$K)	δ	ΔEQ* (mm/sec)
HbO$_2$	77	0.26	2.19
	195	0.20	1.89
FeP(N–MeIm)O$_2$	77	0.26	1.8
	195	0.24	1.4
FeP(N–$\underline{n}$BuIm)O$_2$	77	0.26	1.4
	195	0.23	1.2
FeP(THF)$_2$O$_2$	77	0.33	2.2
	195	0.29	2.1
Fe(TPP)(THF)$_2$	77	0.95	2.7
	195	0.89	2.4

P = $\alpha,\alpha,\alpha,\alpha$-TpivPP
*values relative to Fe metal

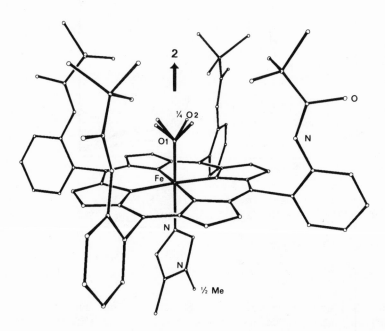

Figure 4. Perspective view of one molecule of the iron dioxygen complex showing the crystallographic 2-fold axis of symmetry and the four way statistical disorder of the terminal O$_2$ oxygen atom of dioxygen.

atropisomer and solution spectral studies indicate reversible oxygenation.[21] Although a crystalline oxygen complex has not yet been isolated from this case, molecular models suggest the coordinated dioxygen would not be disordered.

Other axial bases also support reversible oxygenation of these picket fence iron(II) porphyrins in dilute solution. Such bases include pyridine, piperidine, tetrahydrothiophene and tetrahydrofuran. Typical spectral data for THF solutions is shown in Figure 5. Even though these spectral changes indicate oxygenation, the isolation of the dioxygen complex from THF solutions has proven impossible. The μ-oxo iron(III) dimer is formed! This undoubtedly arises from the increased rates of bimolecular atom transfer oxidations discussed above when the solutions are concentrated so as to precipitate the dioxygen complex.

Fortunately the discovery of facile solid gas reactions for these picket fence porphyrins obviated the difficulties described above, allowing us to isolate some unusual dioxygen complexes. Apparently the crystal lattice of these porphyrins are highly porous permitting rapid kinetic access of gaseous molecules.[3,5] For example, (Figure 6) the parent diamagnetic dioxygen complexes lose oxygen under vacuum affording a high spin deoxy compound which upon exposure to air again becomes diamagnetic as dioxygen is coordinated. Exposure to carbon monoxide affords the carbonyl which has also been prepared in solution. A preliminary structural analysis shows the iron CO unit to be rigorously colinear (Hoard) and the fact that the νCO, 1965 cm^{-1}, is only slightly different from that of carbonyl hemoglobin suggests a structural similarity and calls into question earlier suggestions that this carbonyl might be strongly bent.[22] It is interesting that crystals of the diamagnetic six coordinate complex also react with CO affording a CO derivative. This must involve motion of one of the axial bases to make room for the CO group -- a point which remains to be explored by crystallographic analysis. A more interesting facet of these solid state reactions came upon exposure of the bis tetrahydrofuran complex (high spin iron(II) to oxygen. A reversible reaction ensued in the solid state affording a 1:1 paramagnetic oxygen complex.[3] The room temperature moment of this oxygen complex, 2.4 BM, is suggestive of two unpaired electrons. Mossbauer studies on this oxygen compound down to liquid helium temperature [5] give no indication of a spin change. Thus, the paramagnetic oxygen complex apparently having THF as an axial base may be tentatively considered as a compound of iron(III) and superoxide. This gas reaction is completely reversible. That no net oxidation has taken place is demonstrated by reaction with dry CO affording the diamagnetic carbonyl, νCO 1955 cm^{-1}. A more detailed characterization of this substance awaits X-ray diffraction studies in progress (Hoard), but the possible relationship between this dioxygen complex and that formed by cytochrome P-450 is intriguing.

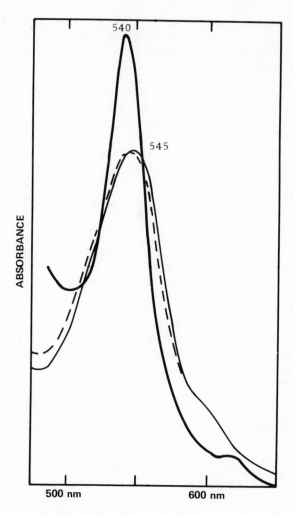

Figure 5. Visible spectrum of 5×10^{-5}M. Fe($\alpha,\alpha,\alpha,\alpha$-TpivPP) in THF: (＿＿＿) under nitrogen, (＿＿＿) under oxygen (1 atm), (-----) under nitrogen after two oxygenation-deoxygenation cycles.

The latter is thought to be diamagnetic in its oxygenated form,[23] but the data are not precise. It is further remarkable that these solid gas reactions do not change the lattice constants of singly crystalline specimens.

    The final issue I wish to discuss concerns the vibrational spectrum of coordinated dioxygen in the picket fence MbO_2 model. Our initial attempts to observe $\nu O\text{-}O$ at ambient temperatures either in the Raman or in the infrared, with the aid of ^{18}O substitution

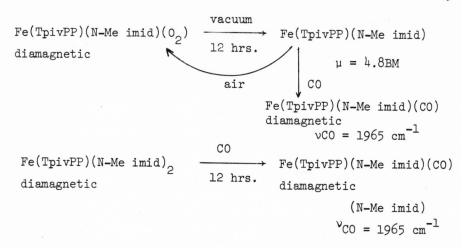

Figure 6. Solid gas reactions at the 25° of complexes of Fe($\alpha,\alpha,-$ α,α-TpivPP)

showed no apparent absorption band.[1] However, collaboration studies (Gray)[4] at low temperatures have revealed a remarkably sharp stretching frequency at -175°C which changes when oxygen is pumped from the solid sample and reappears upon admission of O_2 and is tentatively assigned to νO_2. Isotopic substitution causes this band to shift, but a new absorption is not observed and harmonic oscillator calculations predict the new absorption to be under strong porphyrin bands. The most remarkable aspect of this νO_2 band, other than its amazing temperature dependence, is its frequency at 1385 cm^{-1}! When compared with νO_2 for oxygen, superoxide ion, peroxide, cobalt(III) superoxide complexes and the triangular dioxygen complexes, this 1385 cm^{-1} band clearly implies a higher O-O bond order than is found in any of the other O_2 complexes (Table II). A reasonable assignment for the nature of dioxygen in this oxymyoglobin model would be coordinated singlet oxygen as the observed O-O frequency is about 100 cm^{-1} below that of singlet oxygen -- not an unreasonable value under the influence of modest $d\pi$-$p\pi$ backbonding from iron to oxygen. This frequency is more than 250 cm^{-1} above those for superoxide and its complexes and this shift is far too great to consider these similar models. Caughey[24] recently reported a value for HbO_2 νO_2 as 1107 cm^{-1}. The discrepancy between the frequency we have observed and that reported by Caughey is too large to be explained by hydrogen bonding from the distal imidazole. Gray is reinvestigating Caughey's experiment using the low temperature technique.

A final comment is in order concerning the remarkable tempera-

Table II
Infrared Data

	νO_2 in cm^{-1}
O_2	1555 (R)
$O_2(^1\Delta)$	1483.5
KO_2	1145
Na_2O_2	738
$Fe(\alpha,\alpha,\alpha,\alpha\text{-TpivPP})(\text{N-MeIm})O_2$	1385 (vs at 95°K)
$Fe(\alpha,\alpha,\alpha,\alpha\text{-TpivPP})(\text{N-}\underline{n}\text{BuIm})O_2$	1385 (vs at 95°K)
HbO_2 (D_2O exchanged), Ref. 24	1107 (297°K)
$Co(\text{acacen})(\text{py})O_2$	1123 (vs)
$Ir(Ph_3P)_2(Cl)(CO)O_2$	858 (vs)

ture dependence of our νO_2. The only explanation we have at present is that there are several rotational states of O_2 which are thermally populated at room temperature, each having a slightly different frequency. These may be related to the two forms of coordinated dioxygen as well as to thermal interactions with the tertiary butyl methyl groups (as indicated in the crystallographic analysis). Retrospective examination of the room temperature spectrum reveals a small, broad peak at 1385 cm^{-1} which may have an integrated intensity similar to that observed at low temperatures.

In summary, we have been able to prepare a stable crystalline iron(II) dioxygen complex without the agency of a protein by designing a synthetic porphyrin which confers upon iron some of the major features derived from the protein porphyrin interaction.

ACKNOWLEDGEMENT

I should like to acknowledge the help of many collaborators, associates, and co-workers--in particular, Ward Robinson and Gordon Rodley, University of Canterbury; George Lang, Physics Department, Pennsylvania State University; Lynn Hoard, Cornell University; Harry Gray and Jeffrey Hare, California Institute of Technology; Brian Hoffman, Northwestern University; and Tom Halbert, Jean-Claude Marchon, Craig McAllister, and Tom Sorrell, Stanford University.

REFERENCES

1. J. P. Collman, R. R. Gagne, T. R. Halbert, J. -C. Marchon, and C. A. Reed, J. Amer. Chem. Soc., 95, 7868 (1973).
2. J. P. Collman, R. R. Gagne, C. A. Reed, W. T. Robinson, and G. A. Rodley, Proc. Nat. Acad. Sci., U.S.A., 71, 1326 (1974).
3. J. P. Collman, R. R. Gagne, and C. A. Reed, J. Amer. Chem. Soc., 96, 2629 (1974).
4. J. P. Collman, R. R. Gagne, H. B. Gray, and J. Hare, in submission to J. Amer. Chem. Soc.
5. J. P. Collman, R. R. Gagne, T. R. Halbert, G. Lang, and C. A. Reed, manuscript in preparation.
6. J. L. Hoard, Science, 174, 1295 (1971).
7. J. L. Hoard, M. J. Hamar, T. A. Hamar, and W. S. Caughey, J. Amer. Chem. Soc., 87, 2312 (1965).
8. M. F. Perutz and L. F. Ten Eyck, Cold Spring Harbor Symposia on Quantitative Biology, 36, 295 (1972); M. F. Perutz, Nature, 228, 726 (1970).
9. J. S. Griffity, Proc. Roy. Soc. Ser. A., 235, 23 (1956).
10. L. Pauling and C. D. Coryell, Proc. Nat. Acad. Sci., U.S.A., 22, 210 (1936); L. Pauling, Stanford Med. Bull., 6, 215 (1948); L. Pauling, Nature, 203, 182 (1964).
11. G. A. Rodley and W. T. Robinson, ibid., 235, 438 (1972).
12. G. A. Rodley and W. T. Robinson, Syn. Inorg. Metal-Org. Chem., 3, 387 (1973); J. P. Collman, H. Takaya, B. Winkler, L. Libit, S. K. Seah, G. A. Rodley, and W. T. Robinson, J. Amer. Chem. Soc., 95, 1656 (1973).
13. L. D. Brown and K. N. Raymond, Chem. Comm., 1974, 470.
14. B. M. Hoffman, D. L. Diemente, and F. Basolo, J. Amer. Chem. Soc., 92, 61 (1970); D. Diemente, B. M. Hoffman, and F. Basolo, Chem. Comm., 467 (1970).
15. G. S. Hammond and C. S. Wu, Advan. Chem. Ser., No. 77, 186 (1968).
16. J. P. Collman, D. A. Buckingham, and W. T. Robinson, unpublished results.
17. J. P. Collman and C. A. Reed, J. Amer. Chem. Soc., 95, 2048 (1973).
18. J. P. Collman, J. L. Hoard, G. Lang, L. J. Radonovich, and C. A. Reed, manuscript in preparation.
19. J. L. Hoard and W. R. Scheidt, Proc. Nat. Acad. Sci., U.S.A., 70, 3919 (1973).
20. L. K. Gottwald and E. F. Ullman, Tetrahedron Lett., 3071 (1969).
21. J. P. Collman and J. -L. Roustan, unpublished results.
22. J. O. Alben, W. H. Fuchsman, C. A. Beaudreau, and W. S. Caughey, Biochemistry, 7, 624 (1968).
23. I. C. Gunsalus, J. R. Meeks, J. D. Lipscomb, P. Debrunner, and E. Munck, "Molecular Mechanisms of Oxygen Activation", O. Hayaishi, ed., Academic Press, N. Y., 1974, p. 561.
24. C. H. Barlow, J. C. Maxwell, W. J. Wallace, and W. S. Caughey, Biochem. Biophys. Res. Commun., 55, 91 (1973).

DISCUSSION

DR. MARKS: One problem that occurred to me; I wonder if it is possible that the reason the O-O stretch is so weak at room temperature may be that you are actually photolyzing it off with the infrared spectrometer. I have seen things go to pot with infrared beams, for example, you can knock CO off hemoglobin.

DR. COLLMAN: We thought with raman that would be a problem. We have not exposed the compound to infrared while we are measuring the magnetic moment. The oxygen does go back on very rapidly, and photolysis is a possibility. However, we do see this little nub and it is broad, and I think the integrated intensities are going to be the same.

DR. MARKS: Did Caughey do an O^{18} experiment?

DR. COLLMAN: Caughey has recently reported an oxyhemoglobin spectrum, and he has done this with O^{18}. The quality of those spectra are very, very poor, but what is really distressing to us is his report of an 1107 cm^{-1} band for coordinated dioxygen. Even with hydrogen bonding from the distal imidazole, we can not conceive that the oxygen stretching frequency would be lowered by 280 wavenumbers. Therefore, we suspect this band, and Gray is reexamining that question.

DR. KAESZ: On this O^{18} experiment that you want me to study, could you tell me how he did the exchange?

DR. COLLMAN: We used statistically labeled 50% O^{18}, that was all we could get hold of at the time, and we made up the compound in the usual way, determined that it had oxygen in it, that it had its usual properties-analyzed it, in fact.

DR. KAESZ: Did it have the intensity expected?

DR. COLLMAN: We got a reduced intensity. It was about what you would calculate for the statistical distribution of the O^{16} that was in there. I do not have a good number for the quantization of that. We simply got a reduced band. Using pure O^{18}, by "pure" I mean that it was about 90% O^{18}, we then got just a very tiny band. With the O^{16}, of course, we got the band I showed you.

DR. MARKS: You might expect to see two. You do in a lot of the nitrogen compounds.

DR. COLLMAN: Yes, I know. You might expect to see two. It seems that only the terminal oxygen is sensitive to this substitution. However, the iron-oxygen units seemed to me, naively, that they were acting more or less as a unit in the vibration. Then you might not see two peaks.

DR. TSUTSUI: The X-ray diffraction spectrum was not well refined, is that right?

DR. COLLMAN: It was refined as well as we could with the situation we had.

DR. TSUTSUI: What is the location of the iron? In the plane?

DR. COLLMAN: The iron seems to be in the plane, as near as we can tell.

DR. HALPERN: You referred to the difficulty of making the five-coordinate mono-imidazole complex. What happens in that system if you simply add one methyl imidazole per iron?

DR. COLLMAN: One N-methyl, or one 2-methyl? We used the 2-methyl to grow those single crystals I showed you the structure of. In solution, that is the dominant species.

DR. HALPERN: I am referring to your "picket fence" compound.

DR. COLLMAN: If you add one, you seem to get a mixture of at least 2, and perhaps 3 species. That is, you get some of the diamagnetic compound. It is not a clean situation at all. If you look at the electronic absorption spectrum, it is very difficult, because there are not substantial enough changes, and you do not have a really good spectral analysis of the 5-coordinate compound. It is hard to say what you have. If you try to get it out of solution, of course the least soluble species comes out.

DR. HALPERN: What if you put oxygen in at that stage?

DR. COLLMAN: You get reversible oxygenation.

DR. HALPERN: You drive it back to the one-to-one complex?

DR. COLLMAN: Yes, you drive it to the one-to-one, even when you have 2 or 3 or 4-fold excess. It is very hard to get equilibrium constants out of our system. That is the one big weakness it has. We hope to take care of that soon.

DR. HALPERN: The other question I have is: What about the chemistry of the coordinated oxygen?

DR. COLLMAN: We are looking into that, Jack.

DR. HALPERN: Do you have anything on it at all?

DR. COLLMAN: No comment.

DR. WOOD: Munck and Debrunne at Illinois have done an extensive study of the Mossbauer spectrum of oxygenated heme-proteins. When you look at the Mossbauer spectrum of hemoglobin and myoglobin and cytochrome-P-450 and 1-peroxidase which they have looked at so far, these spectra are virtually superimposable on each other. The question is, is the oxygen intermediate an identical intermediate for all these enzymes, and has anyone got any ideas on how the electrons get in so that these heme proteins catalyze very different reactions.

DR. COLLMAN: But in the P-450, the case I talked about originally, the Mossbauer is different, very much different. In fact, we hope to use that. Our next project is to model P-450. It has very different characteristics. For example, there is a very small temperature dependence of the quadrapole splitting. It is thought to have an axial mercaptan ligand. Incidentally, we have reversible oxygenation with pyridine, and with thio-ethers. The mercaptans are being worked on right now.

DR. OGOSHI: Could you find any significant difference between the parallel and perpendicular isomers?

DR. COLLMAN: We have no evidence which indicates a difference between those two, with the possible exception of the temperature dependence of the vibrational spectrum. I should emphasize that we have not so far been able to measure the vibrational spectra in solution. This is a pity, because we have with the 3-1 isomer a single orientation we believe, in solution, but because of overlapping solvent peaks, and so forth, we have not succeeded.

DR. OGOSHI: What about using iron isomers to elucidate the Fe-O infrared bands? This would give considerable information about the oxygen bonding study.

DR. COLLMAN: Yes, I think that is a good idea. We can not afford even the O^{18}, but at some time, we would like to do that, or have somebody do it. It is just really not our bag. If somebody would wish to do this carefully, we would be glad to collaborate.

DR. OGOSHI: Dr. Gutterman has prepared some divalent? iron-carbon complexes, I think involving two tetrahydrofurans. The electronic spectra of this compound might offer some information about yours.

DR. COLLMAN: We have prepared that compound also, and we have looked at it very carefully in the Mossbuaer. We find that there are two kinds of compounds there. They depend upon the THF that is in the molecule. We get different magnetic properties from the workers you describe. We can pump one THF out of that molecule and we can correlate this with the Mossbauer spectra at 4°K.

DR. OSBORN: Have you made the ESR?

DR. COLLMAN: An attempt has been made to do that at Northwestern by Hoffman. Since there are 2 unpaired electrons, you would be fortunate to find an ESR peak. The experiment is negative-no peak was found for the paramagnetic oxygen complex. Of course, our complexes with an axial N-methyl amidizole are diamagnetic.

DR. OSBORN: Do you see any contact shift?

DR. COLLMAN: We frankly have not taken a careful look at nmr yet, and it needs to be done. There are solubility problems, however.

ORGANORHODIUM AND IRIDIUM PORPHYRIN COMPLEXES A MODEL FOR VITAMIN B$_{12}$

Hisanobu Ogoshi[*], Jun-ichiro Setsune, Takashi Omura and
Zen-ichi Yoshida
Department of Synthetic Chemistry, Kyoto University,
Kyoto, 606, Japan

The cobalt-porphyrin has been considered to be most plausible
model for vitamin B$_{12}$. However, cobalt-porphyrin complex has never
been extensively utilized as a model of cobalamin owing to relatively low stability of alkylcobalt(III)porphyrin and inability of the
porphyrin ligand to stabilize the Co(I) state in aqueous medium.[1,2]
The rhodium-porphyrin in various oxidation states[3] has been found to
be an attractive model to elucidate reaction behavior of vitamin B$_{12}$
as well as the other model such as cobalt complex of dimethylglyoxime.[2] This paper reports two kinds of new rhodium(I)porphyrin complexes, chemical behaviors of Rh(I)-porphyrin and Rh(III)-porphyrin
and iridium porphyrin complexes.

NEW RHODIUM(I)-PORPHYRIN COMPLEXES

The existence of the Rh(I) and Ir(I) complexes represented by
[porphyrin·M(I)CO·Cl]$^-$, (M = Rh, Ir), has been postulated as a precursor of those trivalent metal complexes.[4] We have separated new
Rh(I)-porphyrin complexes with two metal ions bonded to a porphyrin
by treatment of octaethylporphyrin with [Rh(CO)$_2$Cl]$_2$. The structure
of this complex is quite different from that of Rh(I)-porphyrin proposed by Sadasivan and Fleisher.[4] The complex 2' represented by mi-
croanalysis shows a strong tendency to liberate HCl in organic solvent. As a matter of fact, thrice recrystallization from CH$_2$Cl$_2$
gave μ-octaethylporphynato-bis[dicarbonylrhodium(I)] 2. The X-ray
analysis of 2[5] shows that the two Rh(I) atoms are coordinated to the
porphyrin above and below the macrocyclic plane, similar to Re(I)
complex[6] as is shown in Figure 1. Each Rh atom, however, has square-
planar coordination; two nitrogen atoms of the four pyrrole rings
and two carbonyl carbon atoms are coordinated to one Rh atom [Rh$_1$-

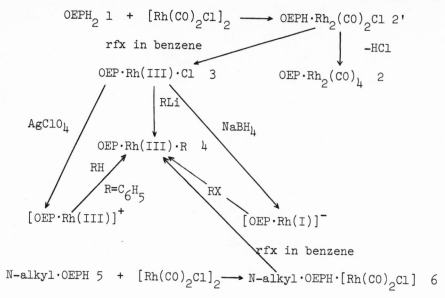

$OEPH_2$ 1 + $[Rh(CO)_2Cl]_2$ ⟶ $OEPH·Rh_2(CO)_2Cl$ 2'

rfx in benzene

$OEP·Rh(III)·Cl$ 3 $OEP·Rh_2(CO)_4$ 2

−HCl

RLi

$AgClO_4$ $NaBH_4$

$OEP·Rh(III)·R$ 4

RH

$R=C_6H_5$ RX

$[OEP·Rh(III)]^+$ $[OEP·Rh(I)]^-$

rfx in benzene

N-alkyl·OEPH 5 + $[Rh(CO)_2Cl]_2$ ⟶ N-alkyl·OEPH·$[Rh(CO)_2Cl]$ 6

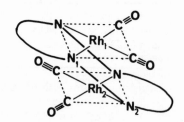

Figure 1.   Coordination sphere in 2

N_1 2.078 (8), Rh_2-N_2 2.087 (7), $Rh-C_1$ 1.837 (10) and $Rh-C_2$ 1.856 (11) Å] and they are coplanar to within ± 0.005 Å. This is the first ex-ample of metalloporphyrin systems in which two monovalent metal a-toms form square-planar coordination with the porphyrin. The Rh a-tom deviates, by 0.105 Å, from the basal plane towards the midpoint of the two nitrogen atoms coordinated to another Rh: Rh_1.....N_2 3.058 (8) Å and Rh_2.....N_1 3.080 (7) Å. There might be some inter-action between these nitrogen and the monovalent rhodium atoms.

Treatment of N-methyloctaethylporphyrin with $[Rh(CO)_2Cl]_2$ in benzene afforded dark violet crystals 6 formulated as $[N-CH_3·OEPH]-[Rh(CO)_2Cl]_2$[3a]; The 220-Hz nmr spectrum shows the striking up-field shifts of N-H at τ 13.74 and $N-CH_3$ at τ 15.90. Eight different me-thyl signals of the peripheral ethyl groups are well resolved as is shown in Figure 2. The ir spectrum in KBr has four strong absorp-tions at around 2000 cm^{-1}. Although crystal structure of 6 has not been established yet, these spectroscopic results suggest that the

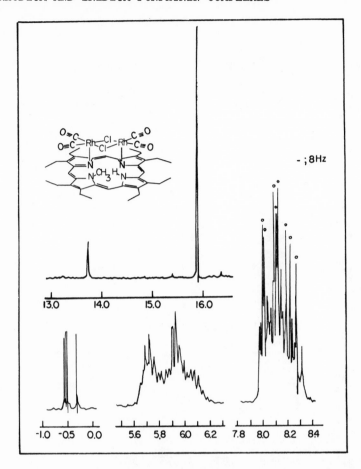

Figure 2.   The 220-MHz nmr spectrum of 6 in CDCl$_3$ using TMS as internal standard.   The symbol 0 indicates the central peak of each
triplet.

symmetry of the porphyrin ring is markedly reduced compared with metal free porphyrin and usual metalloporphyrins.   New iridium(I)-
porphyrin complexes have been obtained for the reaction of 1 or 5
with [Ir(CO)$_3$Cl]$_2$.   The structure of the iridium porphyrin complex 7

1 + [Ir(CO)$_3$Cl]$_2$ $\longrightarrow$ OEP·[Ir(CO)$_3$]$_2$ 7 + OEPIr(III)·CO·Cl 8

5 + [Ir(CO)$_3$Cl]$_2$ $\longrightarrow$ N-CH$_3$·OEP·H[Ir(CO)$_3$Cl]$_2$ 9

seems to be identical with that of Re complex.[5]   The visible spectrum of 9 is almost the same as that of 6.

FORMATION OF THE RHODIUM-CARBON BOND

Reaction of Aquochlororhodium(III) Complex Alkyl and Aryllithium

The treatment of chlororhodium(III) porphyrin 3 with alkyl and aryllithium in dry ether afforded alkylrhodium(III) and arylrhodium (III) porphyrin complexes in moderate yield respectively. Substitution with R group was confirmed by the nmr spectrum which shows remarkable up-field shifts of R group due to the anisotropic effect

$$OEP \cdot Rh(III) \cdot Cl \cdot 2H_2O \quad 9 \xrightarrow{RLi} OEP \cdot Rh(III) \cdot R \quad 10$$

of the porphyrin ring current (Table 1). An alternative method using Grignard reagent (RMgX; X=Br,I) resulted in exchange of the chlorine atom of 3 with halogen(X) of RMgX.

$$OEP \cdot Rh(III)Cl \cdot 2H_2O \quad + \quad RMgX \longrightarrow OEP \cdot Rh(III) \cdot X \quad + \quad RMgCl$$

Table 1
NMR Spectra of $R \cdot Rh(III) \cdot OEP$ in $CDCl_3$

R	1	2	3	4	5	6	7,8	9
$-CH_3$	16.47							
$-C_2H_5$	15.55	14.95						
$-C_3H_7$	15.61	14.97	12.04					
$-C_4H_9$	15.58	15.04	11.93	11.08				
$-C_5H_{11}$	15.56	15.02	11.95	10.75	10.44			
$-C_6H_{13}$[#]	15.58	15.04	11.92	10.80	10.11	9.95		
$-C_9H_{19}$[#]	15.68	15.14	11.99	10.83	10.19	9.59	9.15	9.36

R	o	m	p	$-CH_3$
$-C_6H_5$	—	5.48	4.96	—
$-C_6H_4CH_3P$	7.96	5.64	—	9.07
$-C C-C_6H_5$	5.16	3.85		—

[#] : 220 MHz, $CHCl_3$ standard, all others,

: 100 MHz, TMS standard

Reaction of Rhodium(I)-Porphyrin with Alkylhalide

The chlororhodium(III)-porphyrin 3 can be easily reduced to the

monovalent complex by using $NaBH_4$ in alcoholic alkaine solution.
When alkylhalide was added to the solution containing the Rh(I) complex, the alkylrhodium(III) complex was obtained in high yield. Recently James and Stynes have reported on the synthesis of the rhodium
(I)-porphyrin complex formulated as $H[Rh(I)\cdot(porphyrin)]\cdot 2H_2O$.[3b]
However, we have never separated the similar type of complex in our
system. The visible spectrum of the solution shows quite different

$$3 \xrightarrow[\substack{NaOH \\ C_2H_5OH}]{NaBH_4} [OEP\cdot Rh(I)]^- \xrightarrow{RBr} OEP\cdot Rh(III)\cdot R \quad 10$$

pattern from those of the trivalent rhodium-porphyrins. This fact
suggests the presence of the monovalent rhodium-porphyrin. The rhodium-carbon bond is probably performed by attack of electrophile on
the Rh(I) atom.

Reaction of Rh(I)-Porphyrin with Unsaturated Hydrocarbons

Treatment of the solution containing Rh(I)-porphyrin with olefins substituted with the electron-withdrawing group gave the alkyl-
rhodium(III)-porphyrins 11 in good yield. The nmr spectra of 11
show the Rh-C bonding on β-carbon atom. Addition reaction may proceed through π-complex intermediate as has been proposed in cobaloxime.[2] It is noted that the reaction with phenylacetylene gave only

$$[Rh(I)\cdot OEP]^- \xrightarrow[X=CN, CO_2Et]{CH_2=CHX} OEP\cdot Rh(III)\cdot CH_2CH_2X \quad 11$$

cis isomer 12 in spite of considerable steric repulsion between the
porphyrin ring and the phenyl group. Trans addition of the proton

$$[Rh(I)\cdot OEP]^- + HC\equiv C\cdot C_6H_5 \longrightarrow [Rh(III)\cdot OEP] \qquad 12$$

may be responsible for exclusive formation of the cis conformer.
The cis conformation has been confirmed by the coupling constant of
the vicinal protons (J_{cis} = 7.5 Hz). Further support has been obtained from comparison of the nmr spectrum of 12 with that of the
trans isomer 13 prepared from the reaction with trans-bromostyrene.
The nmr spectrum of 13 shows relatively large vicinal coupling constant (J_{trans} = 13.0 Hz).

Reaction of Rh(I)-Porphyrin with Small Ring Compounds

Treatment of Rh(I)-porphyrin complex with ethylene oxide and
ethylene imine gave 2-hydroxyethyl and 2-aminoethyl·Rh(III)-porphyrin
14 respectively. The reaction behavior of the Rh(I)-porphyrin in the

$$[\text{Rh(I)}\cdot\text{OEP}]^{-} \xrightarrow[\;\overset{X}{\overbrace{\text{CH}_2-\text{CH}_2}}\;]{} [\text{Rh(III)}\cdot\text{OEP}]\quad 14$$

where the product has the group XH–CH$_2$–CH$_2$–CH$_2$– attached.

ring cleavage is quite similar to those of colalamine and cobaloxime
systems. Ring opening of small ring may undergo effectively through
catalytic process with aid of release of strain energy. It has been
found that the catalytic activity of the Rh(I)-porphyrin seems to be
not so potential to cleavage the C-O bond in tetrahydrofuran.

Nucleophilic attack of the Rh(I) atom to cyclopropyl derivatives
affords the organorhodium(I)-porphyrin resulted from cyclopropan
cleavage in which the electron-withdrawing group stabilize a reaction
intermediate. Thus the reaction of the Rh(I)-porphyrin with cyclo-
propylmethylketone and ethyl cyclopropanecarboxylate yielded 3-sub-
stituted-propyl-Rh(III) 15 complex respectively. Formation of the
Rh-C bond occurs at the γ-position with respect to substituent. No
cyclopropane cleavage has been found for cyclopropane and phenylcy-
clopropane. Similar trend has been also noticed in the cases of the

$$[\text{Rh(I)}\cdot\text{OEP}]^{-} \xrightarrow{\;\triangleright\!\!-X\;} \underset{|}{\text{CH}_2\text{CH}_2\text{CH}_2\text{X}}\;[\text{Rh(III)}\cdot\text{OEP}]\quad 15$$

reaction with olefinic compounds. These facts suggest that the
transformation through the π-complex to the σ-complex intermediate
requires the electron-with-drawing group to stabilize the negatively
charged site.

On the other hand, the reaction of the Rh(I) complex with cyclo-
propylcyanide afforded cyanoisopropyl-Rh(III) complex 16 resulted
from the Rh-C bond formation on the β-carbon. When allylcyanide
was added to the solution, the same alkyl·Rh(III) complex was obtained. Both reactions may undergo through hydrogen shift at the
transition state involving equilibrium of the α- and π-complex intermediate. The cyclopropane cleavage is probably initiated by the
interaction between d-orbital of Rh(I) atom and the π-type orbital
of cyclopropane ring.

Oxidative Alkylation

In the related system to metalloporphyrins, Grigg and his co-workers have reported that the alkylation of nickel corrole ambident anions with alkyl halides occurs at the pyrrolic nitrogen atom and the moderate heating of N-alkyl nickel corrole causes alkyl migration from the pyrrolic nitrogen atom to the β-carbon atom of the pyrrolic ring.[7] A similar migration has been reported for the palladium corrole.[7] On the contrary, our result is the first case of alkyl migration from a nitrogen atom to a metal ion. The alkyl migration may proceed concertedly with oxidation of rhodium(I) to rhodium(III). The N-CH_3 bond fission seems to be facilitated by the aid of a low-valent rhodium ion. The reaction of N-ethyloctaethylporphyrin with [Rh(CO)$_2$Cl]$_2$ affords the similar rhodium(I) complex to 6, which is easily oxidized to (OEP)Rh(III)CH_2CH_3 17.

IRIDIUM-PORPHYRIN COMPLEXES

The reaction scheme are briefly summarized as follows:

These products have been confirmed by the spectroscopic measurements and microanalyses.

EVALUATION OF RING CURRENT IN THE ORGANORHODIUM(III) PORPHYRIN

Introduction of alkyl and aryl groups by means of the various

methods have been clearly verified by the nmr spectra. Those substituents attached to rhodium atom are subjected to large shielding
effect due to the anisotropy of the macrocyclic ring current as is
shown in Table 1. The up-field shift has been evaluated by using
Johnson-Bovey's equation.[8] Agreement with the observed shift is not
satisfactory. Therefore, in order to draw the isoshielding map of
the present system, we have modified their equation, which is obtained by using the observed values and least square method with the
aid of computor. Figure 3 demonstrates the isoshielding map.

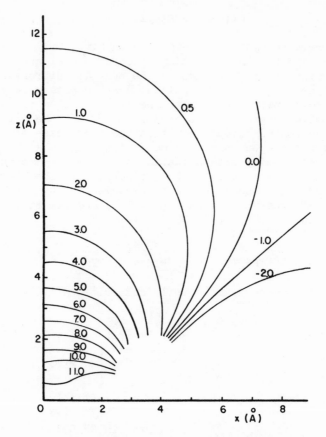

Figure 3. Isoshielding map. The coordinate X and Z denote the ver-
tical distance from the four-fold axis and the distance from the porphyrin ring along the four fold-axis.

SUMMARY

    The rhodium(I) and rhodium(III) porphyrin complexes have been
proved to be an accessible model for vitamin B_{12} (B_{12} and B_{12s}).

The formation of the Rh-C bond can be performed through addition, substitution and ring cleavage by the analogy of vitamin B_{12} system. Catalytic ring cleavage of the cyclopropane derivatives in the present system was noticed.

REFERENCES

1. D. A. Clarke, D. Dolphin, R. Grigg, and A. M. Johnson, J. Chem. Soc. (C), 881 (1968).
2. G. N. Schrauzer, Accounts Chem. Res., 1, 97 (1968).
3. (a) H. Ogoshi, T. Omura, and Z. Yoshida, J. Am. Chem. Soc., 95, 1666 (1973); (b) B. R. James and D. V. Stynes, ibid., 94, 6225 (1972); (c) Z. Yoshida, H. Ogoshi, T. Omura, E. Watanabe, and T. Kurosaki, Tetrahedron Lett., 1077 (1972).
4. N. Sadasivan and E. B. Fleischer, J. Inorg. Nucl. Chem., 30, 591 (1968).
5. A. Takenaka, Y. Sasada, T. Omura, H. Ogoshi, and Z. Yoshida, J. Chem. Soc., Chem. Comm., 792 (1973).
6. D. Cullen, E. Meyer, T. S. Srivastava, and M. Tsutsui, J. Am. Chem. Soc., 94, 7603 (1972).
7. R. Grigg, A. W. Johnson, and G. Shelton, Justus Liebigs Ann. Chem., 746, 32 (1971). M. J. Broadhurst, R. Grigg, G. Shelton, and A. W. Johnson, Chem. Comm., 231 (1970).
8. C. E. Johnson, Jr., and F. A. Bovey, J. Chem. Phys., 29, 1012 (1958).

DISCUSSION

DR. TSUTSUI: Your formulation of your dinuclear rhodium tetra carbonyl is diamagnetic, is that correct?

DR. OGOSHI: Yes.

DR. TSUTSUI: Concerning the rhodium oxidation state, the one side is I, and the other side is II, and you should have an unpaired electron. But the NMR came out so nicely, it looks like it is diamagnetic.

DR. OGOSHI: It is difficult to discuss the electronic structure of the rhodium. We are trying to get some crystallographic studies, but we have had great trouble in making single crystals so far, but in the case of iridium complex, I think it is hopeful, at the present time.

DR. HALPERN: Have you attempted to study any rhodium(II) complexes in this series?

DR. OGOSHI: Yes, we tried, but surprisingly it is not so easy. I did some experimental work for the divalent rhodium in chloroform, and we obtained a chloro trivalent rhodium complex. I think we may extend these divalent complexes.

DR. HALPERN: The problem with rhodium(II) in the dimethylglyoxime series is not that it cannot be made, but that it dimerizes with a very strong rhodium-rhodium bond, and it is almost impossible to a-chieve monomeric complexes in that system.

DR. OGOSHI: You are right, I think we have found some precipitation during our reactions.

UNUSUAL METALLOPORPHYRINS[1]

Minoru Tsutsui[*] and C. P. Hrung

Department of Chemistry, Texas A&M University, College
Station, Texas 77843

The use of metal carbonyls for the insertion of metal ions into porphyrins was first introduced by Tsutsui and co-workers[2,3] in
1966. This method has developed itself to be a useful and unique
technique in the synthesis of new metalloporphyrin complexes within the last decade. In addition to a number of previously reported
metalloporphyrins, the reaction of metal carbonyls and metal carbonyl halides with neutral porphyrins has led to the syntheses of
new metalloporphyrin complexes of chromium, molybdenum, technetium,
ruthenium, rhodium, rhenium, and iridium.[4] Except for the chromium
and molybdenum porphyrin complexes, carbonyl groups are retained by
the metals in the new metalloporphyrin complexes.

By reaction of dirhenium decacarbonyl, $Re_2(CO)_{10}$, or ditechnetium decacarbonyl, $Tc_2(CO)_{10}$, with mesoporphyrin IX dimethyl ester,
H_2MPIXDME, in refluxing decalin under argon, Tsutsui and co-workers
have successfully prepared two unusual rhenium organometalloporphyrins[5,6], $(H-MP)Re(CO)_3$, I, and $MP[Re(CO)_3]_2$, II, a pair of technetium organometalloporphyrins,[7,8] $(H-MP)Tc(CO)_3$, III, and $MP[Tc-
(CO)_3]_2$, IV, and a mixed rhenium technetium organometalloporphyrin,[9]
$(OC)_3ReMPTc(CO)_3$, V, (Figure 1). A single crystal X-ray diffraction
analysis of μ-[meso-tetraphenylporphinato]bis[tricarbonylrhenium(I)-
][10], $TPP[Re(CO)_3]_2$, VI, (Figure 2), has shown that each rhenium ion
is bonded to three nitrogen atoms and that two rhenium atoms are
bonded to one porphyrin on opposite sides of the plane of the porphyrin molecule.

The metal ions in these complexes, I-VI, sit out of the plane
of the porphyrin molecule. The monorhenium and monotechnetium organometalloporphyrin complexes, I and III, where the porphyrin moiety acts as a tridentate ligand, resemble Fleischer's proposed "sit-

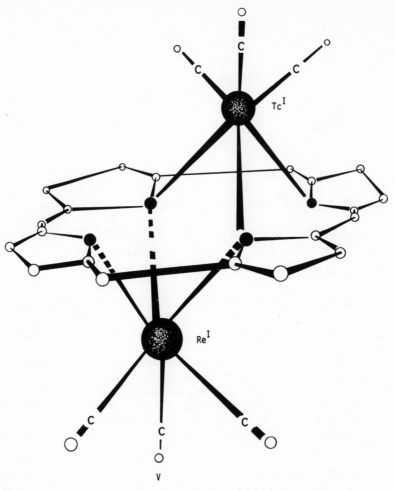

Figure 1.   Schematic diagram of $(OC)_3ReMpTc(CO)_3$, $\underset{\sim}{V}$; The alkyl substituents on the porphine ring were omitted for clarity.

ting-atop complex"[11,12] and are good models for the intermediates
in the insertion of a metal ion into porphyrin.[13] The dirhenium,
ditechnetium, and the mixed rhenium technetium organometalloporphyrin
complexes, $\underset{\sim}{II}$, $\underset{\sim}{IV}$, $\underset{\sim}{V}$, and $\underset{\sim}{VI}$, where the porphyrin moiety acts as a
hexadentate ligand, are examples of the first isolated stable homo-
and hetero-dinuclear organometalloporphyrin complexes.[9] The mono-
rhenium porphyrin complex, $\underset{\sim}{I}$, reacts with $Re_2(CO)_{10}$ or $Tc_2(CO)_{10}$ in
refluxing decalin to form rhenium porphyrin complex,[5] $\underset{\sim}{II}$, and the
mixed rhenium technetium porphyrin complex,[9] $\underset{\sim}{V}$, respectively. Replacement of the pyrrolic proton (N-H) of the monorhenium porphyrin
complex by other metal ions such as Ag^+, Hg^{2+}, and Pb^{2+}, has re-

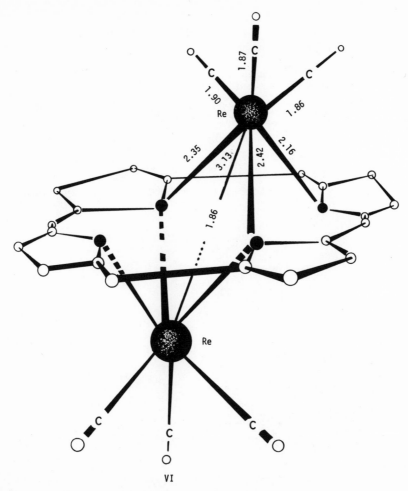

Figure 2.   Coordination sphere showing bond distances around the Re atoms of TPP[Re(CO)$_3$]$_2$, VI.

sulted in unstable complexes.[6]

The temperature dependent nmr spectra of (HMP)Tc(CO)$_3$, (HMP)Re-(CO)$_3$ and (H-TPP)Re(CO)$_3$[14] have been shown to exhibit fluxional be-havior. The metal atoms migrate about the face of the porphyrin ring via an intramolecular mechanism. This is the first example of fluxional behavior of out-of-plane organometalloporphyrins reported.

The monotechnetium porphyrin complex, III, (Figure 3), behaves in a different manner by disproportionating to form a ditechnetium porphyrin complex, IV, and the free porphyrin, H$_2$MPIXDME, by heating

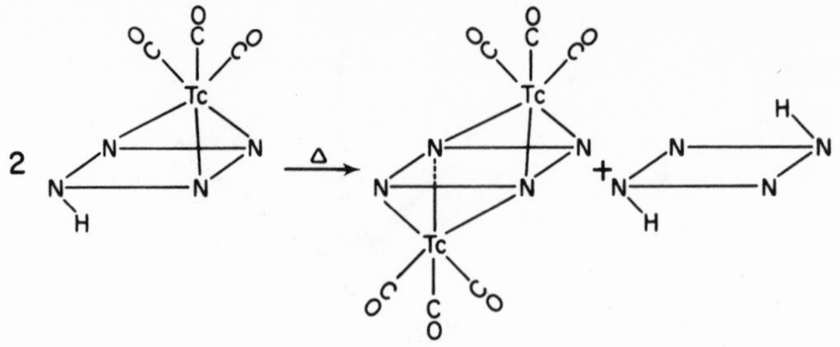

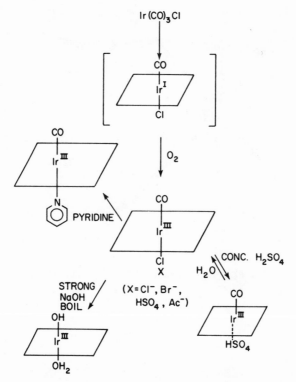

Figure 3. Disproportionation of $(H\text{-}Mp)Tc(CO)_3$, III, by heating.

Figure 4. Reaction scheme for iridium porphyrin prepared via Ir-$(CO)_3Cl$.

in refluxing decalin[8] (Figure 4). This is also the first example of unusual coordination phenomenon reported. Such a reaction was not observed on heating monorhenium porphyrin complex, I, in refluxing decalin.[5,6] It seems that both the rhenium and technetium dimetalloporphyrin complexes are thermodynamically more stable than the monometalloporphyrin complexes, because a reverse reaction of MP[M(CO)$_3$]$_2$ to (H-MP)M(CO)$_3$, (M =.Re or Tc), could not be detected between MP[M(CO)$_3$]$_2$ and H$_2$MPIXDME in refluxing decalin for either the rhenium or technetium dimetalloporphyrin complexes.[5-8]

Two different methods were employed by Fleischer and co-workers in 1967 in preparing the rhodium and iridium porphyrin complexes.[15,16] In one, the freshly prepared metal carbonyl halides, [Rh(CO)$_2$-Cl]$_2$ and [Ir(CO)$_3$Cl], were allowed to react with the porphyrins in glacial acetic acid solution to form the respective metalloporphyrins (Figure 5). In the second method, the cyclooctene complexes of rhodium and iridium were found to be reactive intermediates useful in the metalloporphyrin formation (Figure 6). In both methods, incorporation of rhodium into the porphyrin was more readily achieved than was iridium. By the reaction of [Rh(CO)$_2$Cl]$_2$ with meso-tetraphenylporphine, H$_2$TPP, in refluxing benzene, two stable organometalloporphyrin derivatives of rhodium, Rh^{III}CO(TPP) Cl and (σ-phenyl)-Rh^{IV}(TPP)·Cl, were separated by chromatography on an alumina column by Fleischer and co-workers[17,18] (Figure 7).

Recently, Yoshida and co-workers were able to prepare two novel dinuclear rhodium(I) organometalloporphyrin complexes[19,20], VII and

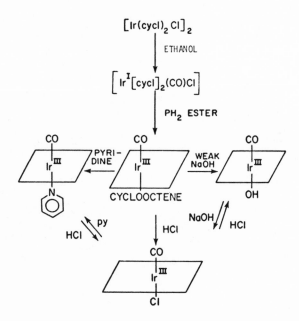

Figure 5. Reaction scheme for iridium porphyrin prepared via iridium cyclooctene complex.

VIII (Figure 8), by modifying Fleischer's reaction conditions for the preparation of Rh^{III}CO(TPP)·Cl and (σ-phenyl)RH^{IV}(TPP)·Cl in refluxing benzene.[17],[18] Octaethylporphyrin, OEPH$_2$, or (N-methyl) octaethylporphyrin reacts with [Rh(CO)$_2$Cl]$_2$ in benzene solution at room temperature under nitrogen atmosphere to produce VII and VIII. From the spectral data and the experimently determined molecular weight, VII was formulated as an acid, H$^+$[OEP·Rh$_2$(CO)$_4$Cl]$^-$, which

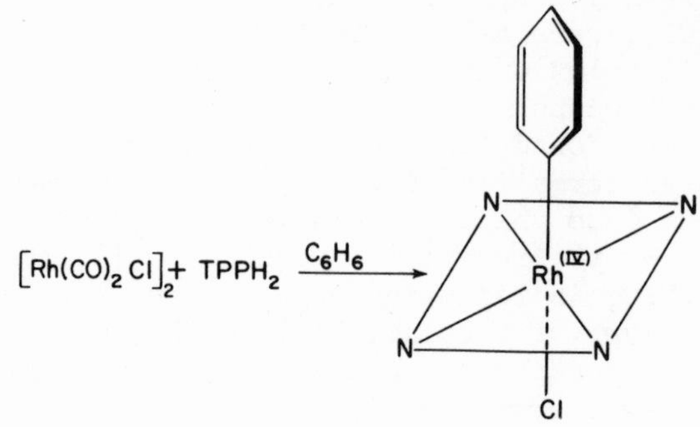

Figure 6. Preparation of new metalloporphyrin complexes of rhodium in refluxing benzene.

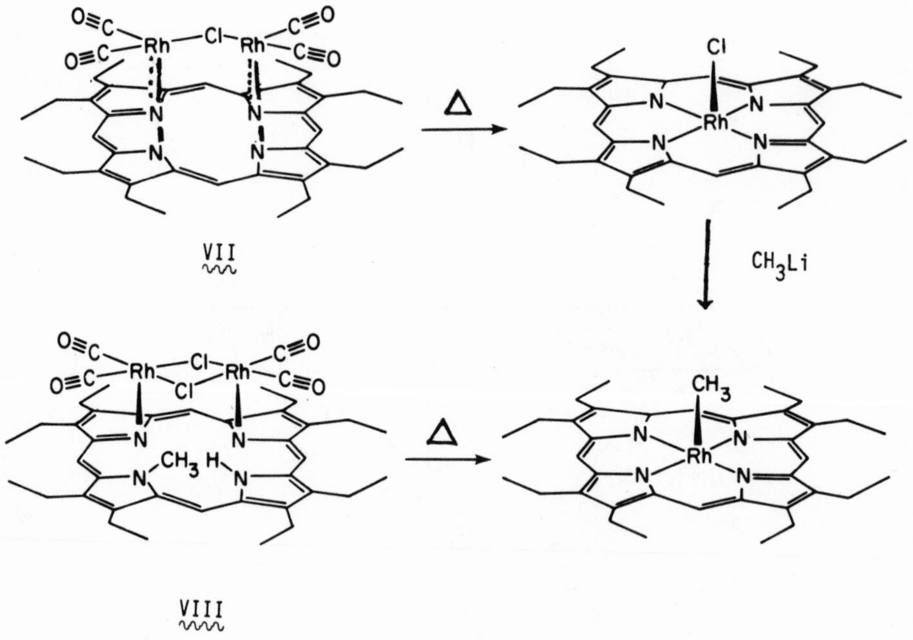

Figure 7. Two unusual metalloporphyrin complexes of rhodium prepared in benzene at room temperature.

contains a Rh-Cl-Rh bridge.[20] The proton nmr and infrared spectral
data indicate that the $[Rh(CO)_2Cl]_2$ moiety is maintained and the N-
H and N-CH$_3$ bonds exist in VIII. Since the Rh-Rh distance in [Rh-
$(CO)_2Cl]_2$ has been reported to be 3.12 Å, and the distance between
the two adjacent nitrogen atoms of planar porphyrin is about 2.9 Å,
it was assumed[19] that the two Rh atoms of the $[Rh(CO)_2Cl]_2$ moiety
are bonded to the two adjacent nitrogen atoms of the porphyrinato
core of VIII, as shown in Figure 8. X-ray analysis of VII, however,
showed the atoms to lie above and below the porphyrin ring and off
of the S$_2$ axis normal to the porphyrin plane (Figure 9).[21] Upon exposure to air, VII was slowly oxidized to form a rhodium(III) chloro
complex of octaethylporphyrin, Rh[III]Cl(OEP)·2H$_2$O, which can further
react with alkyl lithium to give an alkyl-rhodium complex. However,
VIII behaves in a different manner to give the identical alkyl-rho-

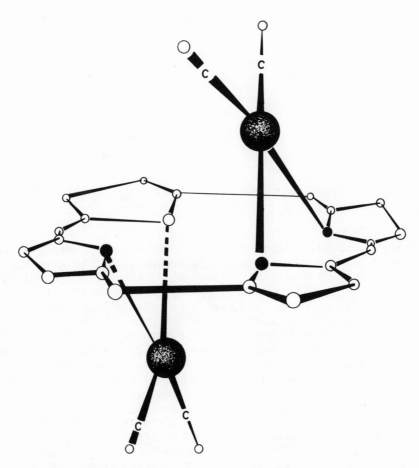

Figure 8. Structure of $[Rh(CO)_2]_2$ porphyrin.

dium complex either by gentle heating in chloroform or chromatography on silica gel (Figure 8). This phenomenon of alkyl migration from a nitrogen atom to a metal ion is reported for the first time. The alkyl migration may proceed concertedly with oxidation of rhodium(I) to rhodium(III). The N-CH$_3$ bond fission seems to be facilitated by the aid of a low-valent rhodium ion.[19] The reaction of (N-ethyl)octaethylporphyrin with [Rh(CO)$_2$Cl]$_2$ yields a rhodium(I) complex similar to VIII, which is also easily oxidized to CH$_3$CH$_2$·Rh^{III}(OEP). The mechanism of metal oxidation and alkyl migration is still unknown.

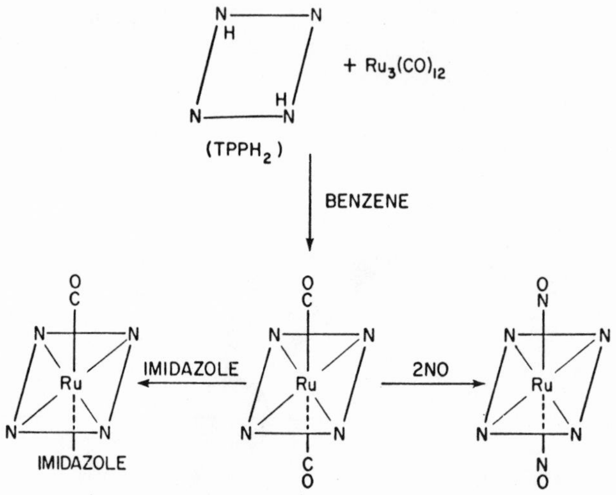

Figure 9.   Reaction scheme for ruthenium(II) porphyrin.

Both ruthenium carbonyl, Ru$_3$(CO)$_{12}$, and ruthenium carbonyl halide, [Ru(CO)$_3$Cl$_2$]$_2$, react with tetraphenylporphine, to give the identical product, monocarbonyl ruthenium(II) tetraphenylporphine,[22,23] TPPRuCO, IX. It was found that IX crystallizes with a molecule of either alcohol or water, and that these weakly bound molecules are trans to the carbonyl group.[23] Recently, a single crystal X-ray diffraction analysis[22,24] confirmed this structure. Imidazole and similar organic bases complex immediately at room temperature with TPPRuCO upon mixing in benzene.[25,26] However, as expected for a low-spin d^6 system, substitution reactions of the monocarbonylruthenium-(II) complex take place slowly and under severe conditions to replace the carbonyl group.[23] Irradiation of degassed benzene or pyridine solutions of monocarbonyl ruthenium(II) aetioporphyrin-I pyridinate with visible or ulta-violet light leads quantitatively to a ruthenium(II) porphyrin photodimer with a metal-metal bond.[27] It is of interest that MPRuCO in benzene solution reacts smoothly with excess nitric oxide to form a dinitrosylruthenium(II) mesoporphyrin complex[28] (Figure 10).

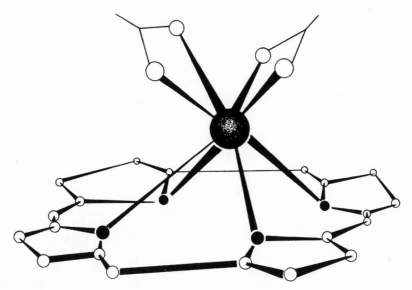

Figure 10.   Scandium porphyrin.

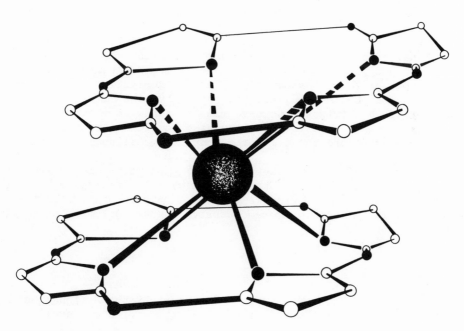

Figure 11.   Structure of U(IV)bis-phthalocyanine.

From the X-ray crystal structure analysis of both rhenium and rhodium out-of-plane metalloporphyrin complexes,[10,21] VI and VII, it would appear that the metal carbonyl moieties prefer to coordinate to adjacent nitrogen atoms rather than alternate nitrogen atoms of the porphyrin ring, causing the metal atoms to lie off the S_2 axis normal to the porphyrin plane. It is of interest that in these unusual metalloporphyrins complexes (Re, Tc, Rh), the porphyrins act as di-, tri-, or hexadentate ligands which are considered to be nonclassical coordination numbers for them. The coordination of each out-of-plane metal atom to two or three nitrogens of the porphyrin ring is predicted by the 18-electron rule with the metal in a low oxidation state $(+1; d^6, d^8)$.

Further use of "unusual" synthetic metalloporphyrins is expected in elucidating the geometries which axial ligands can assume. One approach would be to examine the metalloporphyrin complexes of metals known to exhibit coordination numbers greater than six. Scandium porphyrin[29,30] forms complexes in which axially coordinated acetate or acetylacetonate acts as a bidentate ligand. Considering the size and electronic structure of the metal and the steric requirement of the ligands, it is to be expected that the scandium is situated above the porphyrin plane. Zirconium(IV) and hafnium(IV) form porphyrin complexes[29] each containing two bidentate acetate ligands. It has been proposed[31,32] that the metal atoms are also out of the porphyrin plane, thus coordinating both acetates on the same side of the porphyrin (Figure 11).

X-ray crystal structure analysis of both uranium(IV), and tin-(IV) phthalocyanine complexes[33,34] have shown that both complexes are sandwich type compounds with the metal atoms sitting on the S_2 axis and out of the phthalocyanine plane (Figure 12). Some lanthanide(III), (La, Ce, Nd, Eu, Er, Yb) phthalocyanine complexes[35] were also prepared and proposed to have similar sandwich structures. These out-of-plane metallo-phthalocyanine complexes are close related examples of the above reported unusual metalloporphyrins.

Recently, a "triple decker sandwich" type polynuclear mercuric porphyrin complex[36] has been prepared which illustrates another type of out-of-plane metalloporphyrin complex. This compound illustrates the possibility of forming more extensive stacked polymers of metalloporphyrin (Figure 13).

ACKNOWLEDGEMENT

This investigation was supported in part by the National Science Foundation under Grant No. (GP-28685) and the Office of Naval Research.

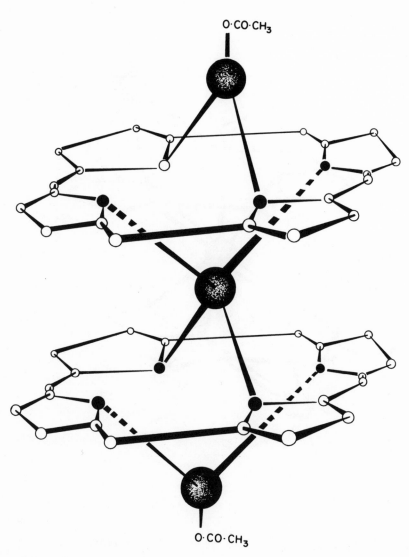

O·CO·CH₃

O·CO·CH₃

Figure 12.   Proposed structure of triple decker mercury porphyrin.

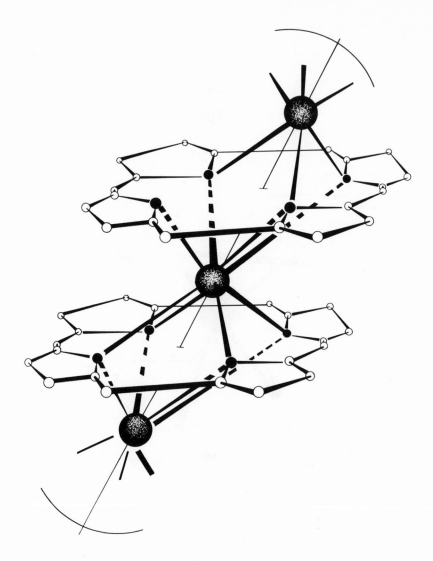

Figure 13.   Proposed structure for stacked polymers of porphyrins.

REFERENCES

1. Unusual Metalloporphyrins.
2. M. Tsutsui, M. Ichikawa, F. Vohwinkel, and K. Suzuki, J. Amer. Chem. Soc., 88, 854 (1966).
3. M. Tsutsui, R. A. Velapoldi, K. Suzuki, F. Vohwinkel, M. Ichikawa, and T. Koyano, ibid., 91, 6262 (1969).
4. D. Ostfeld and M. Tsutsui, Acc. of Chem. Res., 7, 52 (1974).
5. D. Ostfeld, M. Tsutsui, C. P. Hrung, and D. C. Conway, J. Amer. Chem. Soc., 93, 2548 (1971).
6. D. Ostfeld, M. Tsutsui, C. P. Hrung, and D. C. Conway, J. Coord. Chem., 2, 101 (1972).
7. M. Tsutsui and C. P. Hrung, Chem. Lett., 941 (1973).
8. M. Tsutsui and C. P. Hrung, J. Coord. Chem., 3, 193 (1973).
9. M. Tsutsui and C. P. Hrung, J. Amer. Chem. Soc., 95, 5777 (1973).
10. D. Cullen, E. Meyer, T. S. Srivastava, and M. Tsutsui, ibid., 94, 7603 (1972).
11. E. B. Fleischer and J. H. Wang, ibid., 82, 3498 (1960).
12. E. B. Fleischer, E. I. Choi, P. Hambright, and A. Stone, Inorg. Chem., 3, 1284 (1964).
13. R. Khosropour and P. Hambright, J. Chem. Soc., Chem. Comm., 13 (1972).
14. M. Tsutsui and C. P. Hrung, J. Amer. Chem. Soc., 96, 2638 (1974).
15. E. B. Fleischer and N. Sadasivan, Chem. Comm., 159 (1967).
16. N. Sadasivan and E. B. Fleischer, J. Inorg. Nucl. Chem., 30, 591 (1968).
17. E. B. Fleischer and D. Lavallee, J. Amer. Chem. Soc., 89, 7132 (1969).
18. E. B. Fleischer, R. Thorp, and D. Venerable, Chem. Comm., 475 (1969).
19. J. Ogoshi, T. Omura, and Z. Yoshida, J. Amer. Chem. Soc., 95, 1666 (1973).
20. Z. Yoshida, H. Ogoshi, T. Omura, E. Watanade, and T. Kurosaki, Tetrahedron Lett., 11, 1077 (1972).
21. A. Takenaka, Y. Sasada, T. Omura, H. Ogoshi, and Z. Yoshida, J. Chem. Soc., Chem. Comm., 792 (1973).
22. J. J. Bonnet, S. S. Eaton, G. R. Eaton, R. H. Holm, and J. A. Ibers, J. Amer. Chem. Soc., 95, 2141 (1973).
23. B. C. Chow and I. A. Cohen, Bioinorg. Chem., 1, 57 (1971).
24. D. Cullen, E. Meyer, Jr., T. S. Srivastava, and M. Tsutsui, J. Chem. Soc., Chem. Comm., 584 (1972).
25. M. Tsutsui, D. Ostfeld, and L. Hoffman, J. Amer. Chem. Soc., 93, 1820 (1971).
26. S. S. Eaton, G. R. Eaton, and R. H. Holm, J. Organometal. Chem., 39, 179 (1972).
27. G. W. Sovocool, F. R. Hopf, and D. G. Whitten, J. Amer. Chem. Soc., 94, 4350 (1972).
28. T. S. Srivastava, L. Hoffman, and M. Tsutsui, ibid., 94, 1385 (1972).

29. J. W. Buchler, G. Eikelmann, J. Puppe, K. Rohbock. H. H. Schnee-
    hage, and D. Weck, Justus Liebigs Ann. Chem., 745, 135 (1971).
30. J. W. Buchler and H. H. Schneehage, Tetrahedron Lett., 36, 3803
    (1972).
31. J. W. Buchler and K. Rohbock, Inorg. Nucl. Chem. Letters, 8,
    1073 (1972).
32. J. W. Buchler, L. Puppe, K. Rohbock, and H. H. Schneehage, Ann.
    N.Y. Acad. Sci., 206, 116 (1973).
33. A. Gieren and W. Hoppe, Chem. Comm., 413 (1971).
34. D. E. Broberg, Diss. Abst., Int. B, 28, 3204 (1968).
35. S. Misumi and K. Kasuga, Nippon Kagaku Zasshi, 92, 335 (1971).
36. M. F. Hudson and K. M. Smith, J. Chem. Soc., Chem. Comm., 515
    (1973).

DISCUSSION

DR. OGOSHI: In the dirhenium tricarbonyl porphyrin, the unco-
ordinated pyrolle ring should be tilted, shouldn't it?

DR. TSUTSUI: We have done no refinement of x-ray diffraction
data on this yet, so I cannot say definitely, but I expect you are
right.

DR. OGOSHI: In the case of the technetium carbonyl porphyrin,
is it possible to drive it into the center of the porphyrin, to
make it 4-coordinate?

DR. TSUTSUI: Technetium and rhenium tricarbonyl porphyrins are
quite different from your rhodium carbonyl porphyrin which in your
case went into the center of the porphyrin with heating, which makes
it a good model for the mechanism of metal insertion reaction into
porphyrins. The technetium tricarbonyl porphyrins decompose upon
heating without insertion of metal ions, however.

DR. FALLER: Have you tried any electrochemistry on the compound
with the Re-Re bond?

DR. TSUTSUI: Fred Hawthorne suggested that some while ago, but
we have not done any work along those lines yet.

THE CATALYZED REDUCTION OF NITRIC OXIDE BY CARBON MONOXIDE USING SOLUBLE RHODIUM COMPLEXES

Carol D. Meyer, Joseph Reed and Richard Eisenberg[*]

Department of Chemistry, University of Rochester, Rochester, New York 14627

The reduction of nitric oxide by carbon monoxide to yield CO_2 and N_2O is of obvious environmental importance. However, despite its favorable thermodynamics,[2] reaction (1) does not proceed at an appreciable rate in the absence of catalyst.

$$2NO + CO \rightarrow N_2O + CO_2 \tag{1}$$

To date, studies of this reaction have concentrated on the development of solid heterogeneous catalyst systems, both supported and unsupported, and have usually employed temperatures greater than 300°-C.[3] In view of the known redox reactions of Group VIII metal complexes with both CO and NO gas,[4] and of the use of soluble metal complexes as homogeneous catalysts for a range of reactions usually involving organic substrates, we have explored the approach of using such catalyst systems for environmentally important reactions. Recently we reported that ethanolic solutions of rhodium trichloride catalyze the reduction of NO by CO under the surprisingly mild conditions of room temperature and atmospheric pressure.[4a] We have since found that ethanolic solutions of dichlorodicarbonylrhodium(I) anion, $[RhCl_2(CO)_2]^-$, containing small amounts of water and HCl form a highly efficient catalyst system for reaction (1).[4b] These solutions can be repeatedly recharged with NO/CO in a 4:3 ratio without loss of activity, and $[RhCl_2(CO)_2]^-$ appears to be the true catalyst (generated in situ by reduction of $RhCl_3 \cdot xH_2O$ by CO) in the $RhCl_3$/EtOH system.

In a related study, Johnson and Bhaduri[7] have reported reactions (2) and (3), and have suggested the possibility of a continuous redox cycle based on them. Subsequently, Ibers and Haymore[8] have shown that dinitrosyl complexes of the type used in reaction (2) do indeed promote the reduction of NO by CO according to (1).

$$[Ir(NO)_2L_2]^+ + 4CO \rightarrow [Ir(CO)_3L_2]^+ + N_2O + CO_2 \qquad (2)$$

$$[Ir(CO)_3L_2]^+ + 2NO \rightarrow [Ir(NO)_2L_2]^+ + 3CO \quad (L=PPh_3) \quad (3)$$

In view of both the obvious relevance of understanding the combined effects of oxidation state, coordination geometry, and ligand substitution to the design of homogeneous and heterogeneous catalyst systems, as well as the intrinsic interest in the interrelationship of structure and reactivity, we have extended our research to include a variety of phosphine and diolefin Rh complexes, and now wish to report that promotion of the coupled reduction of NO by CO appears to be a phenomenon which is quite general in scope.

The Rh complexes found to show catalytic activity are presented in the Table along with other pertinent data. The reactions were followed for at least 50 hrs, and Figures 1 and 2 depict the formation of CO_2 and N_2O, respectively, as a function of time. No evi-

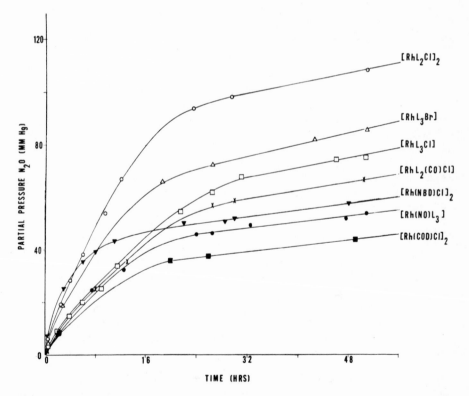

Figure 1. Partial pressure of N_2O vs. time for various "Rh Catalysts". (L = PPh_3; COD = 1,5-cyclooctadiene, NBD-Norbornadiene).

dence for the formation of N_2 or other gaseous products was found.
In all cases the mole ratios of products to starting metal complex
were in the range 4-18 for the first 48 hrs. In addition to the
complexes listed in Table I, we have also observed the coupled reduction of NO by CO using solutions of Rh_2^{4+} in methanol[9,10] and Rh_2-
$(OAc)_4 \cdot 2MeOH$ in methanol/H_2O.[9] However, the reactions with these
systems appear to be nearly stoichiometric, and precipitation of
product complex, possibly corresponding to that previously reported,
is observed in the latter case.[11]

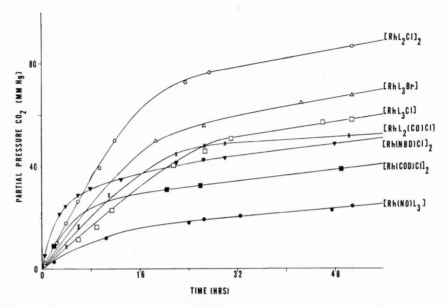

Figure 2. Partial pressure of CO_2 vs. time for various "Rh Catalysts". (L = PPh_3; COD = 1,5-cyclooctadiene, NBD-Norbornadiene).

That the most rapid evolution of product gases in all these
systems occurs during the first 12 hours, paralleling the behavior
of $[RhCl_2(CO)_2]^-$, but in marked contrast to our findings for $RhCl_3 \cdot$
xH_2O for which a substantial induction period is observed,[4] further
supports our hypothesis that a low oxidation state of Rh is needed
to begin the catalytic cycle. While the coupled reduction of NO by
CO catalyzed by $RhCl_3 \cdot xH_2O/EtOH$ or $[RhCl_2(CO)_2]^-/EtOH/HCl$ clearly
follows the stoichiometry of reaction (1)[1,4,11] and the same appears
to be true of the diolefin complexes, substantially more N_2O than
CO_2 is produced initially for the phosphine complexes.[12] This effect
is particularly pronounced for $Rh(NO)L_3$ and $RhClL_3$ for which the i-
nitial ratios of N_2O to CO_2 produced is 14.5 and 13.3, respectively,
and decreases in the order $Rh(NO)L_3 \sim RhClL_3 > [RhClL_2]_2 > RhL_3Br > RhL_3 \sim$

$[Rh(NO)_2L_2]^+$. Although the ratio of N_2O/CO_2 approaches the stoichiometric value of $1:1$,[13] it remains significantly above that value and it is possible that an additional competing reaction involving solely nitric oxide and the complex in solution is also occurring to yield N_2O and an oxidation product such as coordinated nitrite or triphenylphosphine oxide. The reaction between NO and $RhCl(Y)L_2$ (Y=CO, PPh_3) to give such products has been reported previously by Hughes.[3a]

Table I
Relative Reactivities of Soluble
"Rh Catalysts" in the Reduction of NO by CO

Catalyst[a]	Time (hr)	Δp_{NO}[b]	(mm Hg) Δp_{CO}	p_{CO_2}	p_{N_2O}
$[Rh(PPh_3)_2Cl]_2$[c,d]	50.8	247	104	87	109
$[Rh(PPh_3)_3Br]$[c]	50.8	210	86	68	86
$[Rh(PPh_3)_3Cl]$[c]	50.8	183	72	58	75
$[Rh(PPh_3)_2(CO)Cl]$[e]	50.3	179	73	52	67
$[Rh(NBD)Cl]_2$[f]	48.0	163	65	49	58
$[Rh(PPh_3)_3(NO)]$[g]	50.8	151	40	24	54
$[Rh(COD)Cl]_2$[f]	49.0	124	53	39	44
$[Rh(PPh_3)_3]^+BF_4$[h,i]	50.6	87	32	23	27
$[Rh(PPh_3)_2(NO)_2]^+BF_4$[g]	66.0	70	6	10	24

[a] Unless otherwise noted a "Rh catalyst" is 1 mmole of complex dissolved in 100 mls DMSO in a 3ℓ vessel. Systems were degassed, equilibrated to room temperature, and charged with CO and NO gas, respectively, in a 3:4 ratio to an initial system pressure of ca 720 mm. Solutions were well stirred, and reaction monitored by periodic gas sampling analyzed by vpc.[4] Partial pressures were calculated from linear calibration plots for the four gases relating peak height to partial pressure. Estimated errors are ±10% of reported values.

[b] Δp_{NO} and Δp_{CO} are the changes in the partial pressures of the relative gases from t=0 to the time shown for each system; p_{CO_2} and p_{N_2O} are the partial pressures of the product gases at the end of the time period.

[c] J. A. Osborn, F. H. Jardine, J. F. Young, and G. Wilkinson, J. Chem. Soc. (A), 1711 (1966).

[d] NO was added before CO.

[e]D. Evans, J. A. Osborn, and G. Wilkinson, "Inorganic Synthesis", Vol. XI, McGraw Hill Book Co., Inc., New York, N. Y., 1968, p. 99.

[f]J. Chatt and L. M. Venanzi, Nature, 177, 852 (1956).

[g]G. Dolcetti, N. W. Hoffman, and J. P. Collman, Inorg. Chim. Acta, 6, 531 (1972); N. Ahmad, S. D. Robinson, and M. F. Uttley, J. Chem. Soc. (A), 843 (1972).

[h]P. Legzdins, R. W. Mitchell, G. L. Rempel, J. D. Ruddick, and G. Wilkinson, J. Chem. Soc. (A), 3322 (1970).

[i]Solvent was CH_3NO_2.

Significantly, the curves for the generation of product gases in Figures 1 and 2 all approach the same limiting slope, which may indicate structurally similar or identical species as the catalytic intermediates. That the final slopes were truly limiting has been established by evacuating the system after 50 hrs of reaction and recharging with NO/CO to initial pressures. While slow reduction of NO by CO appears to continue indefinitely at a constant rate, the initial rapid formation of products is not duplicated for these systems as it is for the dichlorodicarbonylrhodium(I) anion system,[4] and it appears that the fast initial rates observed for the phosphine and diolefin complexes are associated with stoichiometric promotion of NO reduction. Obviously, the catalytically active species and the factors influencing the reaction are subjects of future study.

Solvent effects appear to be extremely important in these catalyst systems. While ethanolic solutions of $RhCl_3 \cdot xH_2O$ and $[RhCl_2(CO)_2]^-$ are very active in catalyzing the reduction of NO by CO, those of DMSO are catalytically inactive. The immediate color change to bright orange in both instances suggests rapid formation of the substitutionally inert complex $[RhCl_3(DMSO)_3]$ reported by Johnson.[14] Unfortunately, the phosphine containing complexes are insoluble in ethanol, which precludes direct comparison with the previously reported systems. However, product evolution for both $RhClL_3$ and $Rh(NO)L_3$ (L=PPh$_3$) proceeds approximately twice as fast in DMSO as in benzene. Use of the latter solvent also results in a significant (~4 fold) reduction in the initial ratio of N_2O/CO_2, suggesting that oxidation of coordinated DMSO may be a possible side reaction in this solvent. The enhanced catalytic activity observed in the coordinating solvent DMSO may well be due to stabilization of coordinatively unsaturated species formed by dissociation of a phosphine ligand.

$$RhL_3X \rightleftharpoons RhL_2X + L \xrightarrow{DMSO} RhL_2X(DMSO)$$

In particular, our observations that RhL_3Cl and $[RhL_2Cl]_2$ show comparable reactivity (per mole of Rh) in the promotion of reaction (1) in contrast to the established inactivity of the dimer as a hydrogenation catalyst[15] may be explicable in terms of rapid solvation to afford the four coordinate monomer virtually quantitatively.[16] Analogous solvation would be anticipated for the diolefin complexes, although it is apparent from the Figures and the Table that replacement of phosphine ligands with a diolefin chelate alters behavior significantly, perhaps due to the greater π acidity of the latter.

We thank the National Science Foundation (GP-40088X) for support of this research and Dr. Stephen Goldberg for helpful discussions.

REFERENCES

1. Alfred P. Sloan Foundation Fellow, 1972-74.
2. ΔH°_{298} = -91.3 Kcal/mole; ΔG°_{298} = -78.2 Kcal/mole.
3. a. C. W. Quinlan, V. C. Okay, and J. R. Kittrell, Ind. Eng. Chem. Process Des. Develop., 12, 359 (1973); b. J. H. Voorhoeve, J. P. Remeika, and D. W. Johnson, Jr., Science, 180, 62 (1973); c. M. J. Fuller and M. E. Warwick, Chem. Comm., 57 (1974).
4. a. W. B. Hughes, Chem. Comm., 1126 (1969); b. B. R. James and G. L. Rempel, ibid., 158 (1967); c. J. Bercaw, G. Guastalla, and J. Halpern, ibid., 1594 (1971); d. D. Gwost and K. G. Caulton, Inorg. Chem., 3, 414 (1974); e. M. Rossi and A. Sacco, Chem. Comm., 694 (1971); f. J. E. Bercaw, L. Goh, and J. Halpern, J. Amer. Chem. Soc., 94, 6535 (1972); g. D. Evans, G. Yagupski, and G. Wilkinson, J. Chem. Soc. (A), 2660 (1968).
5. a. J. Reed, Jr., and R. Eisenberg, Science, 184, 568 (1974); b. C. D. Meyer and R. Eisenberg, J. Am. Chem. Soc., submitted for publication.
6. a. B. R. James and G. L. Rempel, Chem. Comm., 158 (1967); b. B. R. James and G. L. Rempel, J. Chem. Soc. (A), 78 (1969); c. B. R. James, G. L. Rempel and F. T. T. Ng, J. Chem. Soc. (A), 2454 (1969); d. J. A. Stanko, G. Petrov, and C. K. Thomas, Chem. Comm., 1100 (1969); e. D. Forster, Inorg. Chem., 8, 2556 (1969).
7. B. F. G. Johnson and S. Bhaduri, Chem. Comm., 650 (1973).
8. a. B. L. Haymore and J. A. Ibers, J. Amer. Chem. Soc., 96, 3325 (1974); b. B. L. Haymore and J. A. Ibers, this volume.
9. P. Legzdins, R. W. Mitchell, G. L. Rempel, J. D. Ruddick, and G. Wilkinson, J. Chem. Soc., (A), 3322 (1970).
10. Generated in MeOH by protonation of $Rh_2(OAc)_4 \cdot 2MeOH$ with HBF_4^- as described in Ref. 9.
11. J. Kiji, S. Yoshikawa and J. Furukawa, Bull. Chem. Soc. Jap., 43, 3614 (1970).
12. The only exception to this observation is $[Rh(CO)ClL_2]$.
13. The N_2O/CO_2 ratio increases during the reaction for $[Rh(NO)_2L_2]^+$.
14. B. F. G. Johnson and R. A. Walton, Spectrochimica Acta, 22, 1853 (1966).

15. J. A. Osborn, F. H. Jardine, J. F. Young, and G. Wilkinson, J. Chem. Soc. (A), 1711 (1966).
16. J. Halpern, this volume.

MECHANISMS FOR THE BIOSYNTHESIS AND NEUROTOXICITY OF METHYLMERCURY

John M. Wood

Freshwater Biological Research Institute, College of
Biological Sciences, University of Minnesota, Navarre,
Minnesota

Before the evolution of organic systems on this planet, dynamic
equilibria for inorganic species in the earth's crust were already
important. Upon this inorganic matrix cycles evolved to produce organometallic intermediates many of which proved to be important in
the subsequent catalytic reactions of living systems. It is not
surprising that metals should form the backbone for so many natural
catalytic processes. Witness the importance of zinc in DNA polymer-
ase, and cobalt in the B12-dependent enzyme ribonucleotide reductase.
In this lecture I shall try to review what is known about metabolic
cycles for some elements which are regarded as toxic to higher or-
ganisms, and I shall present evidence pertaining to the principles
for bio-accumulation and neurotoxicity of these elements.

In general terms, elements can be classified according to their
relative toxicity and availability.[1,2] Table 1 shows some of the
elements which I have classified arbitrarily on this basis.

Many elements have been omitted because they cannot be classified in this way. For example, iodine or manganese would fit either
of the first two criteria. However, those elements classified as
very toxic and relatively accessible represent potential public
health problems.

INORGANIC EQUILIBRIA IN BIOLOGICAL SYSTEMS

When inorganic compounds are introduced into biological systems,
it should be recognized that enzyme systems have evolved which are
capable of changing the oxidation states of such compounds.

Table 1

Classification of elements according to their toxicity.
Elements omitted from this table should not be neglected
in the environmental sense. For example, iodine and
manganese are important elements, but they fit more than
one category for the above classification.

Designation	Elements
Non critical elements	Na, K, Mg, Ca, H, O, N, C, P, Fe, S, Cl, Br, F*, Li, Rb, Sr, Al, Si
Very toxic and relatively accessible	Be, Co, Ni, Cu, Zn, Sn, As, Se, Te, Pd, Ag, Cd, Pt, Au, Hg, Tl, Pb, Sb, Bi
Toxic but very insoluble or very rare	Ti, Hf, Zr, W, Nb, Ta, Re, Ga, La,† Os, Rh, Ir, Ru

*
 Some may argue with this designation, but we do add fluoride to
drinking water. †All the lanthanides are very insoluble and some
are very rare.

For example, methanogenic bacteria can reduce As^{+5} to As^{-3}, as
shown in the 8 electron reduction or arsenate to dimethylarsine.[3]
When an element is introduced into an environment rich in microbial
activity, we have to assume that each oxidation state is made available for metabolic interconversions. With mercury, we must consider
the following disproportionation reaction at all times:

$$Hg_2^{+2} \rightleftharpoons Hg^{2+} + Hg^{\circ}$$

This disproportionation reaction can be perturbed by metabolic interconversions of each mercury species. For example, soluble Hg^{2+}
can be reduced to Hg° by an enzyme which is induced, and genetically
transferable via an episomal fragment among species of the Entero-
bacteriaceae.[4] This oxidation reduction reaction uses reduced pyridine nucleotide as the source of electrons.

$$Hg^{2+} + NADPH + H^{+} \longrightarrow Hg^{\circ} + 2H^{+} + NADP^{+}$$

This interconversion of Hg^{2+} to Hg° can be regarded as a detoxification mechanism.

A second example of the power of microbial systems in changing
the oxidation state and chemical properties of an inorganic complex
is seen for the cobalt complex Vitamin B12. For this complex the
following equilibria exist:

$$Co^{III} \xrightarrow{e} Co^{II} \xrightarrow{e} Co^{I}$$

Co^{III} is a good electrophile, Co^{II} a good free radical trap, and Co^{I} the best nucleophile in aqueous systems ever discovered.

These examples offer the basis for our studies on alkyl-transfer reactions from cobalt to mercury, as well as to other metals and metalloids.

ORGANOMETALLIC EQUILIBRIA IN BIOLOGICAL SYSTEMS

Bioaccumulation of reactive inorganic species depends largely upon transport mechanisms from the intestine into the blood stream, and in the case of the potent neurotoxins, partition into the lipid regions (i.e., the brain and central nervous systems). The inter-conversion of inorganic complexes to organometallic complexes forms the essence of this problem, because the introduction of organic components allows a transition from polar to non-polar species. Non-polar species are readily transported and accumulated into those tissues of high lipid content.

The accumulation of methylmercury provides an excellent exam-ple of this process. When methylmercury is ingested, it is converted to methylmercury chloride in the stomach. Methylmercury chloride is quantitatively absorbed into the blood stream where it reacts with sulfhydryl groups which are located at cysteine residues of hemoglo-bin (6)

e.g. $CH_3HgCl + HS\text{-HEMOGLOBIN} \longrightarrow H^+ + Cl^- + CH_3HgS\text{-HEMOGLOBIN}$

In the presence of H^+, methylmercury can fractionate in the membrane regions of the central nervous system.

e.g. $CH_3Hg\text{-S-HEMOGLOBIN} + H^+ \longrightarrow CH_3Hg^+ + HS\text{-HEMOGLOBIN}$

The neurotoxicity of alkylmetals represents a major environmental problem which should be of international concern. Therefore, both the chemical and biological mechanisms of alkylmetal synthesis should be understood in detail, expecially the kinetics and flow of these neurotoxins through aqueous systems. Let us examine some of our knowledge in the area of methylmercury synthesis.

A survey of the methylating agents which are available for me-thyl-transfer reactions in biological systems reveals that these are three coenzymes which are known to be involved in this reaction:

- S-adenosylmethionine,
- N^5-methyltetrahydrofolate derivatives,
- methylcorrinoid derivatives.

S-adenosylmethionine and N^5-methyltetrahydrofolate derivatives trans-

fer their methyl groups to nucleophiles, and so this rules out methyl transfer to Hg^{2+} by these two coenzymes. Methylcorrinoids have
been shown to react with both nucleophiles and electrophiles.[7,8] e.g.

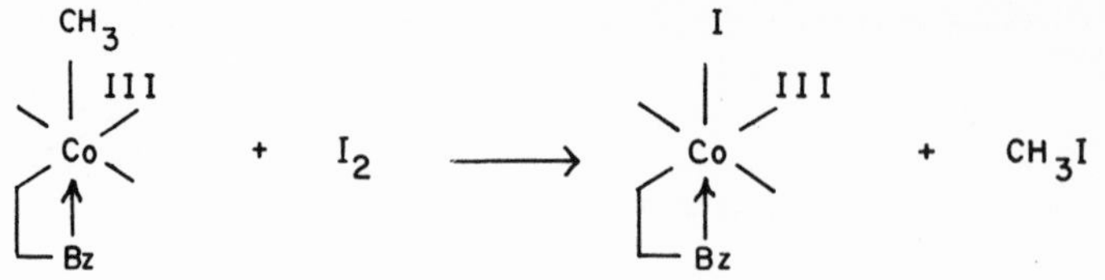

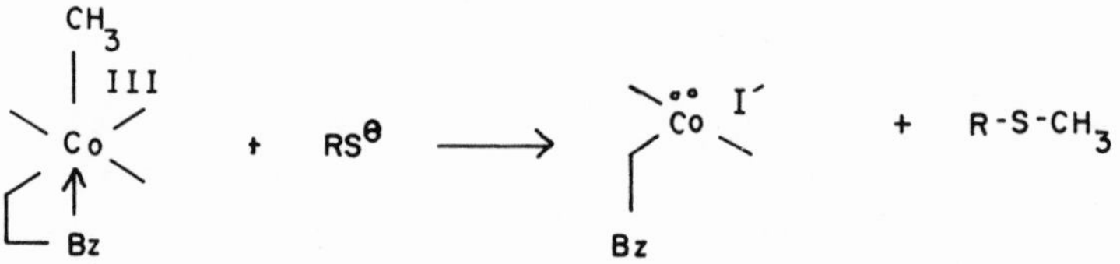

$$\text{CH}_3\text{-Co(III)-Bz} + I_2 \longrightarrow \text{I-Co(III)-Bz} + CH_3I$$

$$\text{CH}_3\text{-Co(III)-Bz} + RS^{\ominus} \longrightarrow \text{Co-I}^{'}\text{-Bz} + R\text{-S-}CH_3$$

Electrophilic attack by I_2 is of environmental interest because both
methylcorrinoids and I_2 are synthesized in the oceans, and the surface of sea water contains significant concentrations of CH_3I.[9,10]

Recent studies have elucidated the details of the mechanisms
for the synthesis of both methylmercury and dimethylmercury from
methyl-B_{12} compounds.[11,12,13]

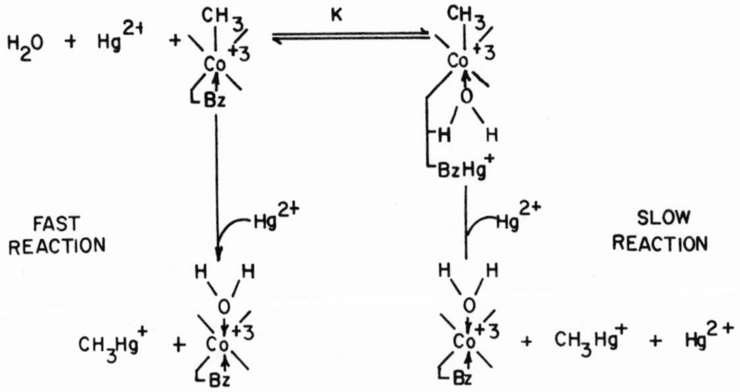

$$H_2O + Hg^{2+} + \text{CH}_3\text{-Co}^{+3}\text{-Bz} \xrightarrow{K} \text{CH}_3\text{-Co}^{+3}\text{-O-H H-BzHg}^+$$

FAST
REACTION

SLOW
REACTION

$$CH_3Hg^+ + \text{Co}^{+3}\text{-Bz}$$

$$\text{Co}^{+3}\text{-Bz} + CH_3Hg^+ + Hg^{2+}$$

The rate of synthesis of methylmercury depends on the equilibrium constant K. Dimethylmercury is also synthesized by an identical mechanism except that CH_3Hg^+ is the reacting species instead of
Hg^{2+}. However, the synthesis of dimethylmercury is about 6,000 times
slower than the synthesis of methylmercury. The pH optimum for the
synthesis of methylmercury either under laboratory conditions or in
the sediments is 4.5 (14). Dimethylmercury is volatile, and, once
in the atmosphere, it is photolysed by UV light to give Hg and me-

thane plus ethane.

$$(CH_3)_2Hg \xrightarrow{h\nu} Hg^o + 2CH_3^o$$

Methyl radicals can abstract hydrogen or couple to give methane plus ethane.

Table 2[11] presents rate data for alkyl-transfer reactions to Hg^{2+} for a number of organocorrinoids. Note the effect of benzimidazole coordination on the kinetics of electrophilic attack by Hg^{2+}.

Table 2

Rate Constants for Decomposition of Organocorrinoids with Hg^{2+}

R	Corrinoid	Medium*	Benzimidazole		Rate constant $(s^{-1} \cdot M^{-1})$
Methyl	Cobinamide	Acetate	—		$1.2 \cdot 10^{-1}$
Methyl	Cobalamin	Acetate	On		$3.7 \cdot 10^{+2}$
Methyl	Cobalamin	Perchlorate	Off		$1.2 \cdot 10^{-1}$
Ethyl	Cobinamide	Acetate	—		$1.0 \cdot 10^{-4}$
Ethyl	Cobalamin	Acetate	On		$1.8 \cdot 10^{-1}$
			Off		$1.0 \cdot 10^{-4}$
Ethyl	Cobalamin	Perchlorate	Off		$1.1 \cdot 10^{-2}$
iso-Propyl	Cobinamide	Acetate	—		$3.8 \cdot 10^{-3}$
n-Propyl	Cobinamide	Acetate	—		
n-Propyl	Cobalamin	Acetate	On	Off	No appre-
n-Propyl	Cobalamin	Perchlorate	Off		ciable
$C^{5'}$-Deoxyadenosyl	Cobinamide	Acetate	—		reaction
$C^{5'}$-Deoxyadenosyl	Cobalamin	Acetate	On	Off	

This reaction has enormous significance in regard to the biosynthesis of methylmercury in the environment. This reaction occurs in sediments,[14,15] many different species of bacteria,[16,17,18] and in B_{12} dependent enzyme systems.[19,20]

In addition to reactions leading to the synthesis of methylmercury, certain microorganisms can detoxify their environment of methylmercury by reducing it to Hg^o plus methane. This detoxification reaction converts the deadly poisonous neurotoxin methylmercury to the much less toxic and more volatile Hg^o (21). In addition to the synthesis of methylmercury and dimethylmercury, methylmercurythiomethyl is synthesized and accumulates in shellfish. From the mercury example, it is clear that biological cycles exist for toxic elements. This mercury cycle is summarized in figure n° 1.

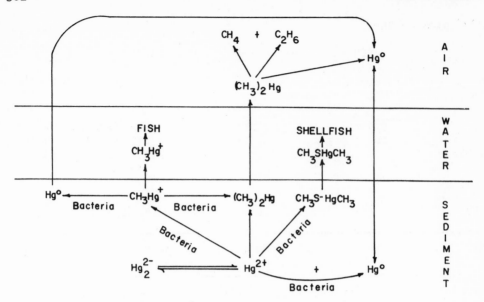

Figure n° 1. The mercury cycle

These interconversions of mercury set up a dynamic system which leads to a steady state concentration of methylmercury in sediments. These steady state concentrations of methylmercury need not reflect the rate of synthesis of methylmercury, and so determination of the concentration of methylmercury in sediments is in this case a meaningless exercise. The parameters which are the most useful to assess a particular mercury pollution situation are the concentration of total mercury in the sediments, and the rate of methylmercury uptake in fish.

From our experience with methylmercury, we can show which other heavy metals can be transformed in a similar way. For example, using the same approach, it can be shown that tin, platinum, gold and thallium will be methylated in the environment. In addition the metalloids arsenic, selenium and tellurium are converted to volatile products of extreme toxicity.[3]

When considering the synthesis of neurotoxins such as methylmercury, one should always recognize that these are natural biosynthetic processes which are influenced by a variety of parameters. For example, the rate of synthesis of methylmercury depends on the concentration of available Hg^{2+}, the microbial population, the pH, temperature, redox potential, and the synergistic or antagonistic effects of other metabolic or chemical processes.

THE NEUROTOXICITY OF METHYLMERCURY

The bio-accumulation of methylmercury into lipids, or hydrophobic regions of the cell, is probably of special significance to the
neurotoxicity of this metal alkyl.[22]

Post mortem examinations of victims of the Minamata disaster,
who suffered neurological disease after eating fish contaminated with
mercury, revealed extensive damage to the central nervous system.
Widespread lysis of cell membranes, especially of neuroglia and gran-
ule cells, was reported.[23,24] Methyl mercury in aqueous systems is
an unreactive molecule, and it is hard to see how it could cause such
irreversible damage. Although methyl mercury reacts readily with
sulphydryl groups, it does not form stable complexes with nitrogen-
containing basis in aqueous solution; in fact it does not complex
with pyridine in aqueous solution.

It is possible that the reaction of methyl mercury with cysteine
residues, which are present at the active sites of several important
enzymes of the glycolytic pathway, could provide a basis for the neurotoxicity of this compound.[25] Concentrations of methyl mercury
greater than catalytic concentrations would have to be present, however, to be effective in enzyme inhibition. As the binding of methyl
mercury to thiols is reversible, relatively large concentrations of
methyl mercury would be required for that enzyme inhibition. Furthermore, inhibition of the enzymes in the glycolytic pathway would have
an indirect and reversible effect on membrane synthesis. Recent experiments by Dr. Segall in my laboratory show that methyl mercury can
react both catalytically and directly with a group of phospholipids
which are important in membrane structure for cells of the central
nervous system.

High resolution nuclear magnetic resonance (NMR) has been used
to assign all the protons of a plasmalogen (L α phosphatidyl ethan-
amine), which has palmitic and stearic acid as alternative residues
to the vinyl ether linkage at the α' carbon, and linolenic as the unsaturated fatty acid at the β carbon (Figure 2).

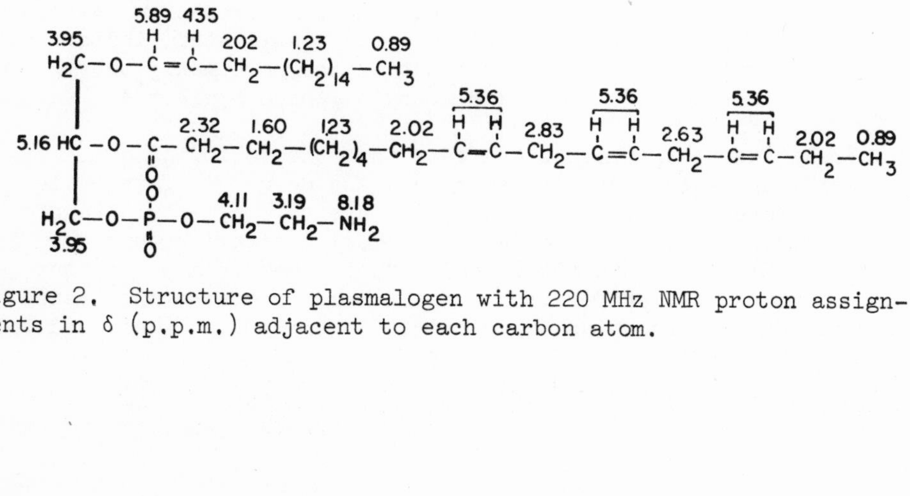

Figure 2. Structure of plasmalogen with 220 MHz NMR proton assignments in δ (p.p.m.) adjacent to each carbon atom.

Methyl mercury chloride is soluble in this phospholipid, and the methyl mercuric ion catalyses rapid hydration and hydrolysis of the
vinyl ether linkage to give a mixture of palmitic and stearic aldehydes plus the linolenic monoglyceride product.

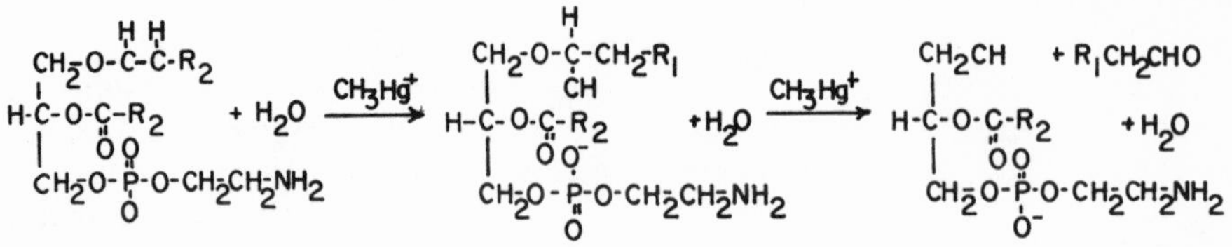

Figure 3. Proposed reaction sequence for methyl mercury-catalyzed
hydrolysis of plasmalogen.

The course of this reaction was followed by 220 MHz NMR, and the aldehyde products were characterized further by mass spectrometry of
their 2,4-dinitrophenylhydrazone derivatives. The relevance of this
reaction to the neurotoxicity of metal-alkyls is probably linked
with membrane lysis, which is specific for those membranes which
contain plasmalogens as a major constituent of the phospholipid
backbone in membrane structure.

The fact that methyl mercury is soluble in phospholipids, and
causes lysis of certain cell membranes in the central nervous system makes it tempting to speculate that this reaction could have
significance in the neurotoxicity of this metal-alkyl. Also, it is
expected that other metal-alkyls should catalyze this reaction by a
similar reaction sequence.

TRANSALKYLATION REACTIONS BETWEEN COBALT, MERCURY AND SELENIUM

Our studies to date indicate that alkyl-transfer reactions between different metals and metalloids occur in aqueous systems in a
predictable sequence. One of the more interesting reactions stems
from the observation that selenium salts detoxify laboratory animals
poisoned with methyl mercury. During this detoxification reaction
it has been shown that dimethylselenide is ventillated, indicating
that methyl groups are transferred out of the more toxic methyl
mercury cycle into the less toxic methylselenium cycle.[25]

Preliminary experiments indicate that the following cycle exists for methyl-transfer between cobalt and mercury. This cycle
offers a number of potential alkylating agents for methyl transfer
to selenium. One possible model for removing methyl groups from the
methyl mercury pool into the selenium pool comes from our recent
studies on methyl-transfer from dimethylmercury to the dehydrated
selenium compound selenium oxychloride. When $SeOCl_2$ is allowed to

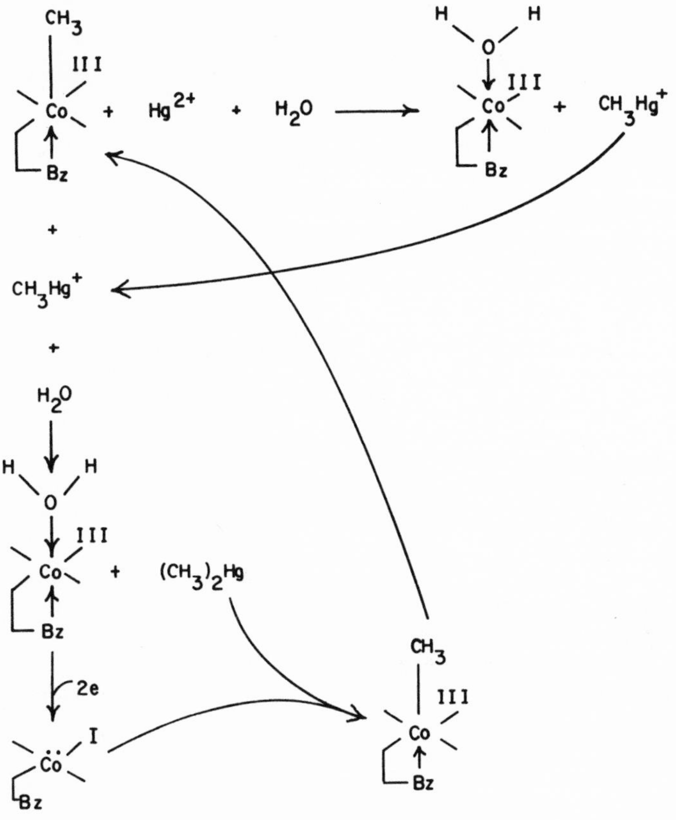

Figure 4

react with dimethylmercury in benzene the major products of this reaction are CH_3HgCl, $CH_3-Se = CH_2$, and $(CH_3)_2Se$. In addition, traces
of $CH_3Se = O$, and $(CH_3)_2 Se-Se-(CH_3)_2$ can be identified by G. C.
Mass Spectrometry. At the present time we are looking into biological mechanisms for the dehydration of selenium salts with a view to
studying mechanisms for alkyl-transfer from dimethylmercury to selenium in both in vitro and in vivo biochemical systems.

ACKNOWLEDGEMENTS

This research was supported by grants from the National Science
Foundation GB 26593X and the United States Public Health Service AM
12599. The contributions of Drs. F. S. Kennedy, M. W. Penley, R. E.
DeSimone, D. G. Brown, H. Segall and W. Chudyk are gratefully acknowledged.

REFERENCES

1. J. M. Wood, Science, 183, 1049 (1974).
2. J. M. Wood, Rev. Intern. Oceanogr. Med., 31:32, 7 (1973).
3. B. C. McBride and R. S. Wolfe, Biochemistry, 10:23, 4312 (1971).
4. A. Summers and S. Silver, J. Bacteriol. (in press), (1974).
5. S. Silver, Nature, (in press), (1974).
6. F. J. Giblin and E. J. Massaro, Trace Substances in Environmental Health, 6, 107 (1972).
7. G. N. Schrauzer, Accounts in Chem. Res., 1, 97 (1968).
8. G. Agnes, H. A. O. Hill, J. M. Pratt, S. Ridsdale, and R. J. P. Williams, Biochim. Biophys. Acta, 252, 207 (1971).
9. J. E. Lovelock, R. G. Maggs and R. J. Wade, Nature, 241, 194 (1973).
10. J. M. Wood, Proceedings of the XVI International Conference on Coordination Chemistry, Dublin, Ireland, August (1974).
11. J. M. Wood and D. G. Brown, Structure and Bonding, 11, 47 (1972).
12. J. M. Wood, M. W. Penley and R. E. DeSimone, Technical Reports Series 137, International Atomic Energy Agency, Vienna, Chapter 4, 43 (1972).
13. R. E. DeSimone, M. W. Penley, L. Charbonneau, S. G. Smith, J. M. Wood, H. A. O. Hill, J. M. Pratt, S. Ridsdale and R. J. P Williams, Biochim. Biophys. Acta, 304, 851 (1973).
14. J. M. Wood, La Recherche (in press), (1975).
15. L. Bisogni, Proceedings of an International Conference on Heavy Metal Pollution, Vanderbilt U., Dec. 4th-7th (1973).
16. L. Landner, Nature, 230, 452 (1971).
17. S. Spangler, Science, 180, 92 (1973).
18. I. Vonk and R. Sijpesteijn, Antonie Von Leeuwenhoek, 39, 505 (1963).
19. J. M. Wood, Advances in Environmental Science and Technology, 2, 39 (1971).
20. J. M. Wood, F. S. Kennedy and C. G. Rosen, Nature, 220, 173 (1968).
21. J. M. Wood, Proceedings of an International Conference on Heavy Metal Pollution, Vanderbilt U., Dec. 4th-7th (1973).
22. H. J. Segall and J. M. Wood, Nature, 248, 456 (1974).
23. T. Takeuchi, Minamata Disease, edit. M. Kuksuna, 141 (Kumamoto University, Japan), (1968).
24. T. Takeuchi, Biological Reactors and Pathological Changes of Human Beings and Animals Under the Condition of Organic Mercury Poisonings 1, 1 (University of Michigan, Ann Arbor), (1970).
25. P. Salvaterra, B. Lown, T. Morganti and E. J. Massaro, Acta. pharmac. tox., 32 (in press), (1974).

DISCUSSION

DR. HALPERN: On the question of the vinyl ether hydrolysis, we concluded from our studies on that reaction that what was happening was electrophilic attack of mercury(II) on the double bond resulting in hydroxymercuration followed by breaking of the other oxygen bond. That occurs very efficiently for the mercuric ion, but from 10^5 to 10^6 times less efficiently with methyl mercury. We would expect it to be extremely inefficient with dimethyl mercury, of course.

DR. WOOD: Yes, our reactions are written with monomethyl mercury chloride, which is very soluble in phospholipid, much more so than in water. I think there is a lot of interesting chemistry to be done, because you can see in the nmr an interaction with the ethanolamine group, and I do not really know what sort of complex we have at the active site in this non aqueous environment.

DR. HALPERN You are not postulating attack of dimethyl mercury, are you?

DR. WOOD: No, we have not tried dimethyl mercury, but I doubt that it would be effective.

DR. HALPERN: I suppose that it would be converted to monomethyl mercury easily enough, though.

DR. WOOD: Yes, it certainly would.

DR. COLLMAN: What foods have high selenium contents?

DR. WOOD: Certain grasses in the Rocky Mountains will give cattle blind staggers from selenium poisoning. There are significant levels of selenium in many plants. It is a required element, as well as being toxic at high levels. The best thing to do is read a review by T. C. Stadtman in Science (March or April, 1974) on the biochemistry of selenium, which beautifully summarizes the sources of selenium.

DR. COLLMAN: In a less facetious question, would sulfur have the same effect?

DR. WOOD: I do not know. That is an interesting question.

DR. TSUTSUI: Are you working on any new chelating agents to promote excretion of methyl mercury?

DR. WOOD: Some work has been done at Dow on the synthesis of a number of thiol resins, which are very good for binding metals. If you eat them with your cornflakes in the morning, they will tie up the methyl mercury and it is not available then for transport into the bloodstream. This works very well, but one of the concerns about it is that it also ties up all the other essential metal ions that one needs, and one has to be very careful about other interactions that occur in the intestinal tract. These studies address themselves to the problem of trying to prevent the methyl mercury from getting into the bloodstream.

DR. MANGO: John, in your classification of the very toxic metals, what was the basis for their classification as such?

DR. WOOD: We put in some metals of known toxicity, like berylium and lead. Beyond that, it was associated with the fact that if you take methyl cobalamin, and you throw in soluble salts of those metals, they displace the methyl group. These are very simple spectrophotometric experiments, and we have been asked by the Food and Drug Administration to the extent of $50,000 to look at these reactions more carefully, especially tin, which they are very concerned about.

DR. MANGO: Are you going to take a phospholipid test, say, bilayer systems, and then treat them with the metal?

DR. WOOD: Well, one of the experiments that Dr. Segal has already done is to make vesicles with lecithin and plasmologen incorporated into the vesicles. You can put a spin label in the vesicle and look at the immobilized spin label spectrum. Then you can add catalytic amounts of methyl mercury and the bilayer becomes very mobile, you get a strongly immobilized spin label spectrum, which says essentially that the membrane is becoming very fluid, and it is beginning to fall apart. Now, it could be that this is a very important mechanism in the mechanism of neurotoxicity of these alkylmetals, or it could be just one of many reactions that occur.

DR. WHITESIDES: Are the vinyl ethers that are the precursors for these plasmologens specifically brain lipids?

DR. WOOD: They are specifically lipids of the central nervous system, yes. You do not find them anywhere else.

DR. TSUTSUI: The methyl mercury catalysis of the lysis is an amazing reaction, which I do not believe you have any mechanism for. If you put a cell in water you get lysis then, too. Is there any other metal ion which catalyzes lysis?

DR. WOOD: I think mercury is quite unique, here.

DR. MARKS: Certainly there are other ways that heavy metal ions can poison you, such as Hg^{2+}. Do you think this kind of behavior is general for any kind of mercury compounds?

DR. WOOD: No. The essence of this problem is that if you have an inorganic metal ion, it is pretty well dialyzed out of living systems. If, however, you have organometallic compounds, such as methyl and dimethyl mercury, these are quantitatively absorbed into the bloodstream, and have a tendency to partition into lipids, and get into the central nervous system. So the whole problem of the neurotoxicity of these things is based on their ability to partition into lipids. Certainly if you got mercuric ion into this plasmologen, the reaction would go much better than with methyl mercury, but I do not think you can get it in there.

DR. COLLMAN: This would also seem to be related to the kinetic stability of the organometallic, of which mercury is a notable example, so that some of the transition metal compounds with no supporting ligand do not form kinetically stable alkyls. I had one other comment; on your freon, I do not suppose you have hooked your sheep up to a gas chromatograph and fed them freon.

DR. WOOD: Oh yes, that has been done with freon, chloroform and all the rest. In fact some dairy scientists at Illinois felt that if they gave cows chloroform, that instead of wasting their energy making methane, they would put the energy into making meat; and they found one of their cows on it's back with it's feet in the air- of

course it inhibits the methionine enzyme and all the other important
enzymes, and so freons are very poisonous.

DR. COLLMAN: I was going to suggest that reaction undoubtedly
is another example of these radical cycle chains John Osborn talked
about.

DR. WOOD: Yes, very much so. I think the most interesting aspect of the freon thing is that it shows you what a good nucleophile
Co(I) is in aqueous systems because it makes such good yields of
freon derivatives.

DR. COLLMAN: I think probably it is not a nucleophile, that it
is doing a chain halogen abstraction.

DR. WOOD: Oh, I see. All right.

DR. HALPERN: The Cobalt(II) alkylation reaction with organic
halides is not a chain reaction, but it is still a radical reaction.

DR. COLLMAN: I was speaking of Cobalt(I), though.

DR. HALPERN: Well, if that went by a radical mechanism, it is
undoubtedly going to Cobalt(II) along the way.

DR. WOOD: It would have to go to Cobalt(II) and make a carbon
radical and then react, and I think that is unlikely. I think you
just have nucleophilic attack on freon by Co(I).

DR. MANGO: The cell continuously manufactures or replaces phospholipid, so there must be some level of the mercury below which it
is safe.

DR. WOOD: Yes, there must be.

DR. MANGO: The next question is, if somebody does have significant brain damage due to mercury, and you were able to flush out
the mercury below that threshhold level, would the brain cells begin
to regenerate.

DR. WOOD: No. Cells in the central nervous system do not regenerate. If you lyse a cell, it is permanent. We may have some damage with low levels of mercury in all of us, but it is at such a
low level that it is not too important to the integrity of the cell
membrane. It might explain why we are grumpy in the morning, and
that sort of thing, but this is all behavioral stuff - it is presumably below the toxic level that people need to worry about.

LIST OF AUTHORS

JAMES P. COLLMAN
    Department of Chemistry, Stanford University,
    Stanford, California 94305

RICHARD EISENBERG
    Department of Chemistry, University of Rochester,
    Rochester, New York 14627

JACK FALLER
    Department of Chemistry, Yale University, 225 Prospect St.,
    New Haven, Connecticut 06520

JAMES FRANCIS
    Union Carbide Research Institute, Tarrytown, New York 10591

ROBERT GRUBBS
    Department of Chemistry, Michigan State University,
    East Lansing, Michigan 48824

JACK HALPERN
    Department of Chemistry, University of Chicago,
    Chicago, Illinois 60637

FREDERICK HAWTHORNE
    Department of Chemistry, University of California,
    Los Angeles, California 90024

MASANOBU HIDAI
    Department of Industrial Chemistry, University of Tokyo,
    Hongo, Tokyo, Japan

YOSHIO ISHII
    Department of Synthetic Chemistry, Nagoya University,
    Faculty of Engineering, Chikusa, Nagoya, 464, Japan

KENJI ITOH
      Department of Synthetic Chemistry, Nagoya University,
      Faculty of Engineering, Chikusa, Nagoya, 464

HERBERT KAESZ
      Department of Chemistry, University of California,
      Los Angeles, California 90024

JITSUO KIJI
      Department of Synthetic Chemistry, Kyoto University,
      Faculty of Engineering, Yoshidahonmachi, Sakyo, Kyoto, Japan

R. BRUCE KING
      Department of Chemistry, University of Georgia,
      Athens, Georgia 30601

MAKOTO KUMADA
      Department of Synthetic Chemistry, Kyoto University,
      Faculty of Engineering, Yoshidahonmachi, Sakyo, Kyoto, Japan

HIDEO KUROSAWA
      Department of Petroleum Chemistry, Osaka University,
      Faculty of Engineering, Yamadaue, Suita, Osaka, 565, Japan

SY LAPPORTE
      Chevron Research, Richmond, California 94801

FRANK MANGO
      Shell Development Company, 3737 Bellaire Blvd.,
      Houston, Texas 77025

TOBIN MARKS
      Department of Chemistry, Northwestern University,
      Evanston, Illinois 60201

S. -I. MURAHASHI
      Department of Chemistry, Osaka University, Faculty of
      Engineering Science, Machikaneyama, Toyonaka, Osaka, Japan

AKIRA NAKAMURA
      Department of Chemistry, Osaka University,
      Faculty of Engineering Science, Machikaneyama, Toyonaka, Osaka,
      Japan

RYOJI NOYORI
      Department of Chemistry, Nagoya University,
      Faculty of Engineering, Chikusa, Nagoya, 464, Japan

HISANOBU OGOSHI
        Department of Synthetic Chemistry, Kyoto University, Faculty
        of Engineering, Yoshidahonmachi, Sakyo, Kyoto, 606, Japan

IWAO OJIMA
        Sagami Chemical Research Center, 4-4-1, Nishi-Onuma,
        Sagamihara, Kanagawa, 228, Japan

ROKURO OKAWARA
        Department of Applied Chemistry, Osaka University,
        Faculty of Engineering, Yamadagami, Suita, Osaka, 565, Japan

JOHN OSBORN
        Department of Chemistry, Harvard University,
        Cambridge, Massachusetts 02138

SEI OTSUKA
        Department of Chemistry, Osaka University, Faculty
        of Engineering, Machikaneyama, Toyonaka, Osaka, 560, Japan

GEORGE PARSHALL
        E. I. DuPont de Nemours and Company, Central Research
        Department, Experimental Station, Wilmington, Delaware 19898

ROWLAND PETTIT
        Department of Chemistry, University of Texas,
        Austin, Texas 78712

CHARLES U. PITTMAN, JR.
        Department of Chemistry, University of Alabama,
        University, Alabama 35486

MEMBO RYANG
        Department of Petroleum Chemistry, Osaka University,
        Faculty of Engineering, Yamadagami, Suita, Osaka, 565, Japan

TAKEO SAEGUSA
        Department of Synthetic Chemistry, Faculty of Engineering,
        Kyoto University, Kyoto, Japan

KAZUO SAITO
        Department of Chemistry, Tohoku University, Faculty
        of Science, Aoba, Aramaki, Sendai, 980, Japan

TOSHIO TANAKA
        Department of Applied Chemistry, Osaka University, Faculty
        of Engineering, Yamadagami, Suita, Osaka, 565, Japan

MINORU TSUTSUI
    Department of Chemistry, Texas A&M University,
    College Station, Texas 77843

YASUZO UCHIDA
    Department of Industrial Chemistry, University of Tokyo,
    Hongo, Tokyo, Japan

GEORGE WHITESIDES
    Department of Chemistry, Massachusetts Institute of Technology,
    77 Massachusetts Avenue, Cambridge, Massachusetts 02139

JOHN WOOD
    Department of Biochemistry, University of Illinois,
    Urbana, Illinois 61801

AKIO YAMAMOTO
    Tokyo Institute of Technology, Research Laboratory of
    Resources Utilization, Ookayama, Meguro, Tokyo, 152, Japan

SUBJECT INDEX